Weather Toys

Weather Toys

Building and Hacking Your Own 1-Wire Weather Station

Tim Bitson

WILEY

Wiley Publishing, Inc.

Weather Toys: Building and Hacking Your Own 1-Wire Weather Station

Published by
Wiley Publishing, Inc.
10475 Crosspoint Boulevard
Indianapolis, IN 46256
www.wiley.com

Copyright © 2006 by Wiley Publishing, Inc., Indianapolis, Indiana

Published simultaneously in Canada

ISBN-13: 978-0-470-04046-7
ISBN-10: 0-470-04046-7

Manufactured in the United States of America

10 9 8 7 6 5 4 3 2 1

For general information on our other products and services or to obtain technical support, please contact our Customer Care Department within the U.S. at (800) 762-2974, outside the U.S. at (317) 572-3993 or fax (317) 572-4002.

Library of Congress Cataloging-in-Publication Data

Bitson, Tim.
  Weather toys : building and hacking your own 1-wire weather station / Tim Bitson.
     p. cm.
  ISBN-13: 978-0-470-04046-1 (paper/website)
  ISBN-10: 0-470-04046-7 (paper/website)
  1. Meteorological instruments--Design and construction. I. Title.
  QC876.B58 2006
  551.5028--dc22

                     2006027619

This book is dedicated to my Mom...
Who has always been there for me.

About the Author

Tim Bitson as been building weather measurement equipment for 30 years, and has been active in the 1-Wire weather community since it started in the late 1990s. He has designed two of the commonly used 1-Wire weather sensors, the "Bitson" barometer and the 1-Wire lightning detector. He has developed weather station code for Basic Stamp, PalmOS, and Tiny InterNet Interface (TINI), just to name a few.

During the day, Tim works as a "rocket scientist" at a major aerospace firm in southern Arizona, which he has been doing for more than 25 years. His specialty is integrating new designs and getting them to work for the first time.

During the evenings and weekends, he enjoys working with electronics, writing software, building neon art, and riding ATVs in the desert. Tim lives with his wife Pam, two sons Kyle and Andrew, one dog, and two cats (at last count).

Tim holds a B.S. degree in Information Technology, and has been programming since the late 1970s.

Credits

Executive Editor
Chris Webb

Development Editor
Ed Connor

Technical Editor
Eric Vickery

Production Editor
Angela Smith

Copy Editor
Kim Cofer

Editorial Manager
Mary Beth Wakefield

Production Manager
Tim Tate

Vice President and Executive Group Publisher
Richard Swadley

Vice President and Executive Publisher
Joseph B. Wikert

Compositor
Maureen Forys, Happenstance Type-O-Rama

Illustrator
Jeff Wilson, Happenstance Type-O-Rama

Proofreading
Jennifer Larsen, Word One

Indexing
Johnna VanHoose Dinse

Cover Design
Anthony Bunyan

Acknowledgments

There were quite a few people who helped make this book a reality. First and foremost, I want to thank my family for their support. My wife, Pam, proofread every chapter, providing helpful suggestions and guidance. Andrew developed many of the graphics, logos, and figures. Kyle helped with the proofreading and software testing. Kyle also took the lightning photo used in Chapter 10. Thank you! I could not have done this without you.

I want to thank the team at Wiley Publishing, especially Chris Webb for having the insight for this book's concept. I also want to thank Ed Connor for guiding a novice writer and for being so patient with my bazillion questions. Thank you, Chris and Ed.

Eric Vickery was involved from the beginning and provided the technical editing. Thanks for your help Eric.

A special thanks goes to Al Felice, Claude Merrill, and Eric Borg for their support while I was writing this book. Thanks guys. Yes, I'll be around more now that it's done.

There are several other individuals that I want to thank for their help and support:

Joan Peterson at Davis Instruments

Dick Norford at LaCrosse Technology

Curt Lillie at Adaptive Microsystems

Tim Sailor for sponsoring the 1-Wire Weather Mailing List

Aitor Arrieta at AAG Electronica

Marco Mularoni at Humirel Inc.

Trent Javi for his tireless devotion to the RxTx Project

Dmitry Markman for his RxTx Mac OS X port and Tiger Installer

Vanessa Bland at Dallas/Maxim

Chuck Prewitt and Shaun Tanner at The Weather Underground

Contents at a Glance

Contents

$\bullet\ \bullet$

Part II: Build a Weather Station

Introduction

If you're reading this, chances are you are either in the book store thinking about buying this book, or are at home thinking about building your own weather station. Either way, I appreciate your interest! I've tried hard to pack it full of insider tips and tricks.

Why Build a Weather Station?

The information explosion we are currently living in provides opportunities to collect data on just about everything that affects our daily life. People have come to expect information that is up-to-the-minute, relevant to us personally, and relates on a local level.

Just a few years ago, our only source of weather data was the evening news, and often the weather presented was focused on the area the news was broadcast from. If you lived miles away, the data was sometimes misleading or inaccurate. As computers and consumer electronics became less expensive, weather data became more localized. Once the Internet hit the scene, you could look up the weather any time of the day for hundreds of cities.

Now it is possible to own and operate your own weather station. Originally, the first consumer weather stations were costly, hard to install, and difficult to maintain. Now that these stations have become more advanced, they are easier to operate, but also less versatile. If you want to perform a function that your weather station doesn't support, there usually isn't much you can do. For the true hacker, the obvious solution is to build your own that does what *you* want it to do.

This book presents the construction details to build and operate your own weather station. Now you can know the weather in your backyard 24-hours a day. You can post your weather data to the Internet so you can retrieve it worldwide and share it with others. You can control appliances based on weather, activate sprinklers, monitor lightning, and more.

Because you build it, it is customized for your specific application. If you want 12 temperature sensors scattered throughout your house or in the yard, you can do it. If you need to turn on a humidifier in your greenhouse when it gets too dry, you can do that too. Options are what this book is all about.

Who This Book Is For

If you have ever considered building and running your own weather station, this book is for you. For about the cost of dinner and a movie, you can get started exploring the world of weather measurement. You can pick and choose the projects you want and build them in the order you want.

Many of the hardware projects featured can be built from scratch or purchased as a kit. If you are more interested in building the software, fully assembled and tested modules are available.

Most of the hardware projects have an accompanying software project. Each software project is shown step-by-step. Even if you have never written a line code before, you should be able to follow along and complete the projects. However, to get the most out of this book, at least intermediate computer skills are recommended. You need to know common computer terms, how to navigate around the various directories, and install software.

If you prefer not to follow along with the software project, a complete weather station software package is available on the companion web site. There are also several third-party software packages that support the weather station described in this book.

How This Book Is Organized

Weather Toys is divided into three parts. Each part presents a different aspect of the whole project.

Part I: The World of Weather

This section introduces you to the world of weather: How is weather measured? What are the units of measurement? Where did some of this stuff come from? These are all questions that are touched upon in the first chapter.

You will also get a quick look at the various types of hobbyist weather stations available for purchase. It is interesting to see what is available commercially as you start to build, and later expand, your project. You'll also get an introduction to the 1-Wire weather station, looking at a couple of the pre-built modules and sensors.

This section wraps up with an introduction to 1-Wire and 1-Wire devices, the building blocks of the weather station you're going to build. You'll learn what it is and how it works. Most importantly, you will learn how to design and install a reliable 1-Wire network.

Part II: Build a Weather Station

This part of the book is where you get your hands dirty. There are eight hardware and software projects, including six that build the basic weather sensors, including:

- Temperature
- Wind speed and wind direction
- Humidity
- Barometric pressure
- Rainfall
- Lightning

You can build one or all of the sensors. If desired, you can have several sensors of the same type. Each sensor is independent of the other, so you can start with just one sensor and add another when you want. Once you have completed your weather station sensors, there's a chapter on installing your weather equipment.

This section also presents the software project used to access the sensors and collect the weather data. You will build SimpleWeather, a Java-based weather program. All the code you need is provided with documentation on how it works. Although SimpleWeather can be used as your main weather program, its primary purpose is to show *you* how to talk to the weather devices, process the data, and then *do* something with it.

Part III: Expanding Your Weather Station

In Part III, you have the option of expanding your weather station with several additional hardware and software projects. From building your own weather station web site to a smart sprinkler timer based on weather, this section provides ideas and projects to take your weather station to the next level. Some of the projects include:

- Build a Weather Web Server
- Turn Appliances On or Off Based on Weather
- Add an LED Sign to Display Weather Conditions
- Build a Smart Sprinkler Timer or Home Thermostat
- Build a Stand-Alone Weather Station to Free up Your PC

This part of the book is a little less detailed than Part II. Even though each project is complete, it is meant to inspire you to come up with new and unique ways to utilize your weather station.

Conventions Used in This Book

Many of the project chapters contain code. Code listings and references in the text are printed in a fixed-width font. Code that you need to enter is presented in bold. For example:

```
// get temperature
temp = ts1.getTemperature();
System.out.println("Temperature = " + temp + " degs F");

// get wind speed & direction
windSpeed = ws1.getWindSpeed();
windDir = ws1.getWindDirection();

// get humidity
humidity = hs1.getHumidity();
System.out.println("Humidity = " + humidity + " %");
```

In the preceding code, the three lines in bold would be the ones you manually type in. There is usually some non-bold code to help you locate where it goes. Often there is a reference line number to help you locate the section of code. Because the actual line numbers can vary depending on how the code is formatted, use the line numbers as a general reference.

The text editor used in the software projects supports lines longer than will fit in the book. These lines are wrapped and indented like this:

```
System.out.println("Wind Speed = " + windSpeed + " MPH " +
                   "from the " + ws1.getWindDirStr(windDir));
```

When entering the code, you can type the entire statement on a single line (preferred) or maintain the line wrap and indent.

All of the source code for this book is provided on a companion web site. Rather than waste pages printing all the code, when walking through the code, portions may be skipped. This is indicated by an ellipsis (…). For example:

```
// get temperature
temp = ts1.getTemperature();
System.out.println("Temperature = " + temp + " degs F");
...
// get humidity
humidity = hs1.getHumidity();
System.out.println("Humidity = " + humidity + " %");
```

The ellipsis indicates that one or more lines of code are missing, and are not relevant in this context. If you need to see the missing code, open the file in the text editor.

In many of the software projects, the running code outputs messages to a window in the development environment. The output listings are also shown in a fixed-width font and the area of interest may be highlighted in bold text:

```
Time = Fri Apr 07 07:05:13 MST 2006
Temperature = 69.9125 degs F
Wind Speed = 0.0 MPH from the  E
Humidity = 22.485159 %
Dewpoint = 30.207876 degs F
Pressure = 30.009905 inHg
Rain = 0.0 in
Lightning = 0 SPM
Updating Log WeatherLog.txt
X10 Housecode = A  Unitcode = 1  Cmd = OFF
X10 Command Sent
```

Directories Are Folders

Some people use the term "folder"; some prefer the term "directory" when describing file path names. I'll try to be consistent and use "folder," but on occasion I might slip in "directory" just to keep you on your toes.

Path and file names are also indicated in a fixed width font such as this:

`C:/Documents and Settings/Tim/Weathertoys`

Sensors Are Not Modules, but Modules May Be Sensors

An electronic part that measures some parameter is a sensor. When you combine the sensor with some electronics to make it work, that's a module. However, because a module can also measure some parameter, it can also be a sensor, right?

As I was writing this book, I tried to standardize and use the term "sensor" for the bare electronic part and the term "module" for a packaged sensor with some electronics, but it just didn't sound right. So throughout this book I'll use the term "sensor" for an electronic device or module when it is referenced in the context of measuring some weather parameter, and the term "module" when discussing a packaged sensor.

There are special icons for Tips and Warnings:

 A Tip is a reminder or an extra tidbit of information related to the project.

 A Warning means that you need to pay extra careful attention to an item. It could be a safety issue or a risk to damaging your project.

What You Need To Use This Book

To get the most out of this book, you need a few things:

- A computer:
    - Windows-based 1GHz Pentium 4 PC or better
    - Windows 2000 or XP
    - At least 256 MB RAM
    - 250 MB free hard disk space
    - At least 1 free serial port (2 preferred; see Chapter 4)
  Or
    - Macintosh G4 or G5

- Running Mac OS X 10.3 Panther or 10.4 Tiger
- At least 256 MB RAM (512 MB preferred)
- 200 MB free hard drive space
- A USB-to-Serial Port adapter (2 preferred; see Chapter 4)

- Common hand tools
- A soldering iron
- Common electrical supplies such as electrical tape, wire cutters, and strippers. Most of these supplies are available at Radio Shack or your local hardware store.

There are a few more tools that come in pretty handy when you start building some of the weather sensors and cabling, and are described in Chapter 3.

What's on the Companion Web Site

There are two companion web sites. The publisher maintains all the source code, tools, and book updates. This can be found at www.wiley.com/go/extremetech.

All of the source code and specialized software tools are also available on my web site at www.weathertoys.net. There are also several additional resources. Here's a quick list of some of what you'll find:

- Complete source code for each project
- Completed projects for each chapter
- Special install files for Mac and Windows users
- 1-Wire tools developed especially for this book
- The source code for the ExtremeTech Weather Server, a complete weather server package for your 1-Wire weather station
- Book updates and errata
- Bonus code and projects not covered in the book
- Links to other readers' weather station web pages
- Live links to the book's resources
- Weather station pictures submitted by other readers

The World of Weather

This section of the book is an introduction to the world of weather and weather measurement. We'll start off by looking at some of the common weather measurements, the units of measure, and how the weather sensors work. You'll also learn a little history about home and hobbyist weather stations.

Next, we'll look at several types of weather stations you can build or buy. There are many weather stations on the market. Some are fully assembled and ready to go, while others are purchased as a kit. You'll also begin to learn about weather stations that are built from the ground up, the real focus of this book.

Finally, I'll finish up this section with an introduction to 1-Wire. I'll start off explaining what it is, how it works, how to hook it up, and tips for making it work reliably. You read about some of the 1-Wire devices and weather sensors, preparing you to start building your station.

Some of this information you may already know, but at the conclusion of this section, we'll both be using the same terms and you'll understand some my techno-speak. Don't worry too much if you don't understand everything presented in these first few chapters. Once you start building your weather station in Part II, most of these topics will begin to make sense.

Measuring the Weather

I've always been fascinated by measuring weather. I still remember back in the 1960s when I was a kid. I would ride my bike several miles to the downtown bank. Up on a pole was the coolest thing — a sign that displayed the temperature digitally. Keep in mind that computers hadn't been invented yet. Heck, integrated circuits weren't even invented yet. So how did they do that? Measure temperature and light up incandescent lights to display the value. How high tech!

It wasn't until about 15 years later that I built my first digital thermometer. It consisted of a simple temperature sensor connected to a digital voltage meter. It cost me about $100 to build. It sat on my desk when I wasn't showing it off to my other nerd friends (the term "geek" wasn't used yet). I would periodically look at it and think to myself, "Wow, the temperature went up 0.1 degrees!!"

Well, over the years things have changed considerably. First, digital integrated circuits became commonplace. Soon, lots of things became "digital": clocks, radios, oven timers, and even thermometers. Then computers hit the scene. Soon "digital" was replaced by "intelligent." Products could now process the data they collected. Microwave ovens know when your food is cooked. Your car knows exactly how much gas to mix with air for the optimal fuel combustion. And, yep, weather data can be collected and processed to control heating, air conditioning, irrigation sprinklers, and thousands of other possibilities.

Climate Is What You Expect, Weather Is What You Get

We've all heard the local weather forecasters on TV talking about the weather. They talk about yesterday's highs and lows, current conditions, and try to predict the weather for the next few days. They show colorful maps that show temperatures for the surrounding area and maybe a national map. If there are storms in the area, they may show a radar map highlighting conditions for the area. But what about the local weather conditions for your area? What about the weather in your backyard?

What is weather? Dictionary.com defines weather as "The state of the atmosphere at a given time and place, with respect to variables such as temperature, moisture, wind velocity, and barometric pressure." So to really know what the weather is, you'll need some way to measure the weather conditions.

In recent years, home and hobbyist weather station equipment has become very popular. I guess it might be part of the information age we live in. Many people want to know about the weather conditions in their immediate area. It could be for commercial reasons. They may have a farm or are raising livestock and need to know exactly what the local conditions are. Others, like me, may be simply fascinated by building and running their own weather station.

A Few Terms Defined

Before I start discussing measurements, I'm going to define a few terms. First off, I'm going to define what I mean by the terms *analog* and *digital*. I realize that most of the techno-geeks that are reading this are rolling their eyes. But just for clarification, I'm going to cover it anyway.

Analog refers to a value that is *continuously variable*. Remember the old-style wall clocks that you had to plug in to an outlet? The second hand moved smoothly around the clock face without any steps or distinct values. This is an example of analog. Sure, there were markings on the face, but you had to interpret the value. This was a real analog clock. Almost all weather measurements are analog. Values change smoothly and continuously. Sure, your digital thermometer converts it to digital, but as far as the sensor sees it, it is analog.

Digital, on the other hand, refers to a measurement that has *discrete*, distinct steps. Referring back to the clock analogy, a digital clock displays the minutes in 1-minute increments. It's 10:15 for a whole minute (unless your digital clock has a seconds display), and then it clicks over to 10:16.

Almost all modern weather stations convert their analog measurements to digital. The number of steps between each digital value is defined as *resolution*. For example, even though wind direction can be any value between 0.0 and 359.9 degrees, my weather station converts it into 1 of 16 possible directions. Its resolution is 360/16 or 22.5 degrees. The temperature sensor's output is in 0.5-degree steps. So its resolution is 0.5 degrees even though the display goes down to 0.1-degree steps.

In the next few pages, you will see the term *linear*. In this context, I'm referring to a sensor whose output is a one-to-one straight line in response to an input. In contrast, some sensors exhibit a *non-linear* or *exponential* response.

As long as I'm defining terms, there's one more to cover. *Accuracy* is, yep you guessed it, how accurate the measurements or sensor is. Why is this important? Some vendors advertise their weather station as being accurate to within ½ degree at room temperature. Yet at higher temperatures, the accuracy is only 5 degrees. That's a big error, especially if you live in the desert. A common mistake is to assume that because the display is digital, it must be right. Don't fall into this trap.

Some of the better weather stations provide *calibration* adjustments to allow you to adjust the accuracy. But now you have to have a reference to compare to, so you'll need to know how accurate the reference is, and it can get pretty complicated. Some of the calibration techniques are discussed in Part II.

Temperature Measurement

By far the most common and important weather measurement is temperature. It affects your everyday life: What clothes should you wear? Should you turn on the air conditioning or the heater? Do you need to run your sprinklers? Is there going to be ice on the road as you drive to work? Are your plants going to freeze tonight?

Scientific types define temperature as the amount of heat an object or the air contains, with absolute zero (no heat) defined as the state where all molecular activity stops. Therefore, heat can be viewed, in a sense, as the amount of molecular activity of an object. Although there are several measurement systems, the three most common are Celsius, Fahrenheit, and Kelvin.

In the 1700s, Swedish astronomer Anders Celsius developed a new temperature scale. He based it on two points. He defined his first point to be 0 degrees; the temperature at which pure water *froze* at sea level. His second point was 100 degrees, and was defined as the temperature at which pure water *boiled* at sea level. The Celsius scale is the standard used today throughout most the world, except for the U.S.

Just a few years before, German physicist Gabriel Fahrenheit was working on his own temperature scale, which was also based on two different points. He defined 0 degrees as the lowest temperature he could generate in his lab using ice and ammonia salts. His other point was 100 degrees and was based on what he believed to be the average body temperature (which he thought was constant). Over the years, the Fahrenheit scale has been revised slightly because the points he chose weren't constant. The two points now used (not surprisingly) are the freezing point of water at sea level (32 °F) and the boiling point of water at sea level (212 °F). On this new scale, body temperature is now 98.6 degrees. The Fahrenheit scale is used only in the U.S. and a few of its territories.

The third and less commonly used scale for weather is Kelvin. Kelvin uses the same degree size as Celsius, but is adjusted so that 0 degrees is absolute zero. This helps with thermodynamic calculations by eliminating negative numbers. You can convert degrees Kelvin to Celsius by adding 273.16. Guess what? You'll be using the Kelvin scale later in this book to calculate dewpoint from the temperature and humidity!

Temperature Conversion between Celsius, Fahrenheit, and Kelvin

°K = °C – 273.16

°C = (°F – 32) ÷ 1.8

°F = (°C x 1.8) + 32

Table 1-1 compares the Kelvin, Celsius, and Fahrenheit scales.

Table 1-1   Comparison of Temperature Scales

Temperature	°K	°C	°F
Absolute Zero	0	–273.16	–459.7
Liquid Helium	4	269.16	–452.5
Liquid Nitrogen	77	–196.16	–321.1
Freezing Point of Water	273.16	0	32
Room Temperature	294.2	21.1	70.0
Hot Day	313.16	40	104
Boiling Point of Water	373.16	100	212

Mechanical Thermometers

Changes in temperature cause many different changes in materials. From high school physics, you learned that heat causes things to expand, whereas cold causes them to contract. Heat also causes an increase in molecular activity.

Most early thermometers used thermal expansion as the basis for temperature measurement. The two examples that come to mind are glass-bulb and dial-type thermometers. Figure 1-1 shows several examples.

We've all seen glass-bulb thermometers. They're used as fever thermometers, inexpensive indoor and outdoor thermometers, and pool water thermometers. They consist of a graduated glass tube with a reservoir of liquid at the bottom, usually mercury or red-dyed alcohol. As the liquid expands, it is forced up through the tube. The temperature is read by comparing the height of the liquid to graduated markings. Most glass-bulb thermometers are evacuated and sealed at the top to prevent the liquid from being spilled or evaporating.

Dial-type thermometers use a coil of wire or metal connected to a pointer. Some of the coils are made up of two different metals that have different coefficients of expansion. As the coil heats up, it expands, causing the pointer to rotate around a graduated dial. They are generally more rugged and have higher temperature ranges than glass-bulb thermometers. Your outdoor barbeque may have one in the cover for sensing the inside temperature. Most of the mechanical home thermostats have a coil thermometer. Instead of a pointer, a small switch is attached to the coil. When the temperature reaches a certain temperature, it causes the switch to make contact, turning on your heater or air conditioner.

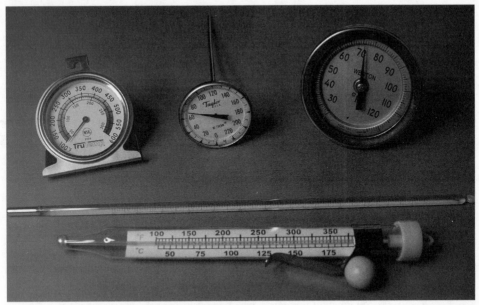

FIGURE 1-1: Glass-bulb and dial-type thermometers.

Mechanical thermometers are great if you're standing there looking at them. But what if you need to log the value? What if you need remote temperature sensing? Ah! What you need is a *temperature sensor*.

Temperature Sensors

Temperature sensors are devices that rely on the electrical changes that occur in the sensor material due to changes in temperature, rather than on mechanical changes. This causes a voltage, resistance, or current change in the output of the device. This can then be measured and displayed, or recorded with a computer. The cool part is that the sensor can be located far from the display. Temperature sensors can be scattered all over a building, with one central monitoring station. This section takes a look at several of the common temperature sensors and how they work.

Thermocouples

Thermocouples are one of the oldest temperature sensors around and are voltage-producing devices. They work on a principle that Thomas Seebeck discovered back in the 1800s. If two dissimilar metals are connected together and heated, they produce a small voltage. This voltage is proportional to temperature: the hotter the connection, the higher the voltage. Thermocouples are classified by the metals they are constructed from. A common type, Type J, is constructed of one wire made from iron and the other wire from a copper/nickel alloy. Table 1-2 lists some of the common thermocouple types.

Thermocouples have a few advantages:

- They are very inexpensive, costing only a few cents to make.

- They are extremely rugged and reliable; they are just two connected wires!

- They have a high and very large temperature range. Because they are metal, they can be used from −100 to over 1000 degrees F!

- Because thermocouples are basically two wires shorted together, they have a very low impedance, which allows them to be placed hundreds of feet away from the measuring device.

- Variation from one thermocouple to another is small. Re-calibration is usually not required when replacing with a same-type thermocouple.

Now for the disadvantages:

- The output voltage is small, usually only tens of millivolts.

- The output voltage is not linear to temperature. Special conversion tables or equations are used to convert the output voltage to actual temperature.

- A reference junction is required when connecting to the measurement system.

Because the output voltage is small, they have poor resolution, usually only 2 to 20 degrees. Thermocouples come in many different ranges, sizes, and packages. Figure 1-2 shows a close-up of a thermocouple junction. For more information on thermocouples, Omega Engineering offers just about any type and style you can imagine. It also has a great online tutorial. See www.omega.com/thermocouples.html.

Table 1-2	Common Thermocouple Types and Ranges		
Type	*Wire 1*	*Wire 2*	*Nominal Range*
J	Iron	Copper-Nickel	0 to 750 °C
K	Nickel-Chromium	Nickel-Aluminum	−200 to 1250 °C
E	Nickel-Chromium	Copper-Nickel	−200 to 900 °C
T	Copper	Copper-Nickel	−250 to 300 °C

Thermistors

Almost as old as thermocouples, thermistors were discovered by Michael Faraday back in the 1800s, and are still widely used today. The term "thermistor" comes from a contraction of the words "thermal" and "resistor." Thermistors are, you guessed it, a device whose resistance changes with temperature. Thermistors are used in weather stations, digital thermometers, temperature-sensing fans, and well, just about everywhere you need to sense temperature.

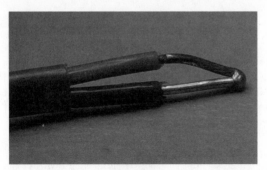

FIGURE 1-2: Close-up of a thermocouple junction.

There are two types of thermistors: Negative Temperature Coefficient (NTC) and Positive Temperature Coefficient (PTC). NTC thermistors' resistance decreases as temperature increases, whereas PTC thermistors' resistance increases as temperature increases. Thermistors come in many sizes, packages, and temperature ranges. They are characterized primarily by two factors: their nominal resistance at 25 °C, and the rate of change of resistance to temperature. Figure 1-3 shows some common thermistors. You can find a great source of information at www.ussensor.com.

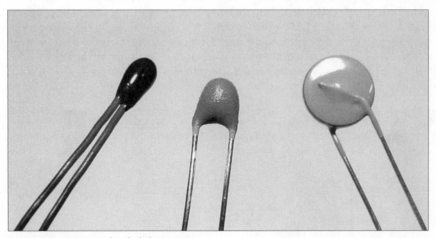

FIGURE 1-3: Common leaded thermistors.

Semiconductor Sensors

Most weather stations, including the one you're going to build in Part II, use semiconductor temperature sensors. These are usually two or three terminal integrated circuit devices that provide a temperature output as a voltage, current, or digital. Many of these devices are also calibrated at the time of manufacture, making them much easier to use. Take a look at some of the more common ones.

Junction Voltage

Although the name sounds high-tech, this is nothing more than a diode (for the real techno-savvy reader, it's technically called a bandgap reference). If you flow a constant current through a diode (or a transistor configured as a diode), the voltage drop developed across it varies with temperature. It turns out that this voltage is fairly linear in the temperature range needed for a typical weather station. Measuring this voltage and applying a simple gain and offset can convert the voltage directly to temperature. Using diodes as a temperature sensor is somewhat outdated. There are devices that measure the junction voltage and scale it for you.

Temperature ICs

This type of sensor is the most common used by hobbyists constructing their own weather stations. Semiconductor manufacturers have taken the junction voltage design and added signal conditioning electronics to provide a calibrated linear output proportional to temperature. There are many types on the market today. Some provide a voltage output, a current output, and some provide a digital output. Here are my current favorites from each category.

The LM34/LM35 are three-terminal voltage output devices. The LM34 provides an output voltage that is scaled to the Fahrenheit temperature, and the LM35's output voltage is scaled to Celsius temperature. Both devices run on 5 to 30 volts. They provide linear +10 millivolts per degree output. A typical circuit is shown in Figure 1-4.

The AD590 is a two-terminal current output device. By simply applying 4 to 30 volts, a current is developed across the device that is proportional to temperature. It is factory calibrated to provide 1 microamp per degree Kelvin. Referring back to Table 1-1, at room temperature, the current flow through the AD590 would be 294.2 microamps. Typically, the output is connected to a resistor to convert the current to a voltage for measurement. Figure 1-5 shows a typical circuit. The main advantage to using a current device is that the device can be located a considerable distance from the measurement equipment with no loss of accuracy or noise problems.

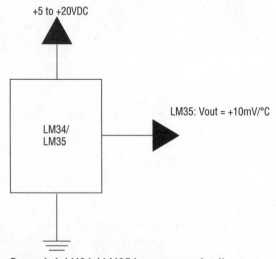

FIGURE 1-4: LM34 / LM35 temp sensor circuit.

Voltage, Resistance, and Current

Many of the sensors you'll read about have a voltage, resistance, or current output. Most weather station applications need a voltage output. To convert current and resistance to voltage, use Ohm's Law:

E = I * R

Where

E = Voltage

I = Current

R = Resistance

For example, suppose you have a device that outputs 5.0 milliamps at 50 °C. If you flow that current through, say, a 1000-ohm resistor, the voltage across the resistor would be E = 0.005 * 1000 or 5.0 volts.

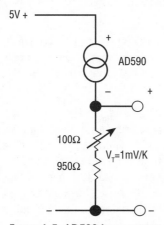

FIGURE 1-5: **AD590 temp sensor circuit.**

The DS18B20 is a direct-to-digital temperature sensor. This means that it measures the temperature and converts the output directly into a digital value. You'll need to interface it to some sort of digital circuit to read the values. In this case, the output is the Dallas Semiconductor/ Maxim IC's 1-Wire interface. This device outputs its temperature in 0.5 °C increments from −55 °C to +125 °C (−67 °F to +257 °F). Its rated accuracy is +/− 0.5 °C. Hmmm… sounds like a great weather station sensor to me! If you've looked ahead in this book, you might have noticed that the weather station you're going to build uses the 1-Wire interface. There's a whole chapter on 1-Wire coming up, so I'm not going to get into the details yet. But here's a hint: you're going to connect to this device and get the temperature digitally with only one wire. Maybe that's how they got the name…

Many digital temperature sensors are on the market today. Most have temperature ranges and resolutions that meet our needs for a weather station. The issue is connecting it to the weather station computer. Most of the interfaces are limited to just a few feet, hardly enough to get the sensor outside. If you're shopping for your own sensor, keep this in mind.

Analog-to-Digital Converters Explained

Suppose you have a sensor that outputs its value in a voltage. It could be temperature, maybe light levels, it doesn't matter. How do you measure that voltage with your computer? You'll need an *analog-to-digital converter* (called an A-to-D in techno-speak). There are many types of A-to-Ds. There are spec'd by the input voltage, conversion time, number of bits in the output, and the output interface.

- **Input Voltage:** The input range of the A-to-D must handle your expected voltage range. Suppose your temperature device outputs −9 volts at its minimum temperature and +8 volts at its maximum temperature. You will need an A-to-D that can handle that range. A-to-Ds come in several different ranges. Common ranges are 0 to +5V, 0 to +10V, and −10V to +10V.

- **Conversion Time:** This is the time it takes to convert the analog voltage to a digital value. It is usually expressed in microseconds to milliseconds, although some can take seconds. Unless you have a lot of sensors to convert, this is not a big concern for weather measurement.

- **Number of Bits:** This defines the resolution of the A-to-D. The higher the number of bits, the better the resolution. For example, suppose you're using an A-to-D that has an input range of −10 to +10 volts, and your temperature sensor outputs −10 volts at −40 degrees, and +10 volts at +180 degrees. That's 220 degrees total range. If you have an 8-bit converter, the resolution is 220/256 or 0.86 degrees. Not bad. However, if you use a 12-bit converter, the resolution is 220/4096 or 0.05 degrees. That's pretty good! Keep in mind that in practice, the sensor's output won't match your A-to-D's input as nicely as this example.

- **Output Interface:** There are many types of outputs from A-to-Ds. Most can be classified as serial or parallel. Serial A-to-Ds output their data on a single line, bit-by-bit, similar to the serial port on your computer. There are also a few control lines to signal when the A-to-D is ready. Common serial interfaces include I2C, SPI/Microwire, and 1-Wire. Parallel A-to-Ds have a data line for each bit, so a 12-bit A-to-D would have 12 lines plus a few control lines. These are generally used when connecting directly to a processor or processing electronics, and aren't too useful for building a weather station.

Humidity and Dewpoint Measurement

Humidity is probably the second most common weather measurement. It affects our comfort, how much water we give our lawns, and how fast water evaporates. It determines how effective evaporate coolers will work. In factories that work with electronics, humidity is monitored closely, because low humidity levels cause an increase in static discharge (ESD), potentially damaging sensitive electronic parts.

Several types of humidity sensors are available on the market: bare sensors, sensor modules, and sensor ICs. Most of the bare sensors require some electronics to convert their output to a usable form. This can be either a circuit board module or a sensor integrated into an IC. Humidity sensors used in weather stations usually contain electronics to convert the sensor's output to a DC voltage so that it can be applied to an A-to-D converter.

Humidity measurement devices are also called *hygrometers*. I'll use the term *sensor* when I'm talking about the device that just converts humidity to an electrical signal, and hygrometer when I'm talking about a device that measures *and displays* humidity.

When weather folks talk about humidity, what they are really talking about is *relative humidity*. Relative humidity (RH) refers to how much water vapor is in the air and is expressed as % RH. Why is it "relative?" It turns out that the warmer the air, the more water it can hold. And vice versa; as it cools, it can't hold as much water. Therefore, RH is relative to temperature.

An example of this is fog. Suppose you have an air mass that has high moisture content (80% or more), maybe it just rained, or this is moist air that blew in off of the coast. In the evening hours, the temperature starts dropping. At some point, as the air cools, it can no longer hold all of the moisture it contains (100%), and so it will condense, forming fog. By the way, the temperature point were the water condenses is call the *dewpoint*, which you learn about in just a second.

Humidity is measured in percent, with 100% RH representing air that is fully *saturated* with water vapor, and 0% humidity representing completely dry air (no water vapor). We experience 100% (or close) quite often; in fog, when it's been raining for a while, and in the shower when it's all steamy. About the driest we experience is a little less 10%, and that's on a dry day in the desert.

Mechanical Hygrometers

Swiss physicist Horace Benedict de Saussure invented the first hygrometer in the late 1700s. He discovered that certain organic substances expanded when exposed to moisture, and contracted when dried out. Guess what he used in his first hygrometer? Human hair! He attached one end of a hair to a fixed post, and the other end was attached to a lever arm and pulled gently by a spring. As moisture increased, the hair stretched and moved the arm. Until the 1960s, the most common humidity sensing material was blonde Swedish women's hair!

These days, using human hair is cost-prohibitive. Most mechanical hygrometers use a very light coiled spring that has a coating of a special moisture-absorbing material on one side. As the material absorbs moisture it expands, causing the spring to rotate. Figure 1-6 shows a common mechanical hygrometer with a close-up of the mechanism in Figure 1-7.

FIGURE 1-6: Mechanical hygrometer.

FIGURE 1-7: Internal hygrometer mechanics.

The psychrometer is another instrument to measure the moisture in the air. It uses the principle of evaporative cooling to measure humidity. As most of us know, when water evaporates, it produces a cooling effect. The amount of cooling is related to how dry the air is. The drier the air, the more cooling takes place.

The psychrometer typically uses two glass-bulb thermometers. One of the thermometers has a small cotton "sock" over the glass bulb, which is saturated with water (the "wet bulb") and the other is left uncovered (the "dry bulb"). Air is forced across the two glass bulbs, either by a fan or by twirling the Psychrometer in the air (a "sling" psychrometer). The wet sock causes the evaporative cooling effect on the wet bulb. After a few minutes, the dry bulb and wet bulb temperatures are read and a table is used to look up the corresponding humidity level. Not very high-tech, but reasonably accurate. I have a sling psychrometer that I use to check the calibration of my weather station (see Figure 1-8) without having to disconnect it from operation.

FIGURE 1-8: Sling psychrometer. Note the cotton sock over one of the glass bulbs.

Humidity Sensors

Up until about 10 years ago, there were just a couple of humidity sensors on the market. Now, like temperature sensors, there are many types of humidity sensors. A quick search on the web returns hundreds of results. Most change resistance or capacitance with relative humidity. Because they can't all be covered here, this section looks at some of the more popular ones.

Capacitive

The most common humidity sensors are capacitive and many use similar technology: A special dielectric material is sandwiched between two plates, forming a capacitor. The dielectric material used absorbs moisture and changes capacitance as a function of the moisture it contains. Typical capacitance ranges from about 160 pF at 0% RH to about 200 pF at 100% RH. One of the leading suppliers of capacitive sensors is Humirel. You can view its products online at www.humirel.com.

Capacitive humidity sensors are inexpensive, and reasonably reliable. The biggest drawback is that capacitance can't be measured directly. It has to be converted to a frequency or a voltage. This requires some support circuitry. You can build your own interface electronics, or many vendors offer support electronics for their sensor. Humirel offers a module that includes the necessary support circuitry on a small PC board along with the sensor. Either way, make sure that the circuit is properly waterproofed, especially if your sensor is mounted outdoors (which is kind of the point, isn't it?). Figure 1-9 shows the Humirel HS1011 capacitive sensors and the HMT1375 Module.

FIGURE 1-9: Humirel capacitive sensors and module.

Resistive

Resistive sensors measure the change in the resistance or impedance of a hydroscopic material. Most common resistive RH sensors use a conductive polymer or salt-coated surface with two electrodes. As the moisture level increases, the treated surface absorbs the moisture, and the resistance decreases. The relationship between RH and resistance is non-linear and requires additional circuitry to linearize the output.

The major drawback to using resistive sensors is that they require an AC voltage for operation. If there is any DC bias, the sensor will eventually polarize and become unusable.

Humidity ICs

Quite a few humidity sensors incorporate the interface electronics right in the package. I guess you could really look at these as mini-modules. Most use a capacitive sensor connected to an internal IC that provides either a voltage or digital output. Although there are quite a few to choose from, the two most popular are the Honey HIH series and the newer Sensirion SHT1x series. Both are shown on Figure 1-10.

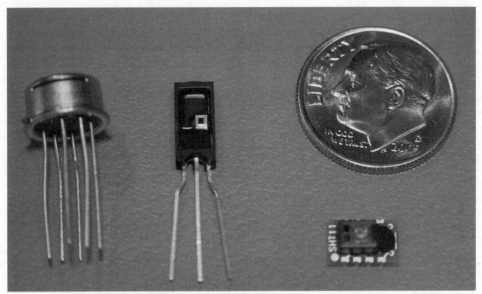

FIGURE 1-10: The Honeywell HIH3602 (left), HIH3610 (middle), and Sensirion SHT11 (right). Note the slit in the can on the Honeywell Sensor to allow air to reach the chip.

The Honeywell HIH series humidity sensor ICs are based on a capacitive sensing element connected to an integrated signal conditioner. They provide a linear voltage output proportional to RH and require between 4.0 and 5.8VDC to operate. They are a ratiometric device, which means that the output voltage is relative to the input voltage. At an input voltage of 5.0V, the output is factory calibrated for 0.8V at 0% RH and 3.9 volts at 100% RH. Accuracy of the HIH series is rated at +/– 2%, which suits most weather station applications. As you'll see in Chapter 7, there are methods to improve this value. The HIH serial devices are also available with a built-in temperature sensor.

The Sensirion SHT series humidity sensors also feature a capacitive sensor. Unlike the Honeywell device, the SHT1x provides a digital output. The capacitive sensor and an internal temperature sensor are connected to an internal analog-to-digital converter. The output is the industry standard I2C interface. These devices are designed to interface directly to a microprocessor for data collection. The SHT features 14-bit output, which provides resolution down to 0.03% RH. Depending on which model you choose, the factory calibration can vary from +/– 4.5% down to +/– 1.8%. The SHT devices operate on a power supply voltage of 2.4 to 5.5VDC.

When selecting a humidity sensor, the interface to your weather station equipment is the key factor. Using a sensor that contains internal electronics will greatly simplify your design. The drawback to all the sensors presented here is the distance you can run the wiring. Like temperature sensors, I2C interfaces are good for only a few feet. Voltage and resistance devices are generally limited to 20 or 30 feet, unless special wiring is used.

Dewpoint

Another important weather measurement is dewpoint. Like humidity, *dewpoint* is a measure of how much moisture is in the air. So why talk about dewpoint when we have humidity?

Relative humidity measurements have a significant drawback. For a given amount of moisture in the air, relative humidity changes with temperature. For example, suppose that the outside air measures 40% RH at 60 °F. Is it dry or is it humid? Forty percent doesn't seem dry. Take the same air, and warm it up to, say, 105 °F (I live in the desert, remember?). The RH would now measure only 9%. That's pretty dry air! What happened? As you learned earlier, the warmer air gets, the more moisture it can hold, and so the relative humidity goes down. Dewpoint, on the other hand, is a measure of the absolute moisture in the air, so it doesn't change with temperature. In this example, the dewpoint was 35 °F in both cases.

Here's another example. Suppose you live where it snows. Outside it measures 25 °F at about 50% humidity. That doesn't sound too dry, does it? So why is your skin chapped and your hands dry? Take the same air, move it indoors, and warm it to 75 °F. The humidity would now be 6% RH. Wow, that's dry! It turns out that dewpoint is a much better indicator of moisture in the air than relative humidity. It's just that most people are used to using humidity and don't really understand dewpoint. Keep in mind this is a simple example, and although the numbers are correct, the humidity in your house can be affected by other factors. Figure 1-11 gives you a feel for what dewpoint temperatures are considered dry, normal, or humid.

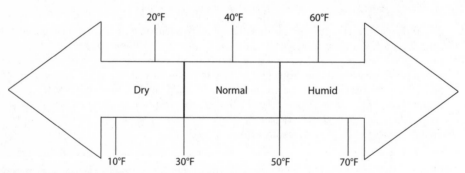

FIGURE 1-11: Dryness perception on the dewpoint scale.

Dewpoint is defined as the temperature at which water will condensate. I'm sure you've had a glass of iced tea or other cold drink collect water on the outside of the glass. That's an example of condensation. The glass temperature is below the dewpoint, and water is condensing on your glass.

Direct dewpoint measurement is complicated. One method is called the "Chilled Mirror Hygrometer." Take small mirror and bounce light off of it. Then start chilling the mirror in small steps. As some point when it gets cold enough, the mirror starts to get foggy (condensation forms) and the reflected light level drops. A temperature sensor mounted directly on the mirror is then checked to see the dewpoint temperature. Most modern dewpoint measuring devices actually measure humidity and temperature, and then calculate the dewpoint. That's the method you'll use in the weather station project.

Wind Speed and Direction

If you stop and think about it, wind is pretty amazing. Conditions in the atmosphere cause the air to *move*. Sometime it's a gentle breeze, sometimes winds can be fierce. As I'm writing this, the U.S. just went through one of its worst hurricane seasons ever. Hurricane Katrina's wind speeds reached more than 150 miles per hour. That's a powerful force.

Wind has two properties we measure: wind speed and the direction it's blowing from. Depending on your location, speed is measured in miles-per-hour (MPH), kilometers-per-hour (KPH), or nautical-miles-per-hour (knots or k). Wind direction is usually measured in degrees, with true north being the reference point of 0 degrees.

Wind Speed

To measure wind speed, you need a device that converts moving air to some electrical output you can measure. This device is called an *anemometer*. There are several types of anemometers; the most common is mechanical, but a few are ultrasonic or thermo-differential.

Mechanical

Typical mechanical anemometers consist of some type of blade that rotates as the wind blows against it. There are two types: omni-directional and directional.

Omni-directional anemometers are designed so that regardless of the wind direction, the force of the wind causes a rotational motion of the blades. Typically, these have three horizontal cupped blades. Wind blowing against the face of the cup imparts more force than wind striking the backside of the other two cups, and causes motion. A typical omni-directional anemometer is shown in Figure 1-12.

Directional anemometers usually have blades that are vertical. The challenge is to keep the blades facing the moving air. If the blades are directed such that the air is blowing from the side, there is no force to rotate the blades. The mechanism used to point the anemometer toward the wind is often a large fin. This mechanism is also used to determine wind direction. An example of a directional anemometer is shown in Figure 1-13. Regardless of the type, mechanical anemometers need to convert the rotational speed to some electrical property we can measure. Most often, this will be a voltage or frequency out.

FIGURE 1-12: Omni-directional anemometer.

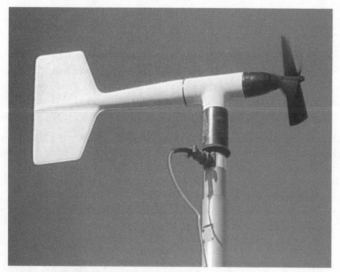

FIGURE 1-13: Directional anemometer.

Voltage

Voltage anemometers connect the rotating blades to a small AC or DC generator inside the housing. This can be as simple as a magnet mounted on the rotating shaft with a fixed coil mounted nearby. The voltage out of the DC generator or frequency out of the AC generator is proportional to how fast the generator is turning. The output can drive a small meter or can be converted to a digital signal using an A-to-D converter.

Frequency

Frequency anemometers produce a series of pulses that increase in frequency as the rotational speed of the blades increases. This is commonly done in two ways: magnetically and optically. The magnetic method attaches a magnet to the rotating blades (usually on a separate rotor in the housing) such that it passes over one or more magnetic reed switches every revolution. Figure 1-14 shows an internal view of a magnet and reed switch anemometer.

The optical method uses a light-emitting diode (LED) and photodiode positioned so the LED is shining on the photodiode. As the rotating blade or disk passes in between the two, it blocks the beam, turning off the photodiode, as shown in Figure 1-15. In either method, the number of pulses are counted in a time period. The greater the number of pulses, the faster the wind is blowing.

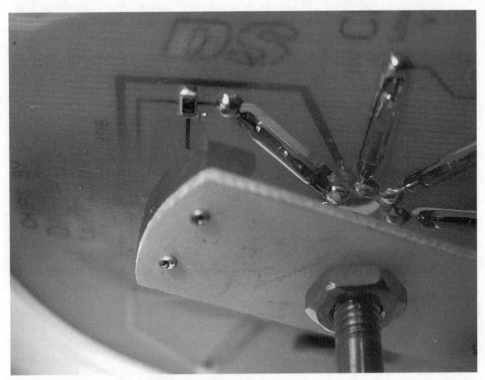

FIGURE 1-14: Close-up of a magnet and reed switch.

FIGURE 1-15: Optical anemometer. The photodiodes are on the circuit board in the foreground. Note the optical disk.

Wind speed measurements with mechanical anemometers are usually collected or averaged over several minutes. If you've been looking ahead, you may have discovered that the weather station you're going to build uses the magnet and reed switch method.

Ultrasonic

Ultrasonic anemometers work on the Doppler principle. Most of us have experienced the Doppler effect as a train blowing its horn has passed us. The horn takes a noticeable shift in pitch just as it goes by. Sound waves change pitch depending on whether the object producing or reflecting them is moving. As the object approaches, the sound waves are compressed, raising the pitch. As the object moves away, sound waves are stretched, and the sound lowers in pitch.

Ultrasonic anemometers use this principle to measure wind speed. A small speaker is mounted in a tube and emits a high-frequency (ultrasonic) tone down the tube. A small microphone mounted at the other end of the tube listens to the speaker. If there is no wind, the sound picked up by the microphone is the same frequency as sent. If there is some wind blowing down the tube toward the speaker, the sound waves are compressed as the microphone picks them up. The sound increases in pitch proportional to the wind speed. Conversely, if the wind is blowing away from the speaker toward the microphone, the sound is "stretched out" and is lower in pitch. By measuring the difference in pitch between the speaker and the microphone, the wind speed can be determined.

Ultrasonic anemometers can also be omni- or uni-directional. Just like its mechanical counterpart, the unidirectional version requires some sort of mechanical positioner to keep the ultrasonic elements pointed toward the wind. Omni-directional units use an interesting method. They have two sensors positioned at 90 degrees from each other. Taking the ratio of the measured wind speed from each sensor and applying a trigonometric calculation yields both the wind speed and direction.

The primary advantage to ultrasonic anemometers is that they can measure wind speed almost instantaneously. The drawbacks are that they tend to be expensive, and don't hold up well when exposed to outdoor weather for extended periods. They also require power to operate, which isn't always available at the weather station site.

Thermo-differential

As wind moves past an object, it "blows" some of the heat away. The faster the wind, the more cooling takes place. This principle can be used to measure wind speed. A few pages back, you read about thermistor temperature sensors. If you flow enough current through a thermistor, it will start to get warm, or "self-heat." Because a thermistor is a temperature-sensing device, you can also measure how hot it is. If you let it heat up and stabilize, then blow some wind across it, it will cool. The drop in temperature is directly proportional to wind speed.

Several different materials are used as heat-sensing elements. For example, some cars use platinum resistance wire in this fashion to measure the amount of air flowing through the intake manifold. Most thermo-differential sensors also use a second temperature sensor as a reference to determine the ambient temperature. That's where the term "differential" comes from. Thermo-differential sensors are generally very robust. However, they require considerable power to operate.

Wind Direction

Wind direction is measured by a wind vane. Wind vanes date back to the 1400s when Leon Battista Alberti, an Italian architect, invented the first mechanical anemometer and wind vane. Robert Hooke later improved the design and is often incorrectly credited as the real inventor. Early wind vanes were nothing more a large flat surface or "fin" attached to a pointer. The assembly was mounted on a pole and allowed to swivel. The fin seeks the position with the least wind resistance, which causes the pointer to face into the wind.

Modern-day wind vanes haven't changed much. Most still use a fin-and-pointer assembly as shown in Figure 1-16. The rotor is now attached to some sort of a sensor that converts the position to electrical signals. The signal is decoded to determine direction. Most weather vanes convert the rotational position to 1 of 16 possible compass points. Table 1-3 lists the 16 compass points and the corresponding degrees.

FIGURE 1-16: Weather station wind vane.

Table 1-3   Wind Direction Compass Points and Degrees

Compass Direction	Degrees	Compass Direction	Degrees
North	0	South	180
North-North-East	22.5	South-South-West	202.5
North-East	45	South-West	225
East-North-East	67.5	West-South-West	247.5
East	90	West	270
East-South-East	112.5	West-North-West	292.5
South-East	135	North-West	315
South-South-East	157.5	North-North-West	337.5

Three common types of wind vane position-sensing are used: resistive, magnetic, and optical. Each type converts the position of the wind vane to an electrical signal that is sensed by the weather station computer.

Resistive

This type of weather vane has the rotating shaft connected to a variable resistor (or potentiometer as it's called in the electronics world.) The variable resistor is connected to power, and produces a variable voltage depending on the resistance. This voltage can be connected to a meter for direct display, or to an A-to-D converter. Resistive wind vanes work well for weather stations. Because they contain only a single passive component, they require no power. A single 2- or 3-wire cable is all that is necessary to connect to a remote computer.

Magnetic

Magnetic wind vanes work similarly to the magnet and reed switch anemometers. Instead of the rotating magnet causing a single reed switch to turn on and off, multiple reed switches are placed in a circular pattern. Depending on the rotational position of the wind vane, the corresponding reed switch would be closed. The number of reed switches in the circular pattern determines the resolution of the wind direction. Typical home/hobbyist weather stations have resolution to 16 compass points or 22.5 degrees. Figure 1-17 shows a close-up view of the 1-Wire weather station circuit board. You can see eight reed switches. The unit is designed so that if the magnet falls between two switches, both are activated. This provides 1 of 16 possible directions.

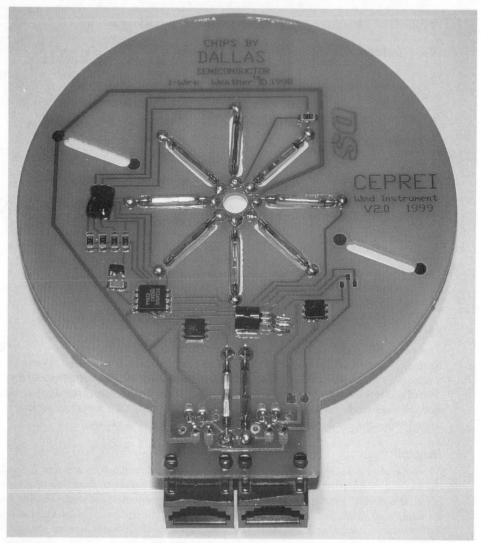

FIGURE **1-17:** 1-Wire weather station circuit board. Note the eight reed switches in the center.

Optical

Optical wind vanes use the same principle as the optical anemometers. Instead of the rotating disk blocking and unblocking light to a single photodiode, four LEDs and photodiodes are used. A pattern is machined or painted on the disk so that the photodiodes produce a unique output for each of the 16 compass points.

Barometric Pressure

Back in the 1640s, the Italian physicist Evangelista Torricelli was experimenting with vacuums. He filled a long tube sealed at one end with mercury and inverted it, forming a vacuum in the sealed end. He noticed that from day to day, the level of the mercury in the tube changed. He theorized that the changes must be caused by variations in the atmosphere, inventing the first barometer.

What is barometric pressure you ask? It is the pressure of the weight of the air above us. It is caused by the earth's gravitational pull on the atmosphere. Without this gravitational pull, our atmosphere would just float off into space. Weather causes minor variations in the pressure, which can be measured with a barometer.

Although we're used to measuring pressure in pounds per square inch (PSI), barometric pressure is usually measured in Inches of Mercury (inHg) or millibars (mb). Table 1-4 shows the relationship between PSI, inHg and millibars.

When you think of measuring pressure, you most likely think of a tire gauge or maybe a pressure gauge on a tank. These types of gauges measure the pressure relative to the surrounding atmosphere. Because the surrounding air is what we want to measure, relative gauges won't work. What we need is an *absolute* gauge.

Table 1-4   Relationship between PSI, inHg and mb

Condition	PSI	inHg	mb
1 PSI	1.000	2.036	68.94
1 inHg	0.04912	1.000	33.86
1 mb	14.51	29.54	1000
Normal Pressure	14.69	29.92	1013
Very High Pressure	15.71	32.00	1084
Very Low Pressure	13.75	28.00	948.2

Mechanical Barometers

Mechanical barometers measure the absolute atmospheric pressure by comparing it to a vacuum. A small metal can that is designed to expand and contract is evacuated. This is called the bellows. One side of the bellows is held fixed in place. The other side is connected to a lever arm to amplify the small movement of the bellows. The lever arm is then attached to a pointer on a dial. As the outside pressure decreases, the bellows expands, causing the pointer to rotate. Figure 1-18 shows a close-up of a barometer mechanism. The round can is the bellows. It is connected to the lever arm, which pushes on the platform in the foreground. The small horizontal wire is actually a small chain that is wrapped around the needle shaft and converts the linear motion to rotational motion.

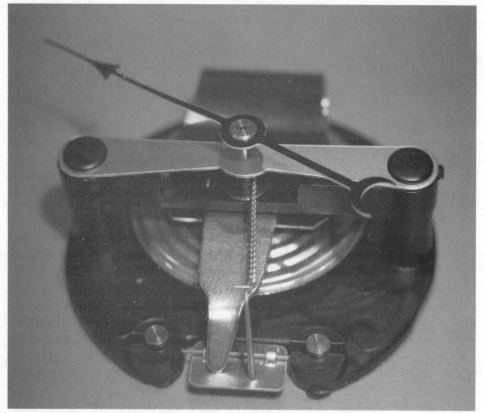

FIGURE 1-18: Mechanical barometer innards.

Barometric Pressure Sensors

Barometric pressure sensors work on a similar principle as mechanical gauges. Instead of a bellows, a small diaphragm is mounted over an evacuated chamber. The diaphragm is manufactured of a special piezo-resistive material that changes resistance in relation to stress. Variations in pressure cause variations in stress on the diaphragm, thereby causing a change in resistance.

There are two primary categories of pressure sensors: bare sensors and sensors that contain integral signal processing electronics. Bare sensors are usually a resistive bridge. DC voltage is applied to two input legs of the bridge, and a DC output is measured across the two output legs.

For weather station applications, having built-in electronics is critical. Because we are measuring extremely small changes in pressure, accuracy is critical in our design. Most pressure sensors are also sensitive to variations in temperature. To minimize this effect, devices with internal electronics contain temperature compensation circuitry, which minimizes the effects. The two most popular pressure sensors used by hobbyists are the Motorola MPC4115A and the Honeywell/SenSym SCX15AN. Both measure 0 to 15 PSI absolute pressure and are shown in Figure 1-19.

FIGURE 1-19: Motorola MPX4115A pressure sensors.

Station Pressure

While I'm discussing barometric pressure, there's one more topic to cover: station pressure versus absolute pressure. Say you're on the beach with your handy barometer and the pressure is normal at 29.92 inHg (or 1013 mb). You start driving inland a few miles and up in the hills. You look at your trusty barometer, and it reads 27.82 inHg (or 941.9 mb). What happened? Did the weather change?

Earlier, I stated that atmospheric pressure is caused by the weight of the air (at least that's how I think of it). So it makes sense that as you rise in altitude, there's less weight from the air above you, and the pressure decreases, until you leave the atmosphere where there is no pressure at all. In this example, you drove up to 2000 feet and the pressure was lower. This is the *absolute pressure*. Table 1-5 shows the effects of altitude on barometric pressure. So how do you compare barometric pressures at different locations? Weather meteorologists have devised a way. They've defined *station pressure* to be an altitude-compensated pressure measurement. When you calibrate your barometer, you will adjust it so that normal pressure reads 29.92 inHg regardless of your altitude. That way, regardless of where you are, pressure readings are all the same and can be compared.

Table 1-5   Absolute Barometric Pressure at Various Altitudes

Altitude (feet)	inHg	Mb
0	29.92	1013
1000	28.86	977.4
2000	27.83	942.3
3000	26.82	908.3
4000	25.85	875.2
5000	24.90	842.2
10000	20.58	696.8

Rainfall

As wind blows across lakes and oceans, it collects moisture. As the moisture rises in the atmosphere, it tends to cool. As you learned earlier in this chapter, the cooler the air, the less moisture it can hold. At some point, the air becomes so cold that the moisture condensates in the atmosphere, forming rain clouds. In a process that scientists aren't exactly sure about, the condensation begins to collect. As enough moisture collects in a drop, it becomes heavy, and eventually falls to earth as rain. Rain has to be at least 0.5 millimeters in size, otherwise it is considered *drizzle*.

Rumor has it that back in 1441, King Munjong of the Choson Dynasty invented the first "standardized" rain gauge. Because the amount of rain that fell in each village determined the potential for each farmer's harvest, he devised a standard rain fall collector and scale. The amount of rainfall recorded with his rain gauge was used to determine how much tax to charge the farmer.

Measuring rain is easy. Stick a bucket in your backyard, drop in a ruler, and then wait for it to rain. Getting the rain data into your computer is another matter. There are several ways to do this, but the most common are the tipping bucket and the drop counter.

Tipping Bucket

About 200 years after King Munjong designed his taxation device, Christopher Wren invented the tipping bucket rain gauge. A small calibrated "bucket" was positioned under a collection funnel. As it rained, the funnel collected the water and directed it into the bucket. When the bucket reached a certain level, it would become unbalanced and tip over, spilling the contents. As it tipped, it punched a hole in a paper tape, thereby recording the rainfall. After the contents emptied, the bucket would fall back into place, starting the collection over again.

The Standard Rain Gauge

The official rain gauge was invented more than 100 years ago. It has been used by official forecasters and weather agencies worldwide. It consists of a glass cylinder with a funnel mounted on top directing the collected water into a smaller glass tube mounted inside the glass under the funnel. The cylinder is 50 cm tall, and the funnel is 20 cm across. The tube under the funnel has a cross-sectional area exactly ⅒ the cross-sectional area of the funnel (6.32 cm) to provide a factor of 10 increase measurement accuracy.

Rain enters the funnel and drips into the lower tube. A scale mounted next to the tube is compared to the water level in the tube. The tube will measure up to 5 cm or 1.97 inches of water. If the water should overflow, the observer empties the inner tube and pours the overflow out of the glass into the tube.

Most of the weather agencies now use computerized tipping-bucket design rain gauges. But now you know the standard!

Over the years, the design hasn't changed much. Most modern designs use two "buckets" to collect water, so while one of the buckets is emptying, the other is filling. Weather station "buckets" generally hold about one tablespoon of water, so they're not really buckets, but that's what they're called. A small magnet is attached to the tipping mechanism. As the bucket tips, the magnet sweeps past a sensing device. This device then triggers a circuit that counts the tips. By counting the tips, you can now determine total rainfall (since you last reset the count) and by tracking the number of counts in a time period, you can determine the rain rate. Figure 1-20 shows a close-up of a tipping bucket rainfall counter.

FIGURE 1-20: Tipping bucket rainfall counter. Can you see the reed switch in the center?

Drop Counter

The drop counter rain gauge also uses a funnel to collect rain. The funnel has a small hole at the bottom that allows water to drip out drop-by-drop. As the drip falls, it briefly touches a pair of wire contacts or electrodes. Because rain has a slight resistance to it, electronics in the rain gauge can count each drop as it touches the electrodes.

Wrap Up

This chapter briefly touched upon many aspects of weather and weather measurement, including the following:

- Temperature
- Humidity and dewpoint
- Wind speed and wind direction
- Barometric pressure
- Rain

You also read about some of the sensors used to measure these parameters. Hopefully, you understand a bit more about the weather and how to measure it. Keep in mind I have only presented just a few samples of many different ways to measure weather. New and better ways are constantly being developed.

If you're thinking about designing your own weather station, hopefully I've given you some insight into what it takes. The biggest hurdle is how to get the data from the outside sensors into your computer. Do you incorporate some electronics in your weather station and run a single cable or do you use a separate line for each sensor? Or just maybe the 1-Wire devices piqued your interest.

As you have guessed, using 1-Wire is a good solution. You've read about the 1-Wire temperature sensor, but how do you convert the other parameters to 1-Wire? Well, you'll see how as you progress through this book. But before you start learning more about 1-Wire, the next chapter looks at a couple of the popular commercial weather stations and shows how they work.

What Kind of Weather Station Can I Build?

In the last chapter, you read about the various types of weather sensors and a little bit about how they worked. In this chapter, you start looking at complete weather stations and how they work. Two of the many commercially available weather stations are reviewed to see their capabilities and features. As you read these reviews, remember, this is not meant to sell you on a prebuilt weather station, but rather to explore the various features and capabilities that can be added to your weather station.

This chapter also introduces you to the 1-Wire weather station, the real focus of this book. You learn what it is and a little bit about how it works, and read about some of the sensors available for 1-Wire.

This chapter wraps up with some of the pros and cons of the 1-Wire weather station. It is important to understand the limitations of 1-Wire before you get started: What are the benefits? What can't you do? However, I'm sure you'll see that the advantages outweigh the disadvantages.

Consumer Weather Stations

If you do an online search, you will find dozens, if not hundreds of weather measurement devices and equipment. Some are simple little indoor-outdoor thermometers, and others are full-blown commercial weather stations. The variety of weather devices available is truly amazing.

If you haven't worked with a weather station before, then by taking a look at some of the complete weather stations available you can learn about many of the features they offer. Some of these features you may want to incorporate into your weather station. On the other hand, you may see features or capabilities that are missing and want to add them. Or maybe you just don't need all the bells and whistles and want something simpler.

This chapter starts by taking a look at two complete weather station packages: the La Crosse Technology WS-3610 and the Davis Vantage Pro2.

The La Crosse Technology WS-3610

The La Crosse Technology WS-3610 weather station presently retails for $339.95. It comes complete with all mounting hardware, instructions, and includes a Windows-based software package.

The WS-3610 consists of two parts: the outdoor weather collection equipment and the indoor base station weather display. By default, the weather data is transmitted from the outdoor equipment to the base station wirelessly. Included in the package is a 30-foot cable that allows a wired connection between the two.

Like most complete weather stations, the WS-3610 measures the following:

- Indoor and outdoor temperature

- Indoor and outdoor humidity

- Barometric pressure

- Wind speed and direction

- Rain

The outdoor equipment package consists of three modules: the main Thermo-Hygro module, the rain collector, and the wind sensor.

Thermo-Hygro Module

The Thermo-Hygro module is shown in Figure 2-1. It contains the outdoor temperature and humidity sensors, the connection to the rain collector, and the connection to the wind sensor. The Thermo-Hygro module also contains the processing electronics for the outdoor section, the transmitter to the base unit, and provides power for all three modules. This unit is powered by two C-size batteries. La Crosse claims that the battery life using alkaline batteries is 2½ years.

The Thermo-Hygro module has a removable radiation shield made of the same plastic as the housing. It slides off, providing access to the wind and rain sensor connectors and battery compartment door.

The connectors used on the Thermo-Hygro module are standard RJ-14 connectors. Although they are protected by the radiation shield, they are not completely waterproof. Wind-driven rain could reach these connectors, possibly causing problems.

Rain Collector

The rain collector is a small rectangular shaped tipping-bucket design shown in Figure 2-2. The top collecting area is 4⅞ × 2⅛ inches providing about 10.4 square inches of collecting area. This is considerably less than the NOAA collection standard of 25 square inches for statistical accuracy. The rain collector is connected to the Thermo-Hygro unit with a 25-foot modular phone type cable.

FIGURE 2-1: The WS-3610 Thermo-Hygro module.

FIGURE 2-2: The La Crosse rain collector.

Wind Sensor

The wind sensor uses a rather unique design. As shown in Figure 2-3, it uses a unidirectional wind speed sensor employing a small impeller. The wind vane attempts to keep the impeller pointed into the wind. Embedded in the impeller blades are small magnets. The magnets pass by a small sensor located on the main body of the sensor triggering the wind speed counter, requiring no electrical connection to the impeller.

The wind vane uses a rotating optical disk with LEDs and photodiodes to provide 16 possible wind directions. The wind module connects to the Thermo-Hygro module with a 25-foot modular cable.

In practical use, the wind sensor suffered from a few problems. The small size of the impeller allowed it to quickly get "clogged" up with snow. Once the sun came out, the snow quickly melted and the wind speed resumed proper operation. There were similar problems with rain: the impeller would get wet and not spin properly. Also, in lighter winds, the wind speed didn't always read correctly. The wind vane lagged in keeping the impeller positioned into the wind, causing low or no wind speed. However, this test was performed in my backyard, so out in the open where there is less turbulence it may work better.

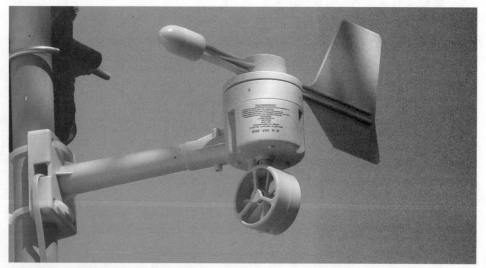

FIGURE 2-3: The wind speed and direction sensor.

The Base Station

The WS-3610 base station has a 3½ by 6½-inch touch-screen display. A switchable backlight is activated when the screen is touched, allowing you to use the base unit in poorly lit areas. Three AA batteries power the base unit.

As shown in Figure 2-4, the screen is divided into seven major sections.

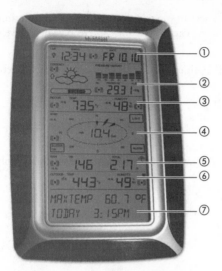

Figure 2-4: WS-3610 base station.

1. This section displays the current time and date. Like many of the La Crosse weather stations, the date and time are set automatically to the WWV time signal transmitted from Boulder, Colorado ("Atomic Time").

2. This section displays the barometric pressure, a pressure history graph for the past 24 hours, a trend indicator, and a weather "forecast" icon. The barometer is adjusted for station pressure by setting it to a known reference, such as a nearby NOAA weather station. In my experimenting, the weather forecaster works exceptionally well, predicting rain or sunny skies by displaying an icon of the forecast.

3. The present indoor temperature and humidity are displayed in this area. Both values can be set to sound an alarm if they exceed a certain threshold.

4. Underneath the indoor temperature is the display for wind speed and direction. The wind speed is shown in the center of a 16-point compass rose. As the wind direction vane moves, the display is updated with a current and average value. Wind speed resolution is 0.1 MPH

5. This section displays the rain count in two parts: the rain in the past 24 hours and the year-to-date totals.

6. Below the rain section is the outdoor temperature and humidity.

7. On the bottom of the display is a section that has a 2-line by 16-character text display. This display is used to display setup information, messages, and storm warnings.

One of the unique features of the WS-3610 is its touch-screen display. By touching the various sections shown in the figure, additional data is displayed in the text display. For example, if you touch the indoor temperature, text "buttons" appear in the text display that activate a minimum or maximum temperature display. Most parameters can display max and min values as well as alarm settings.

The touch screen also has its disadvantages. By design, the extra layer used to detect touches tends to add a "fuzzy" look to the display. It is not as crisp and easy to read as a non-touch screen display, making it hard to read from just a few feet away.

Software

The WS-3610 comes packaged with two software applications: HeavyWeather Pro and HeavyWeather Publisher. Both programs require the use of the included 6-foot serial cable to connect the base station to your computer.

The HeavyWeather Pro application displays most of the data that is displayed on the base station screen plus some additional statistical data as shown in Figure 2-5. Its primary features are as follows:

- Real-time weather display

- Maximum and minimum values

- Data logging to hard drive

- Graphing of selected weather data

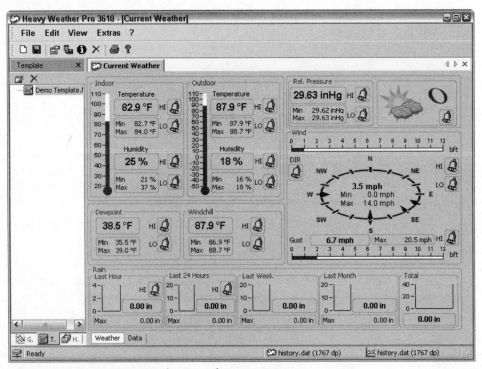

FIGURE 2-5: La Crosse HeavyWeather Pro software main screen.

HeavyWeather Publisher, shown in Figure 2-6, is a novel application that creates a weather graphic that can be added to your local web site (if you have one), or can be automatically sent to a remote site via FTP (File Transfer Protocol). The graphic can be configured to display the time, temperature, humidity, and rain. An icon for the forecast, similar to what the base station displays, can be added to the graphic to show the weather prediction.

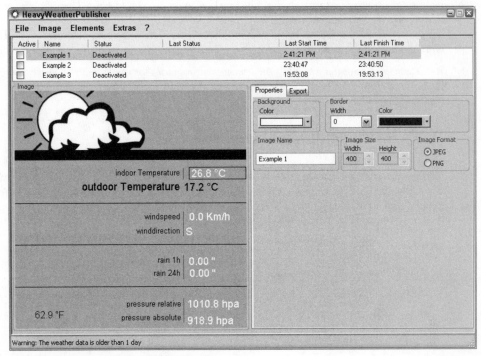

FIGURE 2-6: HeavyWeather Publisher builder screen.

Technical Features

- 433 MHz Wireless Operation
- Line-Of-Sight Transmission Range: 330 feet
- Indoor (base station) Measurement Interval: 20 Seconds
- Display Update Rate (wireless mode) 128 Second at Wind Speeds < 6.2 MPH, 32 Seconds at Wind Speeds > 6.2 MPH
- Stores up to 1750 sets of Data
- Temperature Resolution: 0.1°
- Humidity Resolution: 1%

A complete list of technical data and features are available on the La Crosse Technology web site at www.lacrossetechnology.com/.

Summary

The WS-3610 is a complete package. Everything you need to get started is included in the box: hardware, PC interface cable, and software. Installation is straightforward; however, the mounting hardware isn't very strong. The brackets to mount the Thermo-Hygro modules are inexpensive plastic. During installation, I must have over-tightened one of the included U-bolts and cracked the plastic. It was nothing that a little 5-minute epoxy couldn't fix, but I recommend being extra careful.

La Crosse does not offer any additional sensors or upgrades for the WS-3610. What you get in the box is all there is. Overall, other than the quirky wind speed sensor, the WS-3610 is a solid performer in its price range.

Davis Instruments' Vantage Pro2

The Davis Instruments Vantage Pro2 is the standard by which most other home/hobbyist weather stations are compared. Davis offers several versions of the Pro2: a wired version and a wireless version. The wireless version can be purchased with the option of a 24-hour fan-aspirated solar radiation shield. This review covers the model 6152: the wireless model without the fan-aspirated shield.

The basic weather station measures the standard suite of weather parameters:

- Indoor and outdoor temperature
- Indoor and outdoor humidity
- Barometric pressure
- Wind speed and direction
- Rain

The Pro2 has quite a few different options, including a few additional sensors:

- Solar radiation
- UV sensor
- Leaf wetness sensor
- Soil moisture detector

The Vantage Pro2 has two major parts: the outdoor Integrated Sensor Suite (ISS) and the indoor console. Depending on the model, the two may be connected with an included 100-foot cable, or wirelessly. Using the wireless option, Davis claims the console can receive data from the ISS up to 1000 feet away in a clear area. If this isn't long enough, Davis also offers repeaters for the Vantage Pro2, extending its range to almost 2 miles.

The Integrated Sensor Suite

The ISS consists of four pieces: the Sensor Interface Module (SIM), rain collector, solar radiation shield, and wind vane/anemometer. The SIM, rain collector, and solar radiation shield are integrated as one assembly, as shown in Figure 2-7. The wind vane/anemometer is connected to the SIM with a 40-foot cable. This allows you to place the wind assembly in a different location, such as higher up the mounting pole.

FIGURE 2-7: The Vantage Pro2 Integrated Sensor Suite.

The installation kit for both the ISS and wind assembly comes with two sets of mounting hardware, which allows it to be pole-mounted or screwed directly onto a mounting surface such as a wall or a fence post.

The rain collector measures 6½ inches in diameter, which is about 33 square inches, well exceeding the NOAA spec of 25 square inches. Below the rain collector is the solar radiation shield. The model reviewed did not have the fan-aspirated option. Inside the solar radiation shield is the temperature and humidity sensor. The solar radiation shield is mounted under the rain collector, which provides additional shading. In my tests, this design worked very well, providing accurate temperatures.

The electronics for processing the sensor data and transmitting it to the console are contained in the SIM. A solar panel on the front of the SIM charges an internal (replaceable) 3-volt lithium battery. As shown in Figure 2-8, the SIM has internal connectors to add the optional UV and solar radiation sensors.

FIGURE 2-8: Under the SIM cover.

The wind speed sensor assembly uses an omni-directional 3-cup design as shown in Figure 2-9. The rotor motion triggers a small reed switch located in the base. The wind vane mounts on the top to a 360° rotating potentiometer (variable resistor). If you need to calibrate the wind vane for north, you remove the vane from potentiometer's shaft and reposition it. I suggest you do this before mounting the wind vane; trying to adjust this on a ladder is not easy. You can also adjust the north calibration using the console.

FIGURE 2-9: The wind sensor assembly.

The Console

The console for the Vantage Pro2 sports 16 buttons: 12 to select functions and 4 in a cruciform shape for up/down and left/right navigation. The LED backlit screen, shown in Figure 2-10, measures approximately 6 inches wide by 3½ inches high. It is loosely divided into seven sections.

1. This section displays the time, date, phase of the moon, and the weather forecast icon. Unlike the La Crosse models, the clock does not set itself to WWV. The weather forecast icon is supplemented with a descriptive message in the text display.

2. The outdoor temperature, humidity, and barometric pressure. The temperature has selectable units (°F or °C). The barometer has a simple trend arrow that has five positions to indicate rising or falling pressures.

3. Wind direction is displayed on a 16-point compass rose, with the wind speed displayed in the center. The current wind direction is displayed as an arrow on the compass rose. The wind direction history is shown by a series of hollow arrows.

4. The inside temperature and humidity, along with a re-assignable location, are highlighted in a box near the center of the display. The heat index, dewpoint, or wind chill values can be selected for display in the re-assignable location.

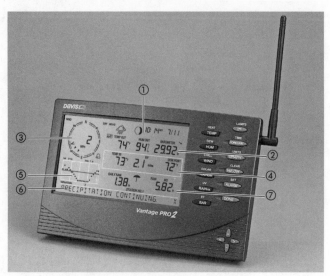

FIGURE 2-10: The Vantage Pro2 Console.

5. On the bottom left of the display is a user-programmable weather graph. By pressing one of the six weather sensor buttons, the Pro2 will display a graph of the selected variable. If the graph button is pressed, the navigation buttons change the scale or time duration of the graph. This feature allows you to see the trend graphically for any of the weather sensors.

6. The rain section has two sets of numbers. One set displays the rate amount, or the total amount for the current rainstorm. The second set can display rain rate, monthly rain, or yearly rain. Between the two sets of numbers is an umbrella icon, which is displayed when the Pro2 is currently detecting rainfall.

7. The bottom of the screen contains a 29-character text "ticker" display. In some modes the ticker displays additional weather information. For example, if the wind button is pressed the ticker displays the 10-minute wind average. Alarm settings and storm warnings are also displayed here.

Davis claims the wireless range is 1000 feet in line-of-sight or up to 400 feet in a building. In actual tests, this claim seems to be valid. The unit received data flawlessly throughout the entire house and in the yard. To judge just how well it worked, I started walking away from the ISS with the console, and I didn't lose the signal until over a half a block away.

Software

The Vantage Pro2 does not come with software, nor does the console have a built-in serial or USB port. Instead, an add-on WeatherLink hardware/software package must be purchased. There are several different WeatherLink models to choose from, with each package performing a different function. Along with each package is a special DataLogger module that plugs into a special connector on the back of the console and interfaces to your PC (see Figure 2-11). Some of the data logger modules have additional features that interface to other equipment. Here's a quick list of some of the WeatherLink models:

- Windows USB Version

- Windows Serial Port Version

- Windows Serial Port Version with APRS (Ham Radio) Output

- Windows Serial Port Version with Alarm (Relay) Output

- Windows Serial Port Version with CAMEO Support

- Windows Serial Port Version with Irrigation Control Interface

- Mac OS X USB

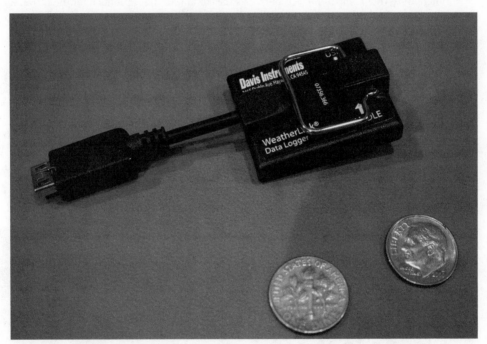

FIGURE 2-11: DataLogger module snaps into the console to provide a USB connection.

The WeatherLink software covers most of the features users would want. Although the graphics are fairly basic, the WeatherBulletin page displays the current weather conditions somewhat similar to the console display as shown in Figure 2-12. The Strip Chart mode allows you to select and plot your weather data strip-chart style. There's also a Report page where you can design you own plots, viewing multiple variables at the same time. One of WeatherLink's strongest features is the numerous ways you can share your data, including the following:

- Build your own web site page exporting graphs, text, and a ticker-tape style display.

- Send your data to Davis Instruments and add it to their list of shared weather stations around the world.

- Submit your data to the Citizen Weather Observation Program (CWOP).

- Share your data with GLOBE, an international weather data sharing program for students.

- Post your data to the Weather Underground, a worldwide weather data clearinghouse.

Overall, the WeatherLink software is a decent package. Although I feel it should be included in the console as standard equipment, the best reason to buy the package is to get the DataLogger interface to your computer. Davis publishes the communications protocol for the DataLogger module, so you can hack your own software. This opens the door to many possibilities.

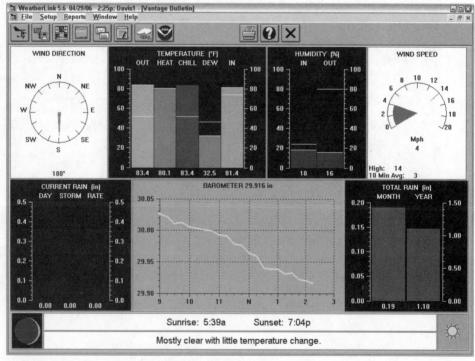

FIGURE 2-12: WeatherLink's WeatherBulletin page.

Accessories

Davis offers several accessories for the Pro2, including the following:

- Add-on daytime fan aspirated solar radiation shield
- Mounting poles and tripods
- Extension cables
- Additional consoles
- Long-range wireless repeaters

Technical Features

- 900 MHz Wireless Operation (US).
- Line-Of-Sight Transmission Range: 1000 feet.
- Display Update Rate: 2.5 seconds to 1 minute, depending on sensor.
- Temperature Resolution: Adjustable between 1° and 0.1°.
- Humidity Resolution: 1%.
- Console Battery Life: up to 9 months wireless, 1 month wired. Includes console AC adapter.
- Optional DataLogger stores up to 6 months' worth of data depending on logging interval.

This section touched upon just a few of the many features of the Vantage Pro2, and it would take many more pages to cover all the options Davis has packed into this unit. A complete list of technical data and features are available on the Davis Instruments web site at www.davisnet.com/.

Summary

The Vantage Pro2 is an awesome weather station offering many features and options. Of course, these features come at a price. Retail price for the Vantage Pro2 Wireless model 6152 reviewed here is $595. The WeatherLink software for Window or Mac with a USB DataLogger interface module is an additional $165. Repeaters to extend the wireless range start at $150. As I'm writing this, the street price of the 6152 with the WeatherLink software and USB module is about $700.

If you are serious about collecting weather data, or need a commercial-quality weather station, the Vantage Pro2 is definitely worth looking at.

The 1-Wire Weather Station

Now that you have had a chance to look at a couple of prebuilt weather stations, here's a look at what is available for 1-Wire weather equipment.

What Is a 1-Wire Weather Station?

The 1-Wire weather station doesn't come in a box with an instruction manual and software. Instead it is a collection of modules and sensors you can build or buy, which connect to your computer with cabling you build. Because 1-Wire is a standard (or rather, a protocol), several suppliers offer 1-Wire weather sensors and modules. And because the specification is public and 1-Wire parts are readily available, you can design and build your own sensors.

Most of the software to collect data from the weather sensors is designed and shared by other 1-Wire weather station hobbyists and experimenters. Over the past few years, the software has become quite mature, offering commercial-grade features at shareware prices. Because the necessary programming libraries are readily available for 1-Wire, there's always the option of writing your own.

The WS-1 1-Wire Weather Instrument

The WS-1 is the device that started it all. Back in the late 1990s, Dallas Semiconductor was looking for ways to show off the capabilities of its new 1-Wire interface. In an article published in the June 1998 edition of *Sensors* magazine by Dan Awtrey, Dallas Semiconductor introduced the 1-Wire Weather Station. It was originally offered as a kit to electronic designers to learn more about 1-Wire, and really wasn't about weather. But because of its low cost and high "coolness" factor, it quickly became the "gotta-have-it" toy for weather hobbyists worldwide. Dallas no longer sells the WS-1 and it is now manufactured by AAG.

The WS-1 measures wind speed, wind direction, and temperature. As shown in Figure 2-13, all three sensors are neatly packaged in a single device. The original WS-1 came complete with a simple weather program that displayed the data from all three sensors. All you had to do was build or buy a longer cable and mount it on a pole, and you were collecting weather data.

Over the years, the WS-1 hasn't changed that much. It still sells for about the same price, but the adapter that interfaces the 1-Wire cable is now sold separately. The internal circuitry has been updated to use some of the newer 1-Wire devices to provide higher reliability.

Note When shopping for your own WS-1, make sure it is the "Version 3" model. The original Version 1 will not work with the software in this book.

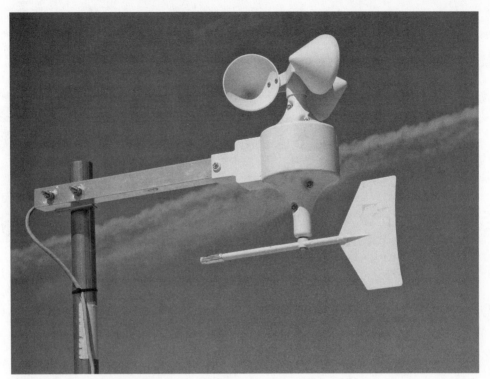

FIGURE 2-13: The 1-Wire WS-1 Weather Station.

Additional Sensors

Dallas published several additional articles that showed how to add humidity and rain measurement to the 1-Wire system. Hobbyists and hackers quickly began hacking new and improved versions, along with designs for barometers, lightning detectors, snow depth measurement, and more.

In the early days, some hackers would design a sensor, others would design the circuit board, and ideas were exchanged, all using email. As the designs solidified and the bugs were worked out, several individuals and companies began selling the designs as kits and prebuilt modules, making it easier for others to jump on the 1-Wire weather bandwagon.

In this book, you will have the option to build many of the weather sensors. Most projects include a schematic, sources for the parts, and in many cases, where to buy the printed circuit (PC) board. If you're an electronics buff, then building your own sensors may appeal to you. However, let me caution you: many of the circuits use surface-mount components. Soldering these parts requires specialized skills and equipment.

For the soldering-challenged folks (like me), you may be interested in buying completed sensors and modules. AAG and Hobby-Boards are two of several companies that offer pre-built and tested 1-Wire devices. As shown in Figure 2-14, these modules are complete with a case and are connected with easy-to-use modular connectors.

FIGURE 2-14: Hobby-Boards and AAG 1-Wire Modules.

Looking back at the features of the consumer weather station features, there were five standard sensors used:

- Temperature (outdoor and/or indoor)
- Humidity (outdoor and/or indoor)
- Barometric pressure
- Wind speed and direction
- Rainfall

These five sensors are available as 1-Wire modules that you can build or buy. In Part II of this book, you add each of these sensors to your 1-Wire weather station. Plus, there are also a couple of specialized sensors and options you can add:

- Lightning detection
- Soil moisture and leaf wetness
- Relay control
- AC power control

Software

The 1-Wire weather station doesn't come with a pre-packaged one-size-fits-all software package. Instead, you have your choice of several free or low-cost software packages. If one doesn't do what you want, you can try another. Several of these software packages are open source, giving you the option to modify it yourself.

As part of the projects in this book, you have the option to build your own complete weather station software package for the 1-Wire weather station from scratch. You will learn step-by-step how to build and run the code to run your station.

Build or Buy?

Now that you have read about a couple of weather stations and their capabilities, you can start to get a feel for what you want in a weather station. As I wrap up this chapter, the final question is should you buy a complete weather station or should you build your own? The fact that you're reading may mean you most likely already know!

The Advantages of Building Your Own

It terms of versatility, the 1-Wire system has many advantages. For example, many of the consumer weather stations only offer two temperature sensors, one indoor and one outdoor. What you if you need a third to monitor your greenhouse, or a fourth to monitor your attic temperature?

Because the 1-Wire protocol and hardware are readily available, you also have the capability to design and build your own sensors. A sampling of some of the many types of specialized sensors 1-Wire hobbyists have built include snowfall, well water height monitoring, solar radiation monitors, and more. In most cases, if you want to measure it, you can find a way.

1-Wire cabling is a simple twisted-pair cable, making it cheap and easy to work with. Most of the 1-Wire sensors are powered directly from the 1-Wire cable. This gives you the freedom to install your weather sensors hundreds of feet apart, allowing you to install the sensor in the best location. For example, you can install the wind instrument above your roof, the temperature sensor over your garden, and the rain gauge away from the trees. You can't do that with a pre-packaged weather station.

Another significant advantage is software. With 1-Wire, you can develop your own software and have the ability to customize it to your needs. Because you build it, you can hack it to your needs. You can use the software project in this book as the launch pad for your own customized software.

The Disadvantages

As you'll learn, the 1-Wire system isn't just plug-and-play. It takes some thought and planning to build a reliable network. Because 1-Wire isn't wireless, cables have to be built and the sensors need to be assembled and tested. You'll have to do some calibration of your modules and if things don't work the first time, you may have to do some troubleshooting.

The biggest reason you should consider buying a complete weather station is the wireless capability. Maybe you're renting a house or living in an apartment and drilling a hole in the wall is just not an option. The wireless capability of some of the weather stations, especially the Vantage Pro2, is amazing. In most cases, once the outdoor unit is installed, all you have to do is install batteries and you're ready to go. If this is your concern, you'll learn how to convert your 1-Wire weather station to "almost wireless" operation; you just need to have an AC outlet near your weather station.

Wrap Up

If you are still not sure that building your own weather station is right for you, look through the rest of this book and see if any of the projects pique your interest. You can ease into the world of 1-Wire weather one step at a time. Start by building the 1-Wire temperature sensor in Chapter 5. Then you can progress to the humidity sensor in Chapter 7. Once you feel comfortable, you can pick and choose the projects you want.

On the other hand, if you're ready to get started, you have one more chapter to read before rolling up your sleeves. In the next chapter you learn a little more about 1-Wire and how it works. You also learn how to build your own 1-Wire cables to connect your sensors, and some of the things to watch out for as you plan your weather station.

1-Wire Exposed

In the last two chapters, you read about 1-Wire devices and a few 1-Wire modules. You've read about how easy they are to connect and how versatile 1-Wire is. So it seems pretty logical that when building your own weather station, 1-Wire is a good choice. So the next step would be to focus a little bit more on 1-Wire.

This chapter begins getting into the nitty-gritty of 1-Wire. After a quick history lesson, you learn how 1-Wire works, and what it can and can't do. After that, the chapter looks at voltages and timing. Finally, you'll see the various types of 1-Wire networks and how to connect them.

This chapter is probably the most technical in the book, but I'll try to present the data so the non-techie will understand it as well.

Introduction to 1-Wire

In the mid 1990s, Dallas Semiconductor was looking for a new way to cheaply and easily add auxiliary memory to microcontroller designs. The existing memory devices required between 3 and 10 lines to read and write data, plus power and ground. As electronic devices got smaller and smaller, so did the number of available pins on processors and microcontrollers. What if the number of pins could be reduced to one?

Well, the engineers at Dallas Semiconductor did just that. They developed a method for reading and writing to a memory device with just one single port pin on a microcontroller. Plus, they added a bonus: That pin would also supply the power needed by the device!

The 1-Wire bus was originally designed for short distances, only 1 or 2 feet. Engineers and electronic hobbyists kept devising new applications that required longer and longer bus lengths. The Dallas engineers responded to these needs by developing better controllers and improved protocols. Today, 1-Wire networks can cover hundreds of feet and dozens of devices. But it isn't just plug-and-play. Careful consideration must be given when designing your network.

in this chapter

☑ How Does 1-Wire Work?

☑ What Kind of Wiring Do I Use?

☑ How do I Connect it?

☑ How Far Can I Go?

☑ What Tools Do I Need?

Note In 2001, Dallas Semiconductor became a subsidiary of Maxim Integrated Products. So from now on, I'll refer to this company as Dallas/Maxim.

Why Is it Called 1-Wire When There Are Really Two Wires?

When you talk about your new 64-bit computer, you're talking about the data bus. Yep, it does have a 64-bit wide data bus, but there are at least a half-dozen or so power, ground, and control lines. By convention, the power and ground lines are normally not counted when describing a data bus. Thus the 1-Wire bus is one wire plus ground.

By the way, the term *bus* refers to the wiring that interconnects the 1-Wire devices. On the other hand, *network* refers to the entire 1-Wire system: the bus, the 1-Wire devices, and the 1-Wire controller.

What's a Device Serial Number?

Each and every 1-Wire device has its own unique serial number lasered into its Read-Only Memory. This acts as the device's node address and allows multiple 1-Wire devices on a single 1-Wire network. When you want to talk to a specific device, you address it using its serial number. The address is composed of 64 bits or 8 bytes. The serial number is composed of three parts: the device's 1-byte family code, the 6-byte address, and a 1-byte CRC. Usually, device serial numbers are written as a 16-digit hexadecimal (hex) value. Figure 3-1 shows the address of a typical device.

3F	00080008F428	10
CRC	Device Address	Family Code

FIGURE 3-1: 1-Wire device serial number format.

Basic Operation

This section describes the fundamental operation of the 1-Wire bus. I'll get into some of the details of voltage levels and communication timing. Because there are many types of 1-Wire devices and applications, it can get quite complex. I'm only going to cover the basics. Don't worry if this seems foreign to you; you won't need to understand it to get your weather station up and running, but it is kind of fun to know about.

Parasitic Power

Most 1-Wire devices are parasitically powered. In other words, they steal their power from the 1-Wire bus. The 1-Wire temperature sensor is a good example. With just one wire (and ground), the device is powered, receives commands, and transmits back the temperature. But where does the power come from? Who does the 1-Wire temperature sensor talk to? That's where the concept of master and slave come into play.

Masters and Slaves

Each 1-Wire network must have one, and only one, master. The master is responsible for all aspects of the 1-Wire network. It provides power for the devices, initiates communications, receives the reply, and interfaces with the host computer or processor. Sometimes the master is the host processor itself, and is connected directly to the 1-Wire bus. In other applications, the host processor talks to a secondary device, such as the DS9097U serial port adapter, which in turn controls the network. A basic 1-Wire (or MicroLAN as Dallas/Maxim calls it) is shown in Figure 3-2.

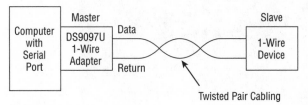

FIGURE 3-2: Basic 1-Wire network.

In the weather station application, the host processor will be your PC. It communicates using weather software (which is covered later) to a serial port, which is accessible outside the computer's case (you probably wouldn't feel comfortable soldering wires directly to your PC motherboard, would you?). The serial port is then connected to the 1-Wire bus through a Dallas/Maxim DS9097U serial to 1-Wire adapter. Inside this adapter, there is a 1-Wire interface integrated circuit that acts as the bus master. It controls the bus, supplies power, and provides two-way communication. The cool part is that the DS9097U is itself a parasitically powered device. It steals its power from your serial port. There are no batteries or power supplies required for most 1-Wire devices. However, when you start adding additional electronics (such as the pressure sensor in the barometer you'll build), there just isn't enough power to run the entire circuit. In those cases, external power is required. Figure 3-3 shows a DS9097U adapter and Figure 3-4 shows the internal circuitry.

Each 1-Wire device is considered a slave. Slaves cannot initiate communication or talk directly to another slave. All communications go directly to the master. You can have many slaves on a bus; the total number of slaves possible is limited by several factors. To give you a feel of the capability of 1-Wire, on a short bus you can have more than 100 devices! In most cases though, I suggest you plan on keeping the number to 20 or less. Once you exceed about 20 devices, loading and other factors begin to degrade the network reliability.

FIGURE 3-3: The DS9097U 1-Wire serial adapter.

FIGURE 3-4: Internal view of the DS9097U.

Voltage Levels and Time-slots

1-Wire buses can operate from between 3.0 and 5.5 volts. However, in your application, potential voltage drop in the cabling requires at least 4.5V at the master. This is usually controlled by the bus master. For example, the DS9097U has an internal 5V regulator that supplies power to the bus.

In its normal state, the bus is *idle* (no communications) and sits at 5V. This provides the power for all the devices on the bus. During communication, data is transmitted on the bus by a series of negative going pulses. If the pulse is below 0.8V, that is considered a logic low. If the bus voltage is greater than 2.2V, that is considered a logic high. Notice that if you transmit a logic low for too long, all the devices on the bus lose power. Not a good thing if you're in the middle of a communication sequence. To prevent this, there is a well-defined timing requirement for all data sent. See the sidebar "For the Techie…" to understand more about 1-Wire timing.

You may be wondering how the slaves stay powered if the bus goes low during a data transfer. Within each 1-Wire device there is a diode and capacitor configured as shown in Figure 3-5. The capacitor is charged while the bus is high. When a logic-0 pulse is sent on the bus, the charge in the capacitor powers the device during this brief period. The diode prevents the charge on the capacitor from discharging back through the bus. As you may have guessed, this can't handle low periods greater than about 1 millisecond, or the device loses power.

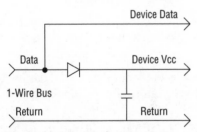

FIGURE 3-5: Parasitic power circuit.

Communication and Device Addressing

As you just learned, the master initiates all communication. Figure 3-6 shows a complete 1-Wire communication sequence. The sequence is divided into three sections: Reset and Initialization, ROM Command, and Function Command. ROM commands are standard commands that are present in all 1-Wire device Read-Only Memories. Function commands can vary from device to device. You will have to look at a specific device's data sheet to see what functional commands it supports.

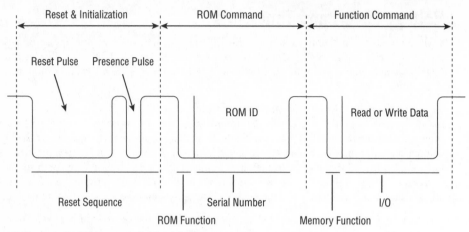

FIGURE 3-6: 1-Wire communication sequence.

For the Techie...

The 1-Wire bus is a wired-AND bus. The master has a fixed pull-up to +5V. The master and slaves can only drive the bus low. Therefore, if one or more devices drive the bus low, the master can't tell. All it sees is a low.

Because the bus master controls all communication, it either sends data to the bus (writes) or gets data from the bus (reads). The data on a 1-Wire bus is sent in "time-slots." There are two types: a "Write time-slot" and a "Read time-slot."

To send data to a device, the master can either send a "write-1" time-slot to send a logic 1 or a "write-0" to send a logic 0. Each time-slot is 60 microseconds wide. To generate a write-1 time-slot, the master pulls the bus low for about 10 microseconds, and then releases the bus allowing it to go high. To generate a write-0 time-slot, the master pulls the bus low and holds it for the full 60 microseconds, and then releases it.

To read data from a device, the master pulls the bus low for a minimum of 1 microsecond and less than 15 microseconds, and then releases it. The currently selected slave device responds by either leaving the bus high for logic 1 or driving the bus low for 45 microseconds for a logic 0. The total read time-slot is 60 microseconds.

If you want to know more, you can find the complete details in the 1-Wire application notes at Dallas/Maxim's web site at www.maxim-ic.com.

Reset and Initialization

All communication sequences start with a logic 0 Reset command issued by the master. The reset pulse must be low for 480 microseconds minimum. This causes all 1-Wire devices on the bus to reset to their initial state. Next, the master releases the bus (logic 1) and 15 to 60 microseconds later, all devices on the bus respond with a logic low presence pulse. If the 1-Wire master doesn't see a device presence pulse, it knows there are not any devices on the bus. After the presence pulse, all devices on the bus start listening for commands.

ROM Command

After the Reset pulse, the master issues a ROM command. There are several standard ROM commands:

- **Search ROM:** As part of the 1-Wire interface, Dallas/Maxim has provided a way to detect 1-Wire devices on the bus and read their unique serial numbers. This command puts all 1-Wire devices on the bus in serial number discovery mode.

- **Read ROM:** If there is only one slave device on the bus, this command is used to read its address.

- **Match ROM:** The Match ROM command followed by the 64-bit address of a slave device addresses a specific device on the bus. This is the most common ROM function. After a device receives its serial number, it begins waiting for a memory function. All other slaves stop listening.

- **Skip ROM:** This command is used to address all devices on the bus.

Memory Function

After the ROM function, the master can send a memory function command. Only the addressed device will respond. This is how the master communicates with a specific slave device. Different 1-Wire devices have different memory functions, but all support at least two commands:

- **Read:** After issuing a Read Function command, the master can read one or more bytes from the addressed device.

- **Write:** After issuing this command, the master can write one or more bytes to the addressed device. Each device has its own set of memory functions. For example, the DS18B20 temperature device has a Convert Temperature command that initiates a single temperature conversion. After the conversion is complete, the master can initiate a new communication sequence, this time with a Read command to retrieve the 2-byte temperature.

1-Wire Devices and Family Codes

Last time I checked, Dallas/Maxim's web site had more than 50 types of 1-Wire devices listed. Rather than go through all the devices, this section focuses on the ones you going are to use to build your weather station.

Previously, I told you about device serial numbers and family codes. As part of the weather station construction steps, you will connect one or more 1-Wire devices and run a utility that will search the bus to find their serial numbers. But how will you know which device is which? You will use the family code that is located in the device serial number. Table 3-1 lists the weather sensor type, device, and family codes for the 1-Wire devices you will use to build your weather station. When you start adding devices to your 1-Wire bus, you can refer back to this page to determine the device type.

Table 3-1   Weather Sensor Family Codes

Sensor	Type	Part Number	Family Code (Hex)
Temperature	Temp Sensor	DS18B20	10
Wind Speed	Counter	DS2423	1D
Wind Direction	Quad A-to-D	DS2420DS2450	20
Humidity	Battery Monitor	DS2438	26
Pressure	Battery Monitor	DS2438	26
Rain	Counter	DS2423	1D
Lightning	Counter	DS2423	1D
Moisture	Battery Monitor	DS2760	30

1-Wire Networks

The most important aspect of designing your weather station is the connection of the devices. Choosing the right connectors, cabling, and network layout are critical. The rest of this chapter focuses on how to set up your 1-Wire network and make it reliable. For best operation, 1-Wire networks should not be free-form ad-hoc designs. Rather, they should be given careful consideration before running the wires. The following section starts by looking at the wiring.

1-Wire Cabling

Simply put, the cable you use to connect your weather sensors is the most important aspect of your system. Poor quality cabling is the number one reason for intermittent problems. I've helped dozens of 1-Wire weather hobbyists whose weather station wasn't working reliably, and in most cases it was the cabling.

Wire has many properties. There's resistance in the wire, capacitance between conductors, and inductance of the line. In Figure 3-6, you saw nice clean square waves. In reality, these signals are not this clean. The combination of resistance, capacitance, and inductance all factor into distorting the 1-Wire waveform. If the waveform becomes too distorted, the master or slave can no longer read the bus correctly, resulting in communication failure.

The theoretical limit to a 1-Wire bus is 750 meters or 2460 feet. That's almost half a mile! Notice I said theoretical. That's one master, one slave, and perfect cable. In practice, you shouldn't plan on exceeding about 200 feet. Why? First off, you'll have a hard time finding "perfect" cable. Then you're going to add several imperfect connectors. Next you'll add several slaves with additional electronics powered off the bus, and finally, you're going to install most of this stuff outdoors, which introduces several other factors. The end result is potential problems exceeding this recommendation.

So it's time to look at the cable. You will use two types of cable: Long lengths to run to your outdoor weather and short lengths to interconnect devices.

Long-Length Cable

Ultimately you are going to install most of your weather station equipment outside and want to connect it to your computer inside. Maybe you're mounting on the roof or maybe you're mounting it on your 50-foot radio tower. That usually requires a considerable length of cable. The de-facto standard cable for these lengths is Category 5 unshielded twisted pair. You can find it at most computer and hardware stores. It is the same cable used in computer networks. It's commonly just called CAT5 or CAT5e.

CAT5 cable comes in several flavors and colors. For your application, get the UV Protected Outdoor flavor (but you can choose your own color). If you strip off the outer insulation, you'll discover that there are actually four pairs of wires as shown in Figure 3-7. You will only use one of the four pairs. By convention, this will be the Blue/Blue-White pair.

FIGURE 3-7: CAT5 Cable 4-Twisted Pair Cable.

When using CAT5 cable, the capacitive coupling to the other pairs can have a significant effect on your 1-Wire bus. To minimize this, you shouldn't connect any of the remaining pairs as shown in Figure 3-8. Some of the 1-Wire sensors you will build in Part II have an external power option that is supplied through the connector. If you use the other pairs in your connection, you would be inadvertently connecting the pairs to power or ground.

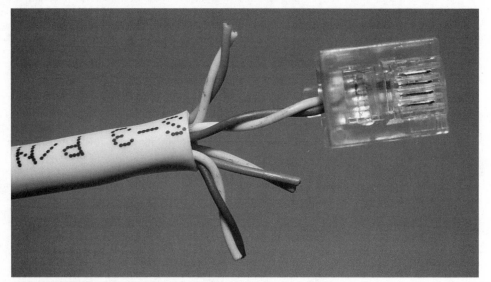

FIGURE 3-8: Using a single pair with an RJ-14.

Note   On shorter runs of 50 feet or less, using the additional pairs to provide power may work. If you need to run power through your bus, go ahead and give it a try. If you experience communication problems, try removing the power connections to see if it clears up.

Why Twisted Pair?

Whenever a wire is sticking up in the air, it picks up noise (Electro-Magnetic Interference or EMI) such as noise from fluorescent lights, radio and TV transmissions, and noise from electrical appliances. The longer the wire, the more susceptible it is. In a typical two-wire circuit like you'll have, if the noise is coupled equally in both wires, the noise tends to cancel.

If the wires aren't twisted, one of the wires is slightly closer to the source of the noise, so it picks up more noise and blocks the noise somewhat from the other wire. By twisting the wires uniformly, the noise gets induced equally, thus providing better cancellation.

Short-Length Cable

CAT5 cable can sometimes be a little difficult to work with when connecting directly to your weather station sensors. Because it has four pairs, it's somewhat big in diameter and pretty stiff to bend. You'll read about it in more detail in Part II, but a common practice is to use CAT5 to run to a junction box (or passive hub) near the weather station, then use more flexible cable to run to each sensor. For these shorter-length cables, I've found two that seem to work well. Most larger hardware stores carry CAT3 2-Twisted Pair Wire (mine's Carol part number C4413) and Radio Shack sells two-twisted pair phone cable model number 279-460 (don't get the four Conductor Modular phone cable, it's not twisted). I prefer the Radio Shack cable because it is stranded, which makes it more flexible and easier to solder.

Keep in mind that these cables are not as good as CAT5. The longer the combined length, the more susceptible you are to potential 1-Wire bus problems.

Connectors

At some point, you're going to have to connect your cable to your 1-Wire serial port adapter and weather sensors. This section looks at some of the common connectors used.

RJ-11/RJ-14

Originally, 1-Wire weather sensors used RJ-14 modular-style telephone connectors because they were used on the DS9097U and the 1-Wire Wind Instrument. These work fine indoors, but hobbyists learned that these connectors didn't do so well outdoors if they were not protected from moisture.

 Note What does RJ stand for? Per Wikipedia.com, RJ stands for Registered Jack. These jacks (and plugs) are registered as part of the United States Code of Federal Regulations. RJ-11, RJ-14, and RJ-12 plugs all have six positions. The RJ-11 uses only the middle two, the RJ-14 uses the middle four, and the RJ-12 uses all six.

RJ-45

The RJ-45 connector is similar to its RJ-11 cousin, just a bit bigger. You've seen an RJ-45 on the end of your network cable. They are designed for eight conductors. Because the contacts are a bit bigger than the RJ-11, some hobbyists have switched to RJ-45s. The Hobby-Boards weather modules presented in Chapter 2 use RJ-45 connectors. Figure 3-9 shows an RJ-14 side-by-side with an RJ-45. Figure 3-10 provides a reference to the location of pin 1 on the plug. With the connector pins (contacts) facing up, pin 1 is on the left.

Screw Terminals

Screw terminals or barrier strips offer a more secure connection than RJ-11s or RJ-45s. Most have good separation distance between the connectors, which offers superior moisture resistance. Many of the Hobby-Boards modules also include screw terminals.

FIGURE 3-9: RJ-14 (left) and an RJ-45 (right).

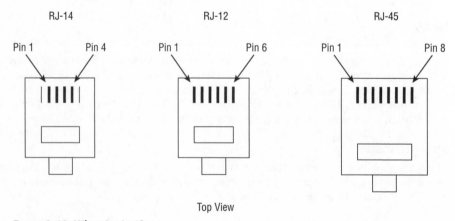

Top View

FIGURE 3-10: Where's pin 1?

Phone Wire Splices

Short of soldering the wire, phone wire splices or crimp connectors are actually one of the best ways to splice 1-Wire network cabling. You simply insert up to three wires and squeeze with a pair of pliers. The crimp pierces the wire and makes a secure connection. There are two types: standard or indoor, and the outdoor weather resistant type that are filled with a silicon dielectric grease. These form a waterproof seal around the wire. Figure 3-11 shows an example of a phone wire splice. The drawback to using splices is that they are permanent. You will have to cut them off to remove them. Telephone splices are available at most hardware stores.

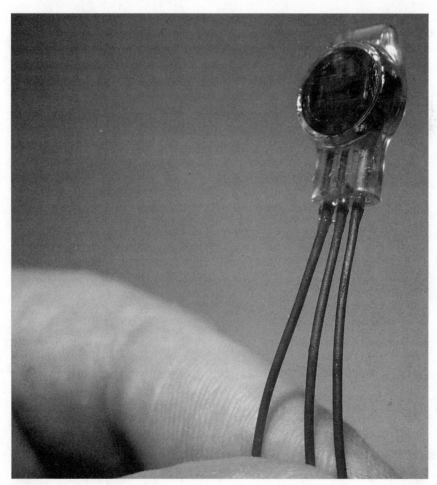

FIGURE 3-11: Phone wire splice.

Connector Pin Assignments and Color Code

There is no "official" standard for 1-Wire connectors or pin assignments. The RJ-14 scheme used today started when Dallas/Maxim released the DS9097. It used pin 2 for data and pin 3 for ground. Most of the 1-Wire modules that use RJ-14s follow this scheme.

As with 1-Wire pin-outs, there is no standard for color-code in the wire, and there are two different wire-coloring schemes. For twisted-pair cable, the "old" style uses solid-color wires, whereas the "new" style uses one solid-color and one white-strip wire of the same color to denote a pair.

Each of the project chapters provides the pin-outs for each of the modules presented. Most modules use the same data and return pins. Figure 3-12 lists the common connectors, pin-outs, and color-codes. Be careful, however, because different modules may use the pins for different functions.

N/C	Black	Red	Green	Yellow	N/C
N/C	Orange	Blue	Blue/White	Orange/White	N/C

		1	2		
	1	2	3	4	
1	2	3	4	5	6

RJ-11 (pins 1–2)
RJ-14 (pins 1–4)
RJ-12 (pins 1–6)

2-Pair	N/C	N/C	Black	Red	Green	Yellow	N/C	N/C
4-Pair	Green/White	Green	Orange/White	Blue	Blue/White	Orange	Brown/White	Brown
	1	2	3	4	5	6	7	8

RJ-45

FIGURE 3-12: Common 1-Wire pin assignments.

Network Types

The other important aspect of your 1-Wire network is the physical layout or topology. Haphazardly connecting your 1-Wire devices can lead to an unreliable network. There are many ways to connect your devices, and this section presents a few of the common topologies.

Linear Networks

The *linear* network is the most common and most reliable. Each device is "daisy-chained" to the next. Many of the 1-Wire modules and kits have two connectors for this purpose. Figure 3-13 shows a block diagram of a PC with the DS9097U and barometric pressure module inside the house, and the temperature, humidity, wind, and rain sensors outside, connected in linear fashion.

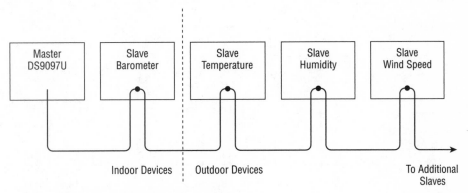

FIGURE 3-13: Linear or "daisy-chain" network.

Star or "Y" Networks

Another common scheme is the star network. Multiple cables from the sensors are all run to a junction box at or near the DS9097U as shown in Figure 3-14. Each line is called a branch. Star networks are more sensitive to cable lengths and loading than the linear network.

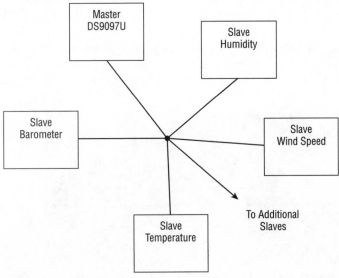

FIGURE 3-14: Star network.

Stubbed Network

In a stubbed network a main line or backbone is run from the master along a path to the last weather device. A short stub is spliced into the backbone and run to each weather sensor as shown in Figure 3-15. Stubbed networks are not as reliable as the linear or star because each stub generates a "reflection" of the bus signal. When stubs of different lengths are used, multiple reflections are generated at slightly different times, which can interfere with normal bus traffic.

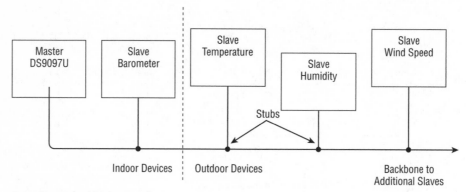

FIGURE 3-15: Stubbed network.

1-Wire Hubs

There are limits to the number of 1-Wire devices and cable lengths used in a bus. Additional electronic devices that are parasitically powered off the bus, such as the humidity IC in the 1-Wire humidity module, can cause additional loading. The folks at Dallas/Maxim have anticipated this, and developed a 1-Wire MicroLAN coupler, which acts as a switchable hub. Under software control, each port is switched on to communicate with devices on that leg, thereby isolating the other legs.

The weather station you will build in Part II does not include a hub. This shouldn't be a problem unless you use very long cables or many stubs. If you find that you need a hub, Hobby-Boards and AAG offer a 1-Wire hub like the one shown in Figure 3-16. See the resources list in Appendix A.

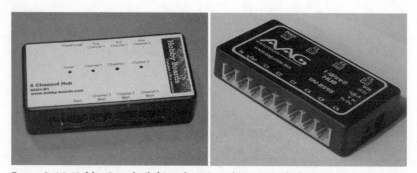

FIGURE 3-16: Hobby-Boards (left) and AAG (right) 1-Wire hub.

What's an iButton?

If you've looked around on Dallas/Maxim's web site you may have noticed in the 1-Wire section there is this thing called an iButton. What is it, you ask? Because many of the 1-Wire devices have only two terminals, the Dallas/Maxim folks came up with a cool idea. Why not package the device in a small sealed container? It would be rugged, and easily transportable. So that's exactly what they did: iButtons are 1-Wire devices in small, coin-cell size packages. Figure 3-17 shows a typical iButton.

You may be wondering what good a portable 1-Wire device is. So far, this book has pretty much focused on weather applications. However, many 1-Wire devices are non-volatile memory ICs. Take a memory device and add a unique serial number, and now you have a custom security device. Think of it as a one-of-a-kind key that has memory. There's a whole web site dedicated to iButtons and their applications. Check out www.ibutton.com.

FIGURE 3-17: iButtons: 1-Wire devices in a can.

Cool Tools

When working with 1-Wire modules and wiring, there are a few tools you just have to have. Other than standard tools such as screwdrivers, pliers, wire cutters, and wire strippers, you will need the following tools.

RJ-11/RJ-14 Crimper

The DS9097U and the 1-Wire anemometer use RJ-14s, and so do all the AAG modules. Because you want to keep your cable lengths as short as possible, making custom cables is a must. RJ-14 crimpers and connectors can be found at most hardware stores and Radio Shack.

I've seen cheapie crimpers for less than 10 bucks. The crimp tool is how you attach the connector to the cable, pushing the little connector pins into the cable.

RJ-45 Crimper

If you are going to use the RJ-45s on the Hobby-Boards modules, you'll also need a good RJ-45 crimper. These aren't quite as common as the RJ-14 crimpers, but I've seen them at Home Depot and Lowe's for around $20 or more. Figure 3-18 shows the crimp tool I use. It has a different set of jaws and can crimp both RJ-14 and RJ-45 connectors.

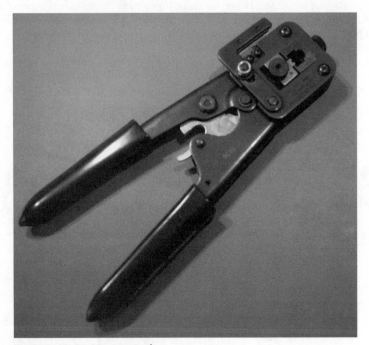

FIGURE 3-18: RJ-45 crimp tool.

1-Wire Test LED

When working with 1-Wire, a common problem is keeping the polarity correct. Here's an easy tool you can build that will help you watch the pluses and minuses. Simply use your new RJ-14 crimper and crimp a two-color LED into the two center pins of an RJ-14 connector as shown in Figure 3-19. The LED will have to be the two-pin type such as the Radio Shack 276-012. You'll need to cut the ends of the pins evenly so they both slide into the RJ-14 evenly. Slide the anode (+) side of the green LED into pin 2 and the other lead into pin 3.

For the LED to light, you'll have to power your 1-Wire bus. You can do it using the OneWireLister program or SimpleWeather, which you learn about in Part II.

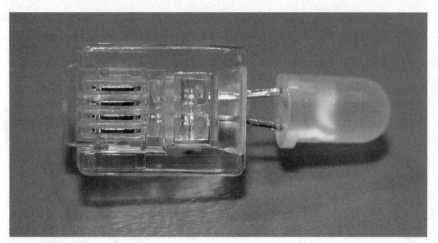

FIGURE 3-19: 1-Wire test LED.

Plug your 1-Wire tester into the DS9097U and run the OneWireLister program. After you select your serial port and click OK, the green LED should light up. As you add new cables and modules, simply plug in the 1-Wire testing and make sure it lights up green. If the LED doesn't light up at all, you'll know you have a disconnected wire somewhere. If the LED lights up red, you know you've got your wires crossed. You can also build the LED tester using an RJ-45 plug. Connect the green LED anode (+) to pin 4 and the cathode (–) to pin 5.

Digital Voltmeter

These days, any serious hobbyist has to have a digital voltmeter to measure voltage and resistance. I've seen them at hardware stores, online, and at Radio Shack for under $20 to over $300. You'll have to decide how much to spend. The test LED described previously will help check the basics, but any serious troubleshooting requires a digital voltmeter.

ESD Warning

Before I end this chapter, I need to caution you about Electro-Static Discharge, or ESD. If you're an old-hat at working with electronics, you probably already have an ESD mat and wrist strap. If not, read on...

We've all experienced a static shock at one time or another. Living here in the desert where the air is usually dry, I get zapped all the time (four times already today!). As you walk across the carpeting or slide out of your polyester-blend fabric chair, literally thousands of volts of electricity are generated. Most of the time, this charge is slowly dissipated and you never notice it. Sometimes this charge dissipates, you touch a metal object, and you get a small shock. That's ESD and it can damage your 1-Wire devices instantly!

Whenever you work with 1-Wire devices or modules, you should always wear an anti-static wrist strap. Anti-static wrist straps go around your wrist and have a wire that gets connected to earth ground. Any static charge that is generated is quickly and safely discharged to ground. This prevents a charge from building up and zapping your electronic devices. Figure 3-20 shows an anti-static wrist strap and cord.

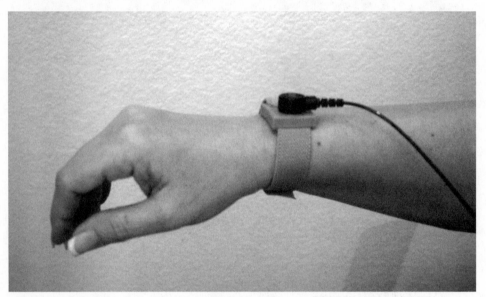

FIGURE 3-20: Anti-static wrist strap. The wire on the right is connected to earth ground.

Wrap Up

By now you should have at least a basic understanding of 1-Wire devices, what the 1-Wire Protocol is, and 1-Wire networks. Once again, just a few pages have covered a lot of ground. There is tons more to learn more about 1-Wire and how it works. If you're interested, pick a 1-Wire device (the DS18B20 Temp Sensor is a good one) and download the full data sheet from the Dallas/Maxim web site. You'll find it is chock full of 1-Wire info. There are also more than a dozen application notes you can download to learn more.

This is the last of the "tutorial" chapters, and I bet you're ready to roll up your sleeves and start building. I hope you have found these first few chapters informative. The goal was to give you enough insider info so that you can plan and build your own weather station without too many problems. Let's get started!

Build a Weather Station

In this section, the real fun begins. You'll build a complete weather station that you can customize to your needs. But what good is a weather station without software you ask? We'll be building that too. Even if you never have written software before, you can follow along, or use the already-built code on the companion website. If you are a seasoned programmer, then you can use the code presented as a launch pad for your own code.

First, we'll start off by setting up the software project, installing an Integrated Development Environment (IDE), and coding the software framework.

Next, I'll show you how to assemble the basic weather sensors. You can build most of the sensors from scratch, buy a kit, or buy a fully assembled version. You also get to pick and choose which sensors you want.

Each hardware project has a corresponding software project that will show you step-by-step how to add the code for each sensor.

Finally, there's a chapter on installing your new weather station. Be sure to read this section for important safety tips before you install your weather station outdoors.

I have teamed up with Hobby-Boards.com to offer you many of the hardware projects presented in this section as kits or fully assembled modules. Hobby-Boards offer these high-quality kits at very reasonable prices. I suggest you scan through this section and pick a few sensors to start out with. A good system to start with is the 1-Wire Weather Instrument, the Temperature and Humidity module, and the DS9097U 1-Wire to Serial Port adapter. This will give you the ability to measure temperature, humidity, wind speed, and wind direction.

On the other hand, if you want to start out as cheaply as possible, order just the temperature sensor IC and the DS9097U adapter and build your own temperature sensor. Regardless of what you choose, at a minimum, you will need at least one sensor and the 1-Wire Serial Port Adapter.

For the do-it-from-scratch hobbyist, Hobby-Boards also offer individual parts and PC boards. Appendix A lists several other sources of 1-Wire parts and accessories.

Building the Software Framework

Before you start building the weather station hardware, I'm going to show you how to set up and code the framework that you will use to run your weather station. Most of the tools and code are available on the companion web site. First, you will download and install the Sun's free Java Software Development Kit (SDK). Next, you'll install NetBeans, a free and really cool Java Integrated Development Environment (IDE). This is where you will write and run your code. There are also a few special Java Libraries needed to access your new hardware. I'll show how to install these too.

The software framework presented in this chapter is the basis for all the software projects in this book. Each chapter builds upon this framework, adding new features one at a time. By the time you reach the end of this book, you will have the intimate details of how the software works, making it easy to modify the code.

If you've never written software before, I encourage you to follow along. Java is an awesome language to learn. All of the development tools and documentation are distributed free at Sun's Java web site. There's also an excellent free online tutorial for learning Java.

What Do I Need?

This is primarily a software project. But all software takes some hardware. Other than your computer, you'll have to have at least one available serial port. Many PCs still have serial ports. If yours doesn't, you'll need to buy a serial port card. They are only about $20 at most computer stores. As long as you're buying one, make sure to get a dual serial port card. Some of the projects in Part III will use the second port.

If you're a Mac user, you will have to buy a USB-to-Serial Port Adapter. All of the projects have been tested with the Keyspan USA-19 series. Do not use the Keyspan USA-28x because the serial ports are RS-422 ports, like the old Macs came with. You need an RS-232 port. Like the PC folks, I recommend you buy two, so when you get to Part III, you can build those projects too.

Before continuing with this chapter, install your new hardware using the manufacturer's installation instructions.

You'll also need a DS9097U 1-Wire Serial Adapter. This is the device that connects the computer to the 1-Wire bus as discussed in the 1-Wire chapter. This can be ordered online from any of the vendors listed in Appendix A.

Parts List

Here's what you'll need to complete this project:

- A computer. You can us a PC running Windows or a Mac with OS X. Most of these projects will also work with a Linux system, but specific instructions are not provided.

- Some of the projects require an always-on Internet connection, preferably a high-speed connection. It also helps to have a high-speed connection to download all of the necessary files and installers.

- 200 MB of available hard drive space. The Java SDK, Java documentation, and development environment require lots of room on your hard drive, especially during initial installation.

- At least one free serial port. You will need a total of two free serial ports to complete all the projects.

- To connect the 1-Wire bus to your computer, you will need a DS9097U 1-Wire to Serial adapter, as presented in Chapter 3.

Why Java?

You may be wondering why this book's projects were written in Java. Other than the fact that it's a really cool language, here's why:

- It runs on almost any computer. There's a version for Windows, MacOS, Linux, Unix, and several embedded operation systems. If you use "pure-java," then your code will run on any Java platform.

- It's free. You can download the entire Java SDK along with complete documentation at no cost. There's even a great online tutorial for free. Start at `java.sun.com`.

- The libraries for accessing 1-Wire devices are provided in Java. Most of the hard work is already done for you!

- There are several development environments for Java. Some are free. Some are great. Some are both!

- Most programmers have a "favorite" language they program in. Mine happens to be Java.

Do I Need to Be a Programmer to Build the Software Project?

Simply put: no. Even if you have never written a line of code, you should be able to follow along. Heck, you just might like it and change careers! You will, however, need at least intermediate computer skills. I'm not going to hand-carry you through every point and click, but I will guide you through each step.

If you're a seasoned Java programmer, you'll still need to install the Java serial and 1-Wire libraries. Once that's done, you can download the code from the companion web site, load it into your development environment, and hack away. Each software project has a pre-written class that you can add to your code.

What Does It Do?

As I stated earlier in the book, weather station hardware is only half the fun. Cool software to collect and log the data is better. Displaying your weather data on the web is great. Having your weather station control your sprinklers or turn on lights is awesome.

You're going to build a Java program called SimpleWeather. It is not intended to be an industrial-strength weather station program. Instead, its primary purpose is to show *you* how to communicate with the weather station sensors, collect the data, and then do something with it. Most of the people I talk to about weather stations want to customize their software. Here's some code that will help you do just that.

To keep things as simple as possible, SimpleWeather won't have a user interface throughout most of this book. In general, you'll launch and run the program from within the development environment. Graphical User Interfaces (GUIs) use lots of code, and tend to obscure the rest of the program. That's not my goal. Also, keep in mind that I'm *not* a whiz-bang super-guru Java programmer. You'll probably see things that could be coded better (a lot better). The code presented here is written for clarity, not speed or compactness. I've tried to use descriptive variables and even sprinkled a few comments in the code.

What Is ETWS? How Does It Compare to SimpleWeather?

If you've poked around on the companion web site, or if you're reading ahead, you may have noticed the ExtremeTech Weather Server (ETWS) and wondered what the heck it is. In the last paragraph I explained SimpleWeather doesn't have a GUI (well, at least not at first). No buttons, no menus. This means that user settings have to be hard-coded in the source files. It also lacks a lot of error checking, and user-specific data has to be hard-coded. That's a bad practice. It's a great learning tool, but maybe you want more. That's where ETWS comes in.

ETWS is a more "user-friendly" version of SimpleWeather. Yep, it's got windows, buttons, and menus. It was actually the original software project for this book, but trying to explain all the code became too hard. So I stripped ETWS down to the bare essentials and created SimpleWeather. However, I strongly encourage you to try the SimpleWeather project. Along the way, you'll learn a lot about how to communicate with 1-Wire devices, how to tweak the code, and the overall program structure.

Using ExtremeTech Weather Server

If you're sure you don't want to follow along with the software project, you can jump right to ETWS. Well, almost. Make sure you follow the next few sections to install Java and the necessary Java libraries. Then you can jump to Appendix B, "Installing the ExtremeTech Weather Server."

Installing Java and Java Docs

The following nine steps walk you through the process of installing Java, the Java Documentation (JavaDocs) package, and some additional Java libraries. There are sections for Windows users and Mac OS X users.

Note Because several of the items you're about to install are constantly being updated, you may want to check the companion web site before continuing to see if there are any revisions to this chapter.

Using Other Versions of Java

The SimpleWeather project will work on Java version 1.4 and higher. If you already have a version of Java installed that meets this requirement, you can skip the Java installation section. Make sure you install the additional libraries.

Step 1: Create Your WeatherToys Directory

At this point, you will need to pick a location on your hard drive where you are going to save downloaded files and build the projects. Maybe you want it on your desktop, or in your documents directory. Regardless of where you put it, create a new folder and give it a name. In this book, I'm going to call it the WeatherToys directory (or folder). Make sure you have read and write privileges to your new directory. From now on, when I refer to the WeatherToys directory, this is the directory I'm referring to.

Step 2: Download the Source Code and Tools

Open your web browser and go to the companion web site. Download `WeatherToys.zip`, and unzip the file in your WeatherToys directory. This will download the entire Weather Toys package, which contains the following:

- The 1-Wire API Java Library
- Source code for all projects
- Copies of the completed projects
- Additional tools
- The ExtremeTech Weather Server software

Step 3: Install Java

The first thing to install is the Java JDK. There are different methods for Mac and Windows users. The Java JDK is a large download, greater than 60 MB.

Note Somewhere between Java 1.4 and Java 1.5, Sun changed the name of its development kit from Software Development Kit (SDK) to Java Development Kit (JDK). It's a bit confusing on the Sun web site and I've tried to keep it straight. Regardless, SDK and JDK mean the same thing.

Mac Users

If you're a Mac OS X user, Apple has pre-installed Java for you. As of this writing, Mac OS X Tiger includes Java 1.4. Most likely by the time you read this, Apple will be shipping Java 1.5. Either way, you are good to go. You can jump down to the section titled "Where's My Stuff At" and look under "Mac Users."

Windows Users

In this step you are going to download and install the Java JDK. Make sure you are logged in as a user with admin (install) privileges and then point your browser to `http://java.sun.com/`. Welcome to the world of Java! New programmers and old hackers frequent this site, and whether they are downloading the latest version of Java or looking for documentation, all things Java can be found here.

There are many flavors of Java. What you want is Java release 2, Standard Edition Java Software Development Kit (J2SE SDK, also called J2SE JDK). Things might have changed since this book was written, so look around if you don't see it at first. On the left side of the web page, click `J2SE`. Now you should be at the J2SE page. Click `Download Java SE (J2SE)`. You will see a page full of downloads. Select `Download JDK 5.0 Update 7` (or whatever the latest update is). Follow the on-screen directions to download and install the JDK.

Step 4: Where Did it Get Installed?

It's important for you to know where your Java stuff is installed. Here's a quick tour.

Mac Users

1. Start off by opening your boot volume.

Mac OS X Boot Disk

For many of the projects in this book, you'll need to know the name of your boot volume, because Mac OS X allows you to name it to almost anything you want. Mine's called Tiger (pretty simple, huh?). It's usually the disk icon in the upper right of your screen, unless you moved it. Take a look and remember its name. So when I refer to the boot volume named Tiger, I'm referring to your boot volume and you should substitute it with your boot volume's name.

---

2. Find and open the `Library` folder, and then look for the `Java` folder. This is your Java library folder.

3. Open the Java Library folder. You should see at least two folders, Home and Extensions, as shown in Figure 4-1. The most important location to remember here is the Java Extensions folder. This is where Apple prefers you to put additional libraries, which you will do soon.

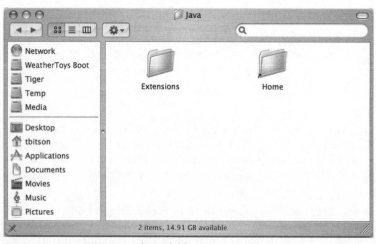

FIGURE 4-1: Mac OS X Java Library folder.

4. Finally, open the Home folder. You'll see something similar to what is shown in Figure 4-2. Don't worry if yours isn't exactly the same because you still have to add a few files. For now, just remember this is your Java Home directory.

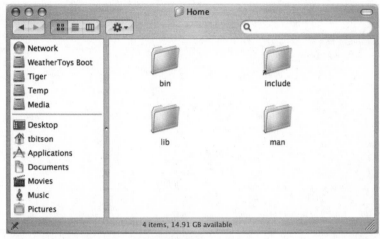

FIGURE 4-2: Mac OS X Java Home directory.

Windows Users

After the Java installer completes, reboot your computer. Log back in if necessary. Now go to C:\Program Files\. Look for the new Java directory and open it. You should see something similar to Figure 4-3.

FIGURE 4-3: Windows Java directory.

The directory jdk1.5.0_07 (yours may be newer) is your *Java Development Directory*. The directory jre1.5.0_07 is your *Java Runtime Directory*.

Open the jdk1.5.0_07 directory. Here you will find several more directories, similar to Figure 4-4.

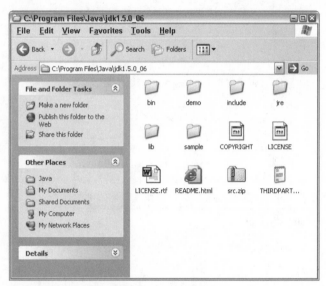

FIGURE 4-4: Windows Java Home directory.

Take a quick look at look at each of these directories:

- **bin** — This is the directory that contains the tools and utilities that will help you develop and execute your programs written in the Java programming language. It also contains useful tools for debugging and developing documentation. In general, you won't use these tools directly. You'll work in your development environment and it will access these tools.

- **jre** — This is a copy of the J2SE runtime environment for use by the JDK. The run-time environment includes a Java virtual machine, class libraries, and several other files that support the running programs in Java.

- **lib** — This directory contains additional class libraries and support files required by the development tools.

- **demo** — This directory is loaded with examples and source code for programming in Java.

- **include** — This directory contains header files that support native-code programming. You won't be using these in this book.

- **src.zip** — This is the Java programming language source files for all classes that make up the Java 2 language. They are for reference purposes only, and most cannot be compiled. You won't be using these either.

Step 5: Setting the Environment Variable — PC Users Only

Windows needs to know where the JDK was installed. The installer may have done this for you. Here's how to check:

1. Open up a command window. If you haven't done this before, choose Start ⇨ Run and then type **cmd** (remember how to do this, because you'll use it again later).

2. Type **javac**.

3. You should see something like Figure 4-5. The command `javac` invokes the java compiler. Because you didn't specify a file to compile, it responded with its usage. If you see this, then your environment variable is correct and you can go to step 6.

4. Right-click My Computer ⇨ Properties. Select the Advanced tab and click the Environment Variables button. This will bring up the Environment Variables window as shown in Figure 4-6. The lower half of the window shows the system variables. Double-click Path to see the path variable. Scroll around the Variable Value text field and look for anything with "java" in it. If you find it, change it to your JDK bin path, which should be `C:\Program Files\Java\jdk1.5.0_0x\bin` (where x is your version number). If you don't see it, add a semicolon at the end of the text and add it to the list. It should look like Figure 4-7 (again, make sure you are using your path; mine probably has a different version number).

5. Click OK three times to close all windows. You'll have to log out and back in again for the changes to take place. Go back to step 1 and repeat this check again to make sure your settings are correct.

```
C:\WINDOWS\System32\cmd.exe                                          _ □ ×
Microsoft Windows XP [Version 5.1.2600]
(C) Copyright 1985-2001 Microsoft Corp.

C:\Documents and Settings\tbitson>javac
Usage: javac <options> <source files>
where possible options include:
  -g                         Generate all debugging info
  -g:none                    Generate no debugging info
  -g:{lines,vars,source}     Generate only some debugging info
  -nowarn                    Generate no warnings
  -verbose                   Output messages about what the compiler is doing
  -deprecation               Output source locations where deprecated APIs are u
sed
  -classpath <path>          Specify where to find user class files
  -cp <path>                 Specify where to find user class files
  -sourcepath <path>         Specify where to find input source files
  -bootclasspath <path>      Override location of bootstrap class files
  -extdirs <dirs>            Override location of installed extensions
  -endorseddirs <dirs>       Override location of endorsed standards path
  -d <directory>             Specify where to place generated class files
  -encoding <encoding>       Specify character encoding used by source files
  -source <release>          Provide source compatibility with specified release

  -target <release>          Generate class files for specific VM version
  -version                   Version information
  -help                      Print a synopsis of standard options
  -X                         Print a synopsis of nonstandard options
  -J<flag>                   Pass <flag> directly to the runtime system

C:\Documents and Settings\tbitson>_
```

FIGURE 4-5: Java's compiler (javac) response in Windows.

FIGURE 4-6: The Environment Variables window.

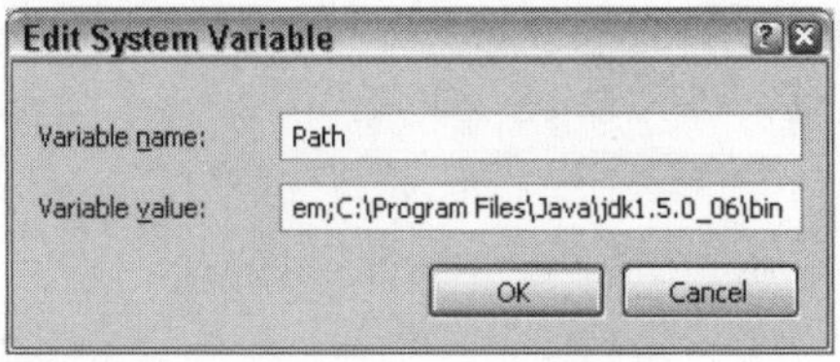

FIGURE 4-7: Adding the JDK bin path.

Step 6: Installing the API Documentation

Mac Users

Apple provides the JavaDocs as part of the Developer Tools package. Even if you have installed the developer tools, the JavaDocs package is a separate install. If you've already installed it, you can skip to the next section. Otherwise, you have two options: install it from the Tiger Install Disk or from Sun's web site.

The preferred way is to install the JavaDocs from the Tiger Install Disk:

> **1.** Insert your Tiger Install Disk.

2. Open X Code Tools ⇨ Packages.

3. Double-Click Java14Documentation to start the install process.

4. When the installer is done, your JavaDocs will be located at `/Library/Java/Home/Docs`.

To install the JavaDocs from Sun's web site:

1. Open your browser and point it to `http://java.sun.com/docs/index.html`.

2. Click the Download J2SE Documentation button and follow the on-screen prompts to download the J2SE API documentation to your hard drive. Save them in your WeatherToys directory, or where all users can access them.

3. The JavaDocs file is zipped. You'll need to unzip it if Safari didn't automatically unzip it for you. Unzip it and save it in your WeatherToys directory. The starting point for JavaDocs is a file named `index.html` located inside the top-level directory. You may want to delete the zipped file when you're done to save disk space.

What's an API?

API Stands for Application Programming Interface. It is a library of high-level routines that allow the programmer easy access to low-level functions. For example, SimpleWeather will use the 1-Wire API. This is a collection of routines that provide easy access to the 1-Wire weather sensors without having to know the intimate details of the sensor. Java's APIs usually come with two parts: the library and documentation.

Windows Users

Open your browser and point it to `http://java.sun.com/docs/index.html`. Click the `Download J2SE Documentation` button and follow the on-screen prompts to download the J2SE API documentation to your hard drive. Save them in your WeatherToys directory.

The JavaDocs file is zipped. You'll need to unzip it. Unzip it to your WeatherToys directory. Inside the top-level directory is a file titled `index.html`.

Take a Look...

Now that you have Java and JavaDocs installed, open the Java Docs folder, double-click the `index.html` file, and explore the world of Java. Make sure you scroll down and check out the Java 2 Platform Specification (`javaDocs/api/index.html`). This lists the thousands of APIs Java provides for you.

NetBeans

When developing code, programmers have to do several things over and over again. They write some code, compile the code, run the code, and (at least for me) debug the code. Writing Java code can be done in any text editor. However, there are some really great text editors specifically designed for editing code. They offer features that programmers like, such as auto-indentation, syntax coloring, version control, and error highlighting, to name just a few.

Next, to compile the code, programmers jump to the command line and type in long, cryptic compiler commands. If the program compiles without error, great; otherwise they go back to the editor and fix their code. Then, it's back to the command line to compile. Maybe it compiles this time, maybe not. (I know several programmers whose code almost always compiles the first time. Not me!) If it compiles, they launch their program and hope the code does what it was supposed to do. If not, back to the editor. You can see where this is going: 'round and 'round. What if the editor, compiler, and debugger were all built into one tool? What if one button started the whole process? Wouldn't that save a bunch of time? Software tool developers thought so too, and created programming applications that contain these features (and many more). They are called Integrated Development Environments, or IDEs. And just like most other programming languages, Java has several to choose from.

Why NetBeans?

Prior to selecting an IDE for this book's projects, I tried several of the free Java IDEs. Some are pretty clunky, some only run on Windows, and some are just hard to use. It boiled down to two Java IDEs: Eclipse and NetBeans. In the end, NetBeans won. Why, you ask? Well, NetBeans seemed to be easier to learn and use. Yes, Eclipse has a few more features, but it seems like it's geared more for bit heads. I like simple. Plus, as I advertised previously, even non-programmers can follow along. So NetBeans it is.

Step 7: Installing NetBeans

In this step, you will install the NetBeans IDE. The NetBeans installer file is large (over 40 MB), so it may take a while.

Mac Users

1. Point your browser to www.netbeans.org and follow the on-screen directions to download the current version of the NetBeans IDE. You don't need to download the optional Application Server.

2. Locate the downloaded file. Double-click the file to expand it.

3. Copy the expanded file to where you want to keep your applications, such as the Applications folder.

4. Double-click the NetBeans application to start it.

5. NetBeans should now start and show the welcome message. Next, go to step 8 where you'll configure a few settings.

Windows Users

1. Make sure you're logged in as an administrator with install privileges. Point your browser to www.netbeans.org and follow the on-screen directions to download the current version of the NetBeans IDE. You don't need to download the optional Application Server.

2. Locate the downloaded file. Double-click the file to start the installer and follow the on-screen instructions.

Step 8: Configuring NetBeans

Next you need to configure NetBeans. The installer has done most of the work for you. All you really need to do is let NetBeans know where you installed the JavaDocs. Here's how:

Mac Users

1. Select Tools ➪ Java Platform Manager.

2. Click the JavaDoc Tab as shown in Figure 4-8.

3. NetBeans may have detected your JavaDocs path and entered it for you. If the path shown is incorrect or missing, do the following:

   a. Click the add ZIP/Folder button.

   b. Navigate to your Java Docs folder you installed a few steps back. Once inside, select the api folder and click the Add ZIP/Folder button. Now NetBeans knows where your JavaDocs are located.

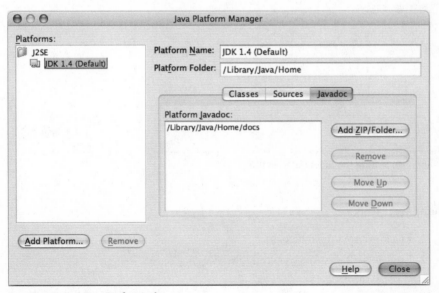

FIGURE 4-8: JavaDoc Prefs window.

4. Click the Close button.

5. Next, you need to adjust an editor setting so your source code will look the same as the book. Select the menu NetBeans ⇨ Preferences.

6. Click the Editor icon.

7. Select the Indentation tab.

8. Set the Number of Spaces per Indent to 2, like in Figure 4-9.

9. Click OK.

10. Quit NetBeans. NetBeans is now configured and you are ready to install the additional libraries.

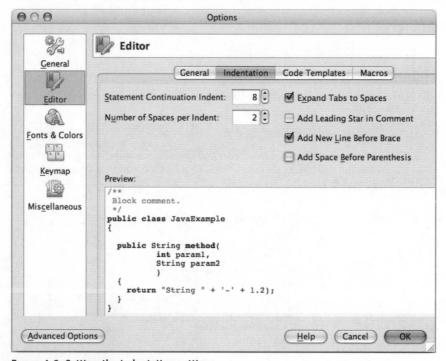

FIGURE 4-9: Setting the Indentation settings.

Windows Users

1. Log in with your user account.

2. Launch NetBeans. You should see the Welcome screen.

3. Select Tools ⇨ Java Platform Manager.

4. Click the Javadoc tab as shown in Figure 4-10.

5. Click the add ZIP/Folder button.

6. Navigate to your Java Docs folder you expanded a few steps back. Inside there, select the `api` folder and click the Add ZIP/Folder button. Now NetBeans knows where your JavaDocs are.

7. Click the Close button.

8. Next, you need to adjust an editor setting so your source code will look the same as the book. Select the menu Tools ⇨ Options.

9. Click the Editor icon.

10. Select the Indentation tab.

11. Set the Number of Spaces per Indent to 2 as shown in Figure 4-9.

12. Click OK.

13. Quit NetBeans. NetBeans is now configured and you are ready to install the additional libraries.

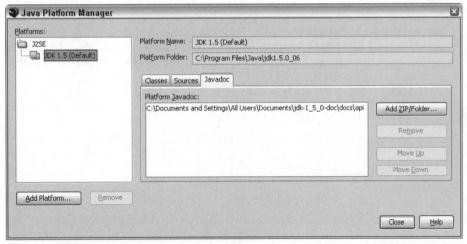

FIGURE 4-10: Setting up the NetBeans Java Docs path.

A Cup of Java...

At this point, you have Java and NetBeans installed. I suggest you take a break from this book, pour yourself a hot cup of coffee (or whatever you drink), and look through the NetBeans IDE Tutorial. You can find it in NetBeans at Help ⇨ Tutorials ⇨ Using NetBeans IDE. At a minimum, learn the following:

- The names of the main windows in the IDE

- How to open and save projects

- How to close a project
- How to look at source files
- How to use the editor

Step 9: Install the Additional Java Libraries

To complete your installation activities, you have two more items to install. These are the Java libraries. What are libraries you ask? They are add-ons or extensions to the Java programming environment. In many cases, they are just simply Java classes that someone else has provided that you can reuse. Some, like the ones you're going to install, allow access to hardware that is not built into the core Java language.

To talk to your 1-Wire weather sensors, you are going to need to access the serial ports (you did install your serial port, right?). The Java folks have developed the Java Comm API. It is a library that you can download and install. Unfortunately, it is only available for Windows and Unix. That leaves Linux and Mac users out in the cold. Fortunately, the good folks at RxTx.org have put together a substitute library that provides serial port functionality, and it works with most operating systems.

You are also going to install the 1-Wire Java Library that provides access to all of the 1-Wire functions. This allows you to read the sensors with just a few lines of code so you don't have to bother trying to figure out how to send 0s and 1s out the serial port in just the right order.

Mac Users

1. Log in to your Mac with your user account.

2. Open your WeatherToys folder and then the folder named MacUsers. There should be two files:

   - RxTxInstaller

   - OneWireAPI.jar

3. Remember your Java Extensions folder you looked at earlier? Good. Copy OneWireAPI.jar to your Java Extensions folder. You may be prompted for your Administrator password.

4. Double-click the RxTxInstaller package to start the RxTx installation. Follow the on-screen prompts. When you're done, you'll have to reboot.

Note You must run the RxTx Installer from the account that will be using RxTx. The installer adds the current user to the uucp group. If more than one user will be using RxTx, each user needs to run the installer. Alternatively, you can add each user to the uucp group if you are familiar with the NetInfo Manager.

Windows Users

Follow these steps to install the RxTx Library:

1. Log in as an administrator.

2. Open your WeatherToys directory. Look inside for a folder named WinUsers. There should be four files:

   - `RXTXcomm.jar`
   - `rxtxSerial.dll`
   - `rxtxParallel.dll`
   - `OneWireAPI.jar`

3. When developing a Java program in NetBeans, it uses the JDK directory. When you execute a Java application, which you will do later, it runs out of the JRE directory. Therefore, each of the Jar files needs to be copied to two places: your Java Development Directory and your Java Runtime Directory. Remember, your Java Development Directory should be at `C:\Program Files\Java\jdk1.5.0_07\bin` and your Java Runtime Directory is at `C:\Program Files\Java\jre1.5.0_07\bin`.

4. Copy `OneWireAPI.jar` to:

   - `Java Development Directory\lib`
   - `Java Runtime Directory\lib\ext`

5. Copy `RXTXcomm.jar` to:

   - `Java Development Directory\lib`
   - `Java Runtime Directory\lib\ext`

6. Copy `rxtxSerial.dll` to your

   - `Java Development Directory\bin`

7. Copy `rxtxParallel.dll` to:

   - `Java Development Directory\lib`

8. Log out and then log back in with your user account.

Congratulations! You now have installed Java, NetBeans, and the necessary library files. Everything should be set up and ready for you to start coding.

Building the SimpleWeather Framework

This section walks through the process of creating a project and the source code. You'll also tell NetBeans what additional libraries you want to use.

Create the Project

1. Launch NetBeans.

2. If NetBeans shows any open projects in the Projects window, close them by right-clicking each project and selecting Close Project.

3. Select File ⇨ New Project.

4. In the New Project dialog, click General in the Category box, and then click Java Application in the Projects box.

5. Click Next. You'll see the New Java Application window, as shown in Figure 4-11.

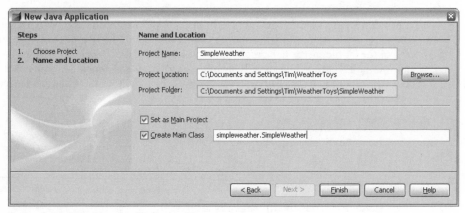

FIGURE 4-11: New Java Application settings.

6. For Project Name, type **SimpleWeather**.

7. For Project Location, click Browse and select your WeatherToys folder.

8. Make sure Set as Main Project is checked.

9. Make sure Create Main Class is checked and type **simpleweather.SimpleWeather** in the text field. Make sure you pay attention to the uppercase letters; Java is case sensitive.

10. Click Finish.

11. NetBeans will now create a new project for you with a default source code file. If NetBeans didn't automatically open your project, go ahead and select File ⇨ Open Project and select your new project. It should look similar to Figure 4-12.

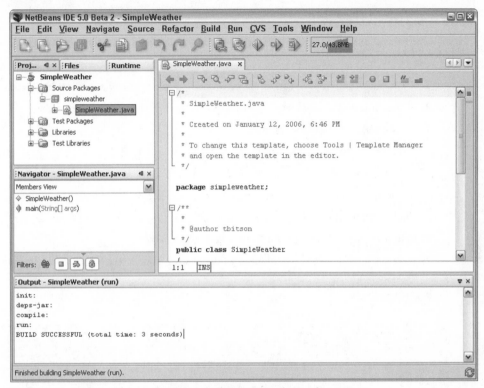

FIGURE 4-12: Project window after creating the new SimpleWeather project.

Java Files

In general, Java requires each class to be in a separate file that has the same name of the class with `.java.` appended to the end. As you build up your project, you will be adding many files to the project and each will be named after the class it contains.

Each class file also belongs to a Java Package. Packages help group classes together and keep them from interfering with each other. By default, NetBeans has created a package named `simpleweather` for you. NetBeans stores your source code files on your hard drive in your projects folder at `Project name/src/Package Name/`. In this case, they're at `/SimpleWeather/src/simpleweather/`.

Pick Your Libraries

The next thing you need to do is tell NetBeans what additional libraries you want to add. Here's how:

1. Right-click the SimpleWeather project in the Projects window and select Properties.

2. In the Project Properties window click Libraries on the left side of the screen.

3. Click the Add JAR/Folder button.

4. Navigate to where you installed your Java libraries:

   ▪ PC Users: `C:\Program Files\Java\jdk1.5.0_05\lib`

   ▪ Mac Users: `Tiger/Library/Java/Extensions`

5. Click `RXTXComm.jar` and then click Choose.

6. Repeat the steps 3 and 4, and this time add `OneWireAPI.jar`.

7. You have just added the two libraries to the project. Your screen should look similar to Figure 4-13.

8. Click OK.

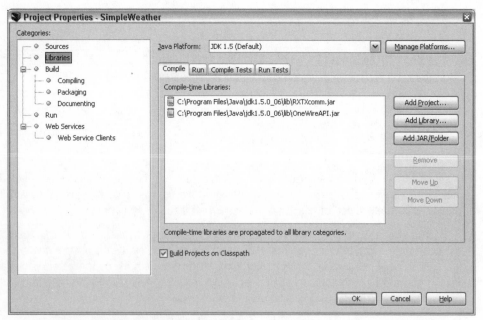

FIGURE 4-13: Project Properties with Libraries added.

Add the Source Code

Hopefully you have had a chance to look at the files from the companion web site. One of the directories contains the project source code. Inside you'll see a directory for each project. Look inside the Chapter 4 directory. You will find a single file and a folder. The file `SimpleWeather.java` is the source file for this project. As you progress through the rest of the projects, you will continue to build on this file, and add some additional ones.

The folder named Chapter 4 Project is a completed project for this chapter. If you have problems getting your project to work, you can refer to this project.

The easiest way to add the source file to your project is to copy it directly to this project's source folder, replacing the existing `SimpleWeather.java` file NetBeans created for you. To do this, copy the `Chapter 4/SimpleWeather.java` file to `Weather Toys/SimpleWeather/src/simpleweather/SimpleWeather.java`, replacing the existing file NetBeans created. Then simply select File ➪ Refresh All Files to have NetBeans update its source files.

Alternatively, you can copy and paste the text directly into the default SimpleWeather file NetBeans created for you. Just make sure you completely replace the original source.

Setting User-Specific Data

Most commercial applications have dialog boxes and preference dialogs to enter your settings. SimpleWeather doesn't. You will break a programming rule and hard-code your settings.

For this project, there is only one setting: the serial port that you will plug your 1-Wire adapter into. So go ahead, find your DS9097U 1-Wire adapter, and put it into one of your serial ports. Make sure you know which port you plug it into. PC users have ports with simple names like COM1 or COM2.

It's a bit harder on a Mac. You'll have to use the manufacturer's utilities to get the port name. If you use a Keyspan USB adapter, the installer put a utility in your applications folder named Keyspan Serial Assistant. Launch it to display the names of all the adapters it found. If you click the disclosure triangle, it will show you the name of the port(s) for that adapter as shown in Figure 4-12.

Note You will have to append a "1" at the end of the name to have the proper name. Don't ask me why. The full name of your port will be something like `/dev/tty.USA19QW2b21P1.1`.

If you really get stuck, I've provided a tool in the WeatherToys download that lists all of the serial ports on your computer. Look in the Tools directory and find `SerialPortLister.jar`. Double-click to launch it. It will display all of the serial ports found on your Mac.

Now that you have your serial port name, open your project, open `SimpleWeather.java` in the editor, scroll down to around line 27, and modify the line that sets the serial port name. For example, if your serial port name is "COM1," it should read:

```
public static final String ONE_WIRE_SERIAL_PORT = "COMI";
```

Don't forget the semicolon at the end.

Set the Main Project

There's one last thing to do: set the main project. NetBeans allows you to have many inter-linking projects. You aren't going to do that for this book, but NetBeans still needs to know which is your "main" project even if there is only one project.

To set it, select File ⇨ Set Main Project ⇨ SimpleWeather.

That's it!

Building and Running SimpleWeather

Are you ready? Let's try it. Hit F6 or Select Run Main Project for the Run menu. Either way, NetBeans will start to compile and you will see the results in the Output window. If things go well, you'll see something like the following:

```
init:
deps-jar:
Compiling 1 source file to C:\Documents and Settings\Tim\My
Documents\Dev\Java\SimpleWeather\build\classes
Note: C:\Documents and Settings\Tim\My
Documents\Dev\Java\SimpleWeather\src\simpleweather\SimpleWeather.j
ava uses or overrides a deprecated API.
Note: Recompile with -Xlint:deprecation for details.
compile:
run:
Starting SimpleWeather 1.0
Native drivers not found, download iButton-TMEX RTE Win32 from
www.ibutton.com
Devel Library
=========================================
Native lib Version = RXTX-2.1-7pre17
Java lib Version  = RXTX-2.1-7pre17
Found Adapter: DS9097U on Port COM1
Resetting 1-wire bus
Time = Fri Dec 16 13:24:41 GMT-07:00 2005

Time = Fri Dec 16 13:25:00 GMT-07:00 2005

Time = Fri Dec 16 13:26:00 GMT-07:00 2005

Time = Fri Dec 16 13:27:00 GMT-07:00 2005
```

This is good! Looking at the last few lines, SimpleWeather is running. So far, it just displays the time once a minute. To stop it, type q in the input window and press Return. You will see this:

```
SimpleWeather Stopped
BUILD SUCCESSFUL (total time: 228 seconds)
```

If SimpleWeather does not start running, look through the output window text to see if there are any clues as to why.

- Couldn't find a library? Follow the instructions in "Step 9: Install the Additional Java Libraries" to pick your libraries again.

- Couldn't find the serial port? Check the name and make sure you entered it correctly as discussed in the "Setting User-Specific Data" section.

- Syntax Error? Check your typing where you changed the serial port name. Is it still in quotes? Is there a semicolon at the end?

Java Objects, Classes, and Methods

Throughout this section and in most of the future projects, I will mention objects, classes, and methods. If you're new to Object-Oriented Programming (OOP) these may be strange terms to you. Here's a quick (very quick!) explanation of OOP. In the real world, we have objects: cars, bikes, ovens, and toasters, for example. Take the car example. We have red cars, fast cars, big cars, and pedal cars. We can group all the different types of cars into a single class: cars.

A class is like a template; it really doesn't exist until you create or instantiate one. Your car is an *instance* of the class cars. Once you instantiate a car, it is now an *object*.

In software, it is also possible to group similar objects into a class. For example, you will create a Temperature Sensor class that can read the temperature from several different types of 1-Wire temperature devices. When you want a new temp sensor, you just instantiate its class.

Java requires that each class be in its own file, and that the name of the file is the same as the class name.

To instantiate an object of the class, you will use the Java keyword *new*. It will create an object of the class and initialize it. You can instantiate as many instances of the temperature class as you need, one for each temperature sensor. When you create it, you pass it some parameters that will make each instance unique.

A class has two main parts: *variables* and *methods*. Variables are pretty much like traditional variables; they hold some numeric value, as well as text and data. Some don't change throughout the object's life; they are called constants or final variables. Methods are the routines that actually do the work: they are the lines of code that program executes.

The Java Tutorial has several great chapters on Object-Oriented Programming, Objects, Classes, and Methods. You can read it on-line at `http://java.sun.com/docs/books/tutorial/index.html`.

Walking Through the Source Code

Take a look at the code for this project. So far, you only have one class file, `SimpleWeather.java`, but that will change in the very next chapter. Follow along as I discuss each section. If there is something you want more information on, check the Java Tutorial at `http://java.sun.com/docs/books/tutorial/index.html`.

Starting at the beginning of the source code, you have a simple header. Headers show the date and version number, along with general comments about the code:

```
/*****************************************************************


   Project Name: SimpleWeather
   File name:    SimpleWeather.java
   Version:      1.0  12/17/05

   Copyright (C) 2005 by T. Bitson - All rights reserved.



   *****************************************************************/
```

Comments in Code

The Java newbie may be wondering what all the asterisks and double slashes in the code mean. That's how programmers embed comments and notes in their code. The comments do not get compiled into the code; they're just in the source code. Comments can be placed anywhere in the code, even in the middle of a line.

There are two types of comments:

- Block comments start with /* and don't end until there's a */. This can cover many lines.

- In-line comments start with // and go to the end of the line only. Here are two examples:

```
/*

    This is a block comment
    It can cover more than 1 line
*/

// This is an in-line comment
```

Next is the package statement. All of the code you will write for SimpleWeather will be in one package. Packages help keep code and variables in one package separate from code in other packages:

```
package simpleWeather;
```

Next, you import the classes that aren't automatically included. You also have to specifically include the classes from the libraries you added:

```
import java.util.Date;
import java.io.*;

import com.dalsemi.onewire.*;
import com.dalsemi.onewire.adapter.*;
import com.dalsemi.onewire.container.*;
```

Because the header and imports don't change much from chapter to chapter, I won't cover them in future chapters. If you want to learn more, consult the Java Tutorial.

This is the definition of your main class. When Java starts running your program, this is the class it looks for:

```
public class SimpleWeather
{
```

Next is where you will define your settings (a.k.a user preferences). Each line sets a constant equal to some value. For example, the following line defines ONE_WIRE_SERIAL_PORT to be equal to the String "COM1". The keyword final means it can't be changed as the program is running. Notice the comments // 1-Wire devices and //sensors. This is where you will add to your program as you progress:

```
// user constants
public static final String VERSION = "SimpleWeather 1.0";
public static final String ONE_WIRE_SERIAL_PORT = "/dev/tty.USA19QW2b21P1.1";

// 1-Wire Devices


// class constants
public static final String ADAPTER_TYPE = "DS9097U";


// class variables
public static boolean debugFlag = false;

private DSPortAdapter adapter;


// sensors
```

After the user constants is the constructor that initializes the class. Constructors automatically get called when the class is created. In this constructor, an instance of the 1-Wire adapter is created using the serial port you supplied in the user constant. If the adapter is found, a 1-Wire bus reset is performed and the program continues. If the serial port is not found or the 1-Wire adapter is not present on the specified serial port, an error message is printed and the program exits:

```
public SimpleWeather()
{
  // get the 1-wire adapter
  try
  {
    // get an instance of the 1-Wire adapter
    adapter = OneWireAccessProvider.getAdapter(ADAPTER_TYPE,
              ONE_WIRE_SERIAL_PORT);
    if (adapter != null)
    {
        System.out.println("Found Adapter: " + adapter.getAdapterName() +
              " on Port " + adapter.getPortName());
    }
    else
    {
      System.out.println("Error: Unable to find 1-Wire adapter!");
      System.exit(1);
    }

    // reset the 1-Wire bus
    resetBus();
  }
  catch (OneWireException e)
  {
    System.out.println("Error Finding Adapter: "+ e);
    System.exit(1);
  }
}
```

After the constructor is the main body of the program where you will be making most of the future additions. It starts off by getting a Date object and declaring a few variables. The code then enters a loop. Inside this loop, it starts by sleeping (pausing) for one second. When it wakes up, it goes and checks the time.

If one minute has passed (the current minute value does not equal the last minute value), the following occurs:

- Update the date (time is contained in the date variable).

- Print the time to screen.

- Update the last minute value.

- Check for keyboard input. Did the user type q to quit? If so, free up the adapter and set the quit flag to true so the loop will quit on the next pass.

- Loop again.

If one minute has not passed, the code simply loops again. Also, notice the comments for //initialize sensors and //get the weather. This is where you will add calls to instantiate the sensor class and the calls to the method that actually gets the weather conditions in future chapters:

```java
public void mainLoop()
{
    Date date = new Date();
    int minute, lastMinute = -99;
    boolean quit = false;
    InputStreamReader in = new InputStreamReader(System.in);

    // initialize sensors


    // main program loop
    while(!quit)
    {
        // sleep for 1 second
        try
        {
            Thread.sleep(1000);
        }
        catch (InterruptedException e) {}

        // check current time
        date.setTime(System.currentTimeMillis());
        minute = date.getMinutes();


        // only loop once a minute
        if (minute != lastMinute)
        {
            System.out.println("Time = " + date);


            // get the weather


            System.out.println("\n");

            // update the time
            lastMinute = minute;
        }
```

```
                // development use only - check for 'q' key press
                try
                {
                  if (in.ready())
                    if (in.read() == 'q')
                    {
                      quit = true;

                       try
                    {
                      adapter.freePort();
                    }
                    catch (OneWireException e)
                    {
                      System.out.println("Error Finding Adapter: "+ e);
                    }
                  }
                }
                catch (IOException e)
                { } // don't care
            }
        }
```

Next is the `main()` method where the program starts running. This method looks at the arguments passed in when the program is launched. If one of those arguments is `-d`, it enables debug mode, which prints diagnostic messages to the screen. You learn more about debug mode in Chapter 7.

After checking the debug mode, `main()` gets an instance of the SimpleWeather class and calls the `mainLoop()` method. If an exception (error) occurs, the program jumps to the `catch()` method, which simply prints the error to the screen and exits the program:

```
public static void main(String[] args)
{
  System.out.println("Starting " + VERSION);

  if (args.length != 0)
  {
    if (args[0].equals("-d"))
    {
      System.out.println("debug on");
      debugFlag = true;
    }
  }

  try
  {
    // get instances to the primary object
    SimpleWeather weatherServer = new SimpleWeather();

    // call the main program loop
```

```
    weatherServer.mainLoop();
  }
  catch(Throwable t)
  {
    System.out.println("Exception: Main() " + t);
  }
  finally
  {
    System.out.println("SimpleWeather Stopped");
    System.exit(0);
  }
}
```

Finally, you have a routine that resets the 1-Wire bus. It calls the reset() method provided in the 1-Wire API. The reset() method returns a value that indicates if the method was successful. If there is a problem, it is printed on the screen and the program continues.

```
private void resetBus() // reset the 1-wire bus
{
    System.out.println("Resetting 1-wire bus");

  try
  {
    int result = adapter.reset();

    if (result == 0)
      System.out.println("Warning: Reset indicates no Device Present");
    if (result == 3)
      System.out.println("Warning: Reset indicates 1-Wire bus is shorted");
  }
  catch (OneWireException e)
  {
    System.out.println("Exception Resetting the bus: " + e);
  }
 }
}
```

Wrap Up

You accomplished many things in this chapter. You installed Java, set up your development environment, and coded the framework for your weather projects. Next you compiled and tested your software. Hopefully things have gone well and you have learned a lot about programming. Don't worry, now that the software project is set up, it will get easier.

If you are a programmer or just curious, now would be a great time to look at the 1-Wire API Java Docs. You just used the DSPortAdapter class and in the next chapter, you are going to use a temperature sensor, which is listed in the OneWireContainer10 class. The entire package can be downloaded from the Dallas/Maxim web site at www.maxim-ic.com/1-Wire.cfm. Click the 1-Wire Software Tools link and then download the 1-Wire API for Java.

Although your software doesn't do much yet, it will. In the next few chapters, you start adding various weather sensors. In fact, the in next chapter you learn how to build and add a temperature sensor and the temperature measurement class to SimpleWeather. You can then start collecting weather data.

Check the Temperature

As you read earlier, temperature is the most common weather measurement, so it makes sense that this is your first weather measurement. Adding a temperature sensor is one of the easier projects in this book and you have several options to choose from.

The 1-Wire weather station introduced in Chapter 3 and several other 1-Wire modules have optional temperature sensors. You can use these in place of the one described in this chapter. However, because building your own is inexpensive and easy, you may want to build it anyway. It never hurts to have an extra temperature sensor for some of the future projects or to use it as an indoor sensor.

This chapter assumes that you have successfully completed the software project in the last chapter and have the SimpleWeather software installed and working. It also assumes that the DS9097U has also been installed. If you are using ETWS, make sure you've read Appendix B and have the ETWS software installed.

Project Overview

This project uses the Dallas/Maxim DS18S20 1-Wire Temperature Sensor. It's a self-powered 1-Wire device. All you need is a single twisted-pair cable and a couple of connecters to connect it to your 1-Wire bus.

Once you have it connected to your 1-Wire bus, I'll show you how to add the Temperature Sensor class to SimpleWeather. Then you'll learn how to use the OneWireLister program to find the sensor's serial number. Finally, you will add the serial number to SimpleWeather and then you're ready to run it!

Options, Options, Options!

Like many of the hardware projects to come, you have several options:

- **Option 1:** Use the 1-Wire Weather Instrument. It has an internal DS18S20 temp sensor. If you choose this option, the hardware is already assembled. All you need to build is a data cable to connect it.

- **Option 2:** You can purchase a complete, assembled temperature sensor module. If you use the Hobby-Boards module, you can add a humidity sensor and solar intensity (light) detector.

- **Option 3:** You can buy a kit that contains the temperature sensor and a PC board. You will have to solder the parts to the PC board. You can add a humidity sensor and solar sensor to the same PC board later.

- **Option 4:** You can buy individual temperature sensors and connect them yourself. This requires a minimal amount of soldering and some basic electrical tools, such as cutters, pliers, and a soldering iron.

All four methods are covered here. If you have purchased a kit or an assembled module that also has other sensors, you are only going to build the temperature module in this project. You'll build and test the other sensors in future projects.

How It Works

The schematic for the temperature sensor is shown in Figure 5-1. This is the in-line version with an RJ-14 plug and jack. The DS18S20 can either be externally powered or parasitically powered from the 1-Wire bus. For your application, you will be powering the device parasitically, so that you don't need a separate +5V power supply. When powered from the 1-Wire bus, Vdd pin 3 must be tied to GND pin 1.

If you are using the Hobby-Boards kit, the only difference is the connector. Hobby-Boards uses the RJ-45, whereas this project uses an RJ-14.

If you're using the WS-1 Weather Instrument, the schematic is a bit different. The DS18B20 Vdd pin 3 is connected to RJ-14 pin 1, the DQ pin 2 to the RJ-14 pin 2, and ground to the RJ-14 pins 3 and 4. The manufacturer supplies an RJ-14 plug that shorts the RJ-14 pin 1 to pin 4 to jumper the DS18B20 pins 3 to 1. I'll cover this in more detail in the next chapter. For now, make sure the RJ-14 jumper plug is installed.

Parts List

There are so many options for this project, it's hard to list exactly what you will need. Here is my best shot! For each of the following options, you'll need a short length of 2-twisted pair data cable. A couple of options were presented in Chapter 3, one of them being CAT3 cable. In this chapter (and future chapters) I'll just refer to this as CAT3 cable.

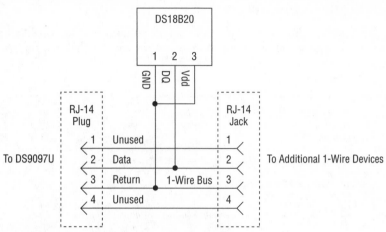

Figure 5-1: Basic temperature sensor schematic.

Option 1

If you are using the 1-Wire Weather Instrument, you will need a data cable to connect the WS-1 to the DS9097U 1-Wire Adapter. The parts required are listed in Table 5-1.

Table 5-1	RJ-14 to RJ-14 Data Cable Parts List for the WS-1	
Qty	*Description*	*Vendor*
2	RJ-14 Modular Plugs	Home Depot
10'	CAT3 2-Twisted Pair Cable	Home Depot, Radio Shack

Options 2 and 3

If you've purchased an assembled module or a kit from Hobby-Boards, you will also need a data cable. Table 5-2 contains the list of parts required.

Table 5-2	RJ-14 to RJ-45 Data Cable for the Hobby-Boards Module	
Qty	*Description*	*Vendor*
1	RJ-14 Modular Plug	Home Depot
1	RJ-45 Modular Plug	Home Depot
10'	CAT3 2-Twisted Pair Cable	Home Depot, Radio Shack

Option 4

If you're building the sensor from scratch, you will need the sensor, a modular phone jack to house the module, and a data cable to connect it to the 1-Wire adapter. Table 5-3 lists the required parts.

Table 5-3	Temperature Module Parts List	
Qty	**Description**	**Vendor**
1	DS18B20 Temperature Sensor	Digikey, Maxim, Hobby-Boards
1	Modular Jack Assembly (Surface Mount)	Home Depot
1'	CAT3 2-Twisted Pair Cable	Home Depot, Radio Shack
1	RJ-14 Modular Plug	Home Depot

Building the Hardware

Once you've collected the necessary parts, you're ready to start building the hardware. Follow the steps for the option you have selected.

ESD Warning: Some of the parts used in this project are ESD sensitive. Be sure to take proper ESD precautions!

Option 1: Using the 1-Wire Weather Instrument

The 1-Wire Weather Instrument uses an RJ-14 connector on the internal circuit board. You will need at a 5- to 10-foot data cable to connect it to the 1-Wire Serial Adapter. Along with the parts listed earlier, you will also need a modular crimp tool (this was discussed in the "Cool Tools" section of the Chapter 3) and a pair of medium-sized diagonal cutters. Here's how to crimp a modular connector on to a twisted pair phone cable:

1. Cut the cable to the desired length, about 10 feet.

2. Remove ¼″ of the outer insulation as shown in Figure 5-2.

3. Trim the end of the wires so they are even.

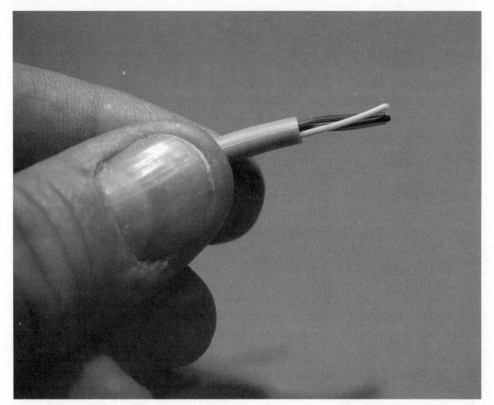

FIGURE 5-2: Strip the outer insulation.

4. Arrange the wires in the correct sequence as shown in Figure 5-3. Refer to Table 5-4 for a refresher on colors and pins.

Table 5-4	RJ-14 Pin/Color Designation	
RJ-14 Pin Number	*"Old" 4-Color Wire*	*"New" 2-Color Wire*
Pin 1 - Left	Black	Orange
Pin 2	Red	Blue
Pin 3	Green	Blue-White
Pin 4 - Right	Yellow	Orange-White

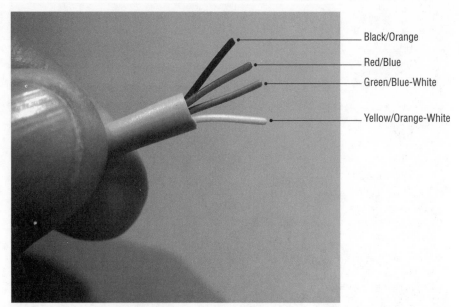

Black/Orange

Red/Blue

Green/Blue-White

Yellow/Orange-White

FIGURE 5-3: Arrange the wires.

5. Group the wires together, keeping them in the correct order.

6. Slide on the modular connector with the pins facing up as shown in Figure 5-4.

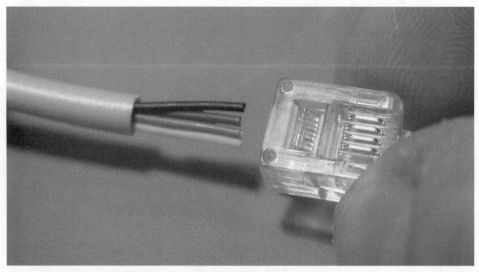

FIGURE 5-4: Slide on the connector keeping the wires arranged.

7. Making sure the wires stay fully inserted, slide the connector into the crimp tool and crimp.

8. Inspect your work. Each wire should still be fully inserted into the connector and in the correct order. Compare yours to Figure 5-5.

Repeat these steps on the other end of your cable. When you're done, you will have a 10-foot cable with modular connectors on each end. If you have a digital ohmmeter, you can check continuity from one end to the other for each pin.

FIGURE 5-5: Completed RJ-14. Notice the outer insulation is captured in the crimp.

Option 2: Building the Hobby-Boards Kit

The Hobby-Boards temperature module kit is simple enough that even beginners should be able to assemble it. There are no surface-mount components, so you don't have to be a soldering expert but you should have some experience soldering circuits. Check the Hobby-Boards web site for parts lists and general instructions for the kits.

Table 5-5 lists what is supplied in the kit and Figure 5-6 shows a photo of the parts you're going to install in this section. If your kit contains other optional components, try to hold off from installing them yet. I know it's hard, but one step at a time.

Table 5-5	Hobby-Boards Temperature Module Kit Components	
Qty	Description	Vendor
1	Temp Sensor PC Board	Hobby-Boards
1	DS18S20 Temperature Sensor	Hobby-Boards
1	Dual RJ-45 Modular Jack Assembly	Hobby-Boards
1	3-Position Screw Terminal Strip	Hobby-Boards

FIGURE 5-6: Hobby-Boards kit parts.

1. **Clean the board.** Wipe the solder-side of the PC board with alcohol to remove any residue or contamination.

2. **Solder the parts.** Insert the parts in the circuit board and solder them carefully.

3. **Clean-up.** Trim the component leads and clean the solder connections with a soft cloth dampened with isopropyl alcohol to remove any flux residue. Over time, this residue will collect moisture and could short out your 1-Wire bus. Use a pair of good quality diagonal cutters to trim excess lead length from the components.

4. **Inspect.** Inspect each connection to make sure there are no shorted parts. Figure 5-7 shows the completed board.

FIGURE 5-7: Completed Hobby-Boards temperature sensor.

Options 2 and 3: Building the Hobby-Boards Data Cable

The Hobby-Boards' products use an RJ-45 connector to connect to the circuit board. You will need an RJ-14 to RJ-45 data cable to connect to your circuit board:

1. Cut a 5- to 10-foot length of CAT3 twisted pair cable.

2. Follow the instructions given previously in Option 1 to crimp on an RJ-14 modular connector to one end of the cable.

3. Remove ½″ of outer insulation off the other end.

4. Order the wires as listed in Table 5-6.

Table 5-6   RJ-45 Pin/Color Designation

RJ-45 Pin Number	"Old" 4-Color Wire	"New" 2-Color Wire
Pin 1 (Left)	No Connection	No Connection
Pin 2	No Connection	No Connection
Pin 3	Black	Orange
Pin 4	Red	Blue
Pin 5	Green	Blue-White
Pin 6	Yellow	Orange-White
Pin 7	No Connection	No Connection
Pin 8 (Right)	No Connection	No Connection

5. Position the RJ-45 connector pin-side up. Carefully insert the wire into the connector, skipping the two holes on the left. As an aid, I take four extra lengths of wire and insert two of them in the two rightmost slots and two in to the two leftmost slots. This helps guide the four wires into the correct slots as shown in Figure 5-8.

6. Carefully insert the RJ-45 into the crimp tool, remove the extra guide wires if you used them, and crimp on the connector.

7. Inspect your work. It should look like Figure 5-9.

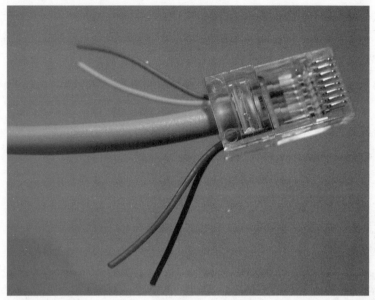

FIGURE 5-8: Using spare wires as a guide. Don't forget to remove them before you crimp!

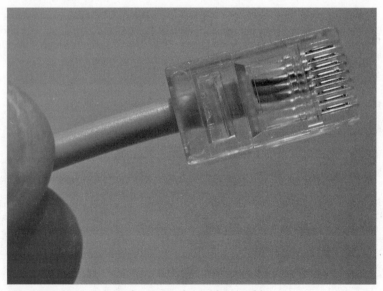

FIGURE 5-9: RJ-45 Crimped to twisted pair phone cable.

Option 4: Building the In-Line Version

Even if you have a temperature sensor on one of your other weather modules, building this is inexpensive and easy, and you'll find having a spare temp sensor useful:

1. Connect pin 1 and pin 3 on the DS18S20 together and twist gently. Bend pin 2 slightly forward so it clears the other pins as shown in Figure 5-10.

FIGURE 5-10: Bend the pins on the DS18S20 to connect 1 to 3.

2. Solder a short length (about 3″) of wire to pin 2, the Data pin. To keep from getting mixed up later, I used a red wire cut from a length of CAT3 cable. Next, solder a short length of wire to the twisted pins 1 and 3 (Return). This time I used a green wire. Sleeve the leads with insulated sleeving or heat-shrink tubing. Yours should look similar to Figure 5-11.

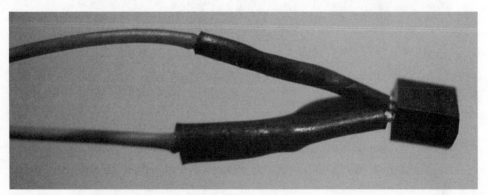

FIGURE 5-11: Attach leads.

3. To build the pigtail, cut approximately a 1-foot length of CAT3 cable. Attach an RJ-14 modular connector to one end as described previously. Strip the outer insulation off the other end so approximately 1 to 2 inches of the twisted-pair wires are exposed. Strip off about 1 inch of insulation from each of the four wires and tin the ends with solder.

4. Attach the modular phone jack's red wire, temp sensor data lead (red), and the pigtail red wire to a screw terminal on the modular jack.

5. Attach the modular phone jack's green wire, temp sensor Return lead (green), and the pigtail green wire to a screw terminal on the modular jack.

6. Attach the modular phone jack's black wire and the pigtail black wire to a screw terminal on the modular jack.

7. Attach the modular phone jack's yellow wire and the pigtail yellow wire to a screw terminal on the modular jack.

8. Attach a wire tie to the outer insulation as shown Figure 5-12. Tuck this inside before you install the cover to act as a strain relief.

9. You can optionally drill several ventilation holes in the cover if you want. Then install the cover. Figure 5-13 shows my completed unit. You have now built a 1-Wire temperature sensor. Actually, it was pretty easy, huh?

FIGURE 5-12: Complete unit without cover.

FIGURE 5-12: Completed in-line temp sensor.

Connect to the 1-Wire Serial Port Adapter

You are now ready to connect your 1-Wire Temperature Sensor to the DS9097U Serial Port Adapter. Plug one end of the data cable into your temperature sensor and the other end into the serial adapter. If you built the in-line version, you can plug it directly into your DS9097U. That's it. You are now ready to add the temperature software module and start measuring temperature.

Add Temperature Capability to Your Software Framework

Let's quickly review what you did in the last chapter. First, you created a SimpleWeather project in NetBeans. It contained one class file, `SimpleWeather.java`. In the SimpleWeather class, you defined a constant for your serial port name. When the SimpleWeather class is instantiated, it tries to find the 1-Wire Serial Adapter using the serial port name you specified. If it was successful, SimpleWeather then enters the main loop, checking the current time. If the current time's minute value changes, it executes the loop. All the first project did in the loop was to print the date and time.

Now you're ready to add the temperature sensor code to SimpleWeather. First, you will add the TempSensor class, a new constant with the device serial number, and a new float (short for floating-point) variable to store the temperature. Next, you will instantiate the class with the device serial number. Finally, you'll call the `getTemperature()` method to get the current temperature and then print it to the screen. Ready? Let's get started.

Add the TempSensor Class

Locate your WeatherToys directory and then find the project source code you downloaded from the companion web site. Next, find the Chapter 5 directory. You will find a single file and a folder. The file `TempSensor.java` is the source file for this project. Copy this file to your SimpleWeather Project directory ⇨ src ⇨ simpleweather folder.

The folder named Chap 5 Project is a completed project for this chapter. If you have problems getting your project to work, you can refer to this project.

Note If you use any of the prebuilt projects from the companion web site, you will most likely get a dialog stating that NetBeans can't find one or more project resources. Follow step 3 in "Find Your Device Serial Number" later in this chapter to remedy this. Keep this in mind in future chapters as well.

Modify SimpleWeather

Launch NetBeans and open your SimpleWeather Project. In the Projects window, double-click the SimpleWeather project, double-click Source Packages, and double-click the SimpleWeather Package to see the list of your source files. If you put `TempSensor.java` in the right spot, then you will see two source files: `SimpleWeather.java` and `TempSensor.java`. Great!

Now you need to add a couple of lines to SimpleWeather. Double-click `SimpleWeather.java` to open it in the NetBeans editor.

Note Remember the steps you took earlier to open a file in the editor? You opened the project (by double-clicking it), then you opened Source Packages, the SimpleWeather Package, and then finally clicked the file you wanted to view. Next time I'll just say open `SimpleWeather.Java` in the editor and you'll know what I mean.

Look through the SimpleWeather code for the `// 1-Wire devices` comment near line 30. Add the following line (the one in bold) under the comment:

```
// 1-Wire Devices
private final String TEMP_SENSOR_ID = "";
```

This will be the serial number of your 1-Wire device. You'll go find it in a second. Finish adding the code first.

Next, you need a variable to hold the temperature. Just a little further down near line 37, find the comment `// class variables` and add 1 more line:

```
// class variables
public float temp;
public static boolean debugFlag = false;
```

Next, create a variable that points to your TempSensor object. At line 45, add this:

```
// sensors
private TempSensor ts1;
```

Next, you need to create a next instance of the TempSensor class you added. Near line 88 in the main loop method, add this:

```
// initialize sensors
ts1 = new TempSensor(adapter, TEMP_SENSOR_ID);

// main program loop
```

Finally in the main loop after the comment on line 114, add three more lines:

```
// get the weather

// get temperature
temp = ts1.getTemperature();
System.out.println("Temperature = " + temp + " degs F");

System.out.println("\n");
```

This is where you actually call the method in the TempSensor class that commands the DS18S20 to measure the temperature and then read it back. The results get stored in the variable temp. For now, you just print it to the screen.

Before you can run your code, you will need to find the serial number of your temperature sensor. Make sure all of you files are saved, and then close the project by right-clicking the project name and selecting Close Project.

Find Your Device Serial Number

In this project and each time you add a new 1-Wire device, you will need to know its unique 16-digit hexadecimal serial number. Because most of the 1-Wire parts are too small to print the number on, Dallas/Maxim has built into each 1-Wire device a way for you to search and display all of the serial numbers for all of the devices on your 1-Wire bus. Pretty cool, huh? In your WeatherToys directory, there is a Tools directory. Inside is a NetBeans project folder named OneWireLister. When run, OneWireLister will search your 1-Wire bus and list all of the devices it finds by serial number. Here's how:

1. If you haven't already done so, plug in your 1-Wire device.

2. If NetBeans isn't already open, launch it. In NetBeans, open the project WeatherToys ⇨ Tools ⇨ OneWireLister.

3. The fist time you run OneWireLister, you may get a dialog complaining that NetBeans can't find one or more project resources. These are the libraries you installed in Chapter 4 in the section "Step 9: Install the Additional Java Libraries." Because I built this project

on my computer, NetBeans is set up to look for where I installed my libraries and not yours. Here's how to fix that:

- Close the dialog.

- Right click the OneWireLister project, and select Resolve Reference Problems from the pop-up menu.

- In the Resolve Reference Problems dialog, click RXTXcomm.jar and then click the Resolve button. Navigate to where you installed *your* RxTx libraries in Chapter 4. Click OK. If necessary, repeat this step for OneWireAPI.jar.

4. Press F6 or select Run ➪ Run Main Project from the menu. OneWireLister will compile and run.

5. OneWireLister will display a dialog asking which serial port you have your 1-Wire bus connected to as shown in Figure 5-14. Select it from the list and click OK.

FIGURE 5-14: OneWireLister 's Serial Port Chooser.

6. OneWireLister will now search for all the 1-Wire devices connected to your 1-Wire Serial Adapter. If all goes well, you should see one or more devices found. If not, go back and check your wiring and that you have your device plugged in. See Figure 5-15 for an example.

7. Even though there isn't an edit menu, you can still copy and paste the serial numbers. Select the serial number that ends in "10" (remember the family codes in Chapter 3?) and press the keyboard equivalent for Copy (Control+C or ⌘+C).

8. Quit OneWireLister by clicking the Close button.

9. Before you close the project, do one more thing: Select Build ➪ Clean and Build Main Project from the menu. You only have to do this once unless you make changes to OneWireLister. I'll show you why you did this in a second.

10. Close the OneWireLister project.

FIGURE 5-15: OneWireLister output.

11. Re-open your SimpleWeather project.

12. Open SimpleWeather.java in the editor.

13. Scroll down to line 31 and paste your serial number in between the quotes after
`TEMP_SENSOR_ID = ""` like this:

```
// 1-Wire Devices
private final String TEMP_SENSOR_ID = "3F00080008F42810";
```

14. Click the Save All button in NetBeans. Your source code is now ready to test. But don't
run it just yet…

Back in step 9, you did a "Clean and Build Main Project." What did this do? Let me show you!
Open your WeatherToys ➪ Tools directory. Now open OneWireLister. Inside, there are several
directories that NetBeans has created. Open the one named dist (which is short for distribution). You will see a single file named `OneWireLister.jar`. Launch the file. What happens?
OneWireLister starts up!

NetBeans has created an executable "jar" file for you. From now on, you don't have to open the
OneWireLister project in NetBeans. Just launch `OneWireLister.jar`. You can even copy it
to a more convenient location, such as your desktop if you wish. You will eventually do the
same with SimpleWeather. Now let's get back and see if you can measure the temperature.

What's a Jar File?

Jar is short for Java Archive. It's the method used to combine and compress Java source and class files. Most libraries are distributed as a single jar file. Certain jar files that contain executable class files and a special manifest file can be executed. Jar files are compressed using the standard Zip format. If you are curious, you can unzip the `OneWireLister.jar` file to see what it is composed of.

Testing the Project

At this point, the SimpleWeather project should be open; if not, open it. Ready? Press F6 (or Run ⇨ Run Main Project). Your project should compile. SimpleWeather will start up, display some initialization information, and then print the date and the temperature. Your output window should look like this:

```
init:
deps-jar:
compile:
run:
Starting SimpleWeather 1.0
Found Adapter: DS9097U on Port /dev/cu.USA19QW2b21P1.1
Resetting 1-wire bus
Time = Tue Dec 27 16:29:02 MST 2005
Temperature = 76.1 degs F

Time = Tue Dec 27 16:30:00 MST 2005
Temperature = 76.1 degs F

Time = Tue Dec 27 16:31:00 MST 2005
Temperature = 76.1 degs F

Time = Tue Dec 27 16:44:00 MST 2005
Temperature = 76.1 degs F
```

You are now displaying the time and temperature! Let your new program run for a while, watching the time and temperature scroll by, marveling at your work. When you have had enough, enter **q** in the input window and press return to stop SimpleWeather.

If your code didn't compile, check your typing. NetBeans will underline errors in red. Look through your code to see if there are any underlines. If so, fix the code. Also, look at the error message NetBeans gave you. Does it provide a clue? If you get really stuck, there is a completed project in the source code folder for this chapter. If you use it, don't forget to change the temp sensor serial number to yours!

Walk Through the Code

The preceding section discussed the modifications to SimpleWeather.java as you went, so there's no need to cover those again. Take a look at the file you added, TempSensor.java.

First, you have the Java classes you need to import. These are the 1-Wire classes that are in the OneWireAPI.jar library that you need for accessing the temp sensor:

```
import com.dalsemi.onewire.*;
import com.dalsemi.onewire.adapter.*;
import com.dalsemi.onewire.container.*;
```

Here's the class declaration and the definition of the class variables. The OneWireAPI groups similar sensors into Java Containers based on the 1-Wire family codes. Because temp sensors are part of family code 10, they are in the OneWireContainer10 class:

```
public class TempSensor
{
    // class variables
    private DSPortAdapter adapter;
    private String deviceID;
    private OneWireContainer10 tempDevice = null;
    private static boolean debugFlag = SimpleWeather.debugFlag;
```

Next is the class constructor. When you instantiate the class, you pass in the adapter you created in the main program, along with the device serial number. Because the serial number is unique for each device, you can create as many temp sensors as you like. The call to OneWireContainer10 returns an instance of the temp sensor for the remaining API methods to use:

```
public TempSensor(DSPortAdapter adapter, String deviceID)
    {
        // get instances of the 1-wire devices
        tempDevice = new OneWireContainer10(adapter, deviceID);
```

Up until now, I have been advertising the 1-Wire temperature sensors as having 0.5°C temperature resolution. Some of the newer devices support resolution down to 0.1°C. This code checks to see if your device supports the higher resolution:

```
        // does this temp sensor have greater than .5 deg resolution?
        try
        {
          if (tempDevice.hasSelectableTemperatureResolution())
          {
            // set resolution to max
            byte[] state = tempDevice.readDevice();
            tempDevice.setTemperatureResolution(tempDevice.RESOLUTION_MAXIMUM,
                        state);
            tempDevice.writeDevice(state);
```

```
      if (debugFlag)
        System.out.println("Temp Device Supports High Resolution");
    }
  }
  catch (OneWireException e)
  {
    System.out.println("Error Setting Resolution: " + e);
  }
}
```

Now comes the method that gets the temperature. The first few lines make sure that the temp sensor exists (is not equal to null) and prints some diagnostic info if the debugFlag is set:

```
public float getTemperature()
{
  float temperature = -999.9f;

  // make sure the temp device instance is not null
  if (tempDevice != null)
  {
    if (debugFlag)
    {
      System.out.print("Temperature: Device = " + tempDevice.getName());
      System.out.print("  ID = " + tempDevice.getAddressAsString() +
                            "\n");
    }
```

Here is the "guts" of the method. It's only about four lines long. That is the joy of having a good library. Here you read the current state of temp sensor with readDevice(), command it to take a temperature reading with doTemperatureConvert(), and then read back the results with getTemperature(). You then convert the output to degrees F. If you desire, comment out the line to leave the temperature in degrees C:

```
    try
    {
      byte[] state = tempDevice.readDevice();
      tempDevice.doTemperatureConvert(state);

      state = tempDevice.readDevice();
      temperature = (float)tempDevice.getTemperature(state);

      // convert to degs F - comment out to get degrees C
      temperature = temperature * 9.0f/5.0f + 32f;
    }
```

Finally, there is a `catch()` method so that if the `getTemperature()` method fails, the program just prints an error message and continues:

```
catch (OneWireException e)
{
    System.out.println("Error Reading Temperature: " + e);
}
}
return temperature;
}
}
```

That's it. Overall, it isn't very complicated. You will follow a very similar process for each of the sensors you add.

Calibration

Because the DS18S20 sensor comes factory calibrated to 0.5°C, there is no calibration required. However, it never hurts to double-check. If you have a known good thermometer, such as a fever thermometer, which is usually calibrated to 0.1° near 100°F, put them side-by-side and let them stabilize for a while. Your readings should be within 0.5°C or 0.9°F.

Wrap Up

Because this was your first hardware project, I walked you through most of the details. As you progress through the following projects, I won't bore you with quite as much detail.

If you want to add more than one temp sensor, all you have to do is repeat the steps in the section "Modify SimpleWeather" using different names for the temp sensor ID, the temp sensor object, and a new variable for reading the temp, such as:

```
// 1-Wire Devices
private final String TEMP_SENSOR_ID = "3F00080008F42810";
private final String TEMP_SENSOR2_ID = "2D0000067382DF10";

// class variables
public float temp;
public float temp2;

// sensors
private TempSensor ts1;
private TempSensor ts2;

// initialize sensors
ts1 = new TempSensor(adapter, TEMP_SENSOR_ID);
ts2 = new TempSensor(adapter, TEMP_SENSOR2_ID);
```

```
// get temperature
temp = ts1.getTemperature();
System.out.println("Temperature = " + temp + " degs F");

Temp2 = ts2.getTemperature();
System.out.println("Temperature = " + temp2 + " degs F");
```

Now that you've completed adding a 1-Wire sensor, you can see that it was really pretty simple. There wasn't a lot of special configuration. All you had to do was get the device serial number and add it to your program. As you add additional sensors, you will follow the same pattern. You'll add another sensor in the next chapter.

Measure the Wind

I n this chapter you are going to build the WS-1 1-Wire Weather Instrument, the device that started the 1-Wire weather phenomenon. It provides temperature, wind speed, and wind direction, all from a single instrument!

The WS-1 is available as a kit. All you have to do is some final assembly and then it is ready to use. As I'm writing this chapter, there are two manufacturers of the WS-1: Texas Weather Instruments and AAG. The Texas Weather Instruments model sells for about $250 and comes with 20 feet of cable. The AAG manufactured unit sells for about $80 and comes with a 6-foot cable. You can do the math, but that extra 14 feet of cable is mighty expensive! You can also purchase the AAG unit at Hobby-Boards.

After making a few modifications to your WS-1, you will add a new class to your software project that provides both wind speed and wind direction capability. Combined with the temperature capability you added in the last project, you can collect data from all three sensors.

Project Overview

In this project, you're first going to make a few minor enhancements to the weather instrument that improve its operation and give it a longer life. Then you will assemble your WS-1 and connect it to your 1-Wire bus.

If you used the WS-1 for your temperature sensor in the last project, then this may be your only 1-Wire device so far. If not, then you can daisy-chain (connect end-to-end in a linear fashion) it to the temperature sensor you have already built.

Options

Because the WS-1 is purchased as a kit, you don't have many options during assembly. You will have to decide if you want to solder in a jumper, but that is covered later in this chapter. You can also start thinking about where you want to install the WS-1. It will need to be in an open area where the wind won't be blocked by houses or trees. I also suggest that it be at least 8 feet above the ground, the higher the better. If you are going to use the built-in temp sensor, you also need to make sure it's not mounted over something that could affect the temperature, such as an asphalt driveway. You will read more about installation in Chapter 11.

---

in this chapter

- ☑ About the 1-Wire Weather Instrument

- ☑ How It Works

- ☑ Upgrades and Modifications

- ☑ Final Assembly

- ☑ Add the Software

- ☑ Test It

- ☑ Check the Wind!

If you don't already have your WS-1 in hand, decide which vendor you want to purchase your WS-1 from. The Texas Weather, AAG, and Hobby-Boards Internet addresses are listed in Appendix A. You might get lucky and find one on eBay, but if you do make sure it is a version 3 unit. The older version 1 (there was no version 2) model uses a different method for wind direction and won't work with this project.

Parts List

All the hardware and supplies to complete this project are listed in Table 6-1. This list assumes a pole-mount configuration, so it includes the mast-mount U-bolt. If you plan on mounting the WS-1 differently, this may not be necessary.

Table 6-1   WS-1 and Mounting Parts List

Qty	Description	Vendor
1	WS-1 1-Wire Wind Instrument	Hobby-Boards, AAG
1	10-Foot Data Cable	Make
1	3-in-1 Oil, Motor Oil (just a few drops)	Hardware Store
1	2-inch Jumper Wire	Make
1	Lead or Aluminum (metal) Tape	Hardware Store
1	"U"-type Mast Mounting Bracket	Hardware Store

What's in the Box?

Referring to Figure 6-1, inside the WS-1 Weather Instrument kit you should find the following items:

- Main Body
- Wind Speed Rotor Assembly
- Wind Vane Assembly
- 10/32 Nuts, Qty 2
- 1-inch Washer
- Aluminum 1-inch Square Mounting Tube
- 1½-inch 6/32 Screw and Locknut
- Tube End Cap

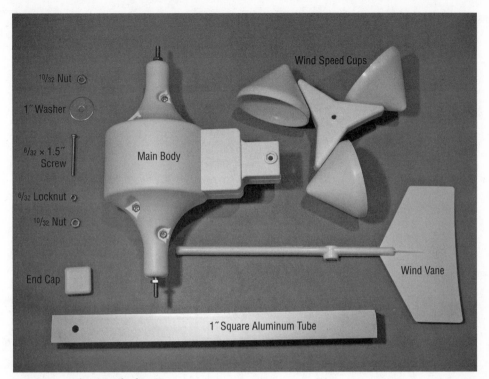

FIGURE 6-1: What's in the box?

How It Works

Looking at Figure 6-2, the schematic is divided up into three sections: the temperature sensor, wind speed, and wind direction.

The temperature section contains a DS18S20 and operates the same as discussed in Chapter 5. The only significant difference is that the DS18S20's Vdd pin is tied to the RJ-14 connector pin 1. This allows the external power option of the DS81S20. However, in most cases, including your weather project, you'll be using the DS18S20 in parasitic power mode. Therefore, RJ-14 pin 1 must be tied to ground on RJ-14 pin 4. The vendor supplies an RJ-14 jumper plug just for this purpose. Unfortunately, your application may require the use of both RJ-14s for daisy-chaining the 1-Wire bus. One of the options presented in this chapter is to solder a jumper across these pins to eliminate the need for the external jumper.

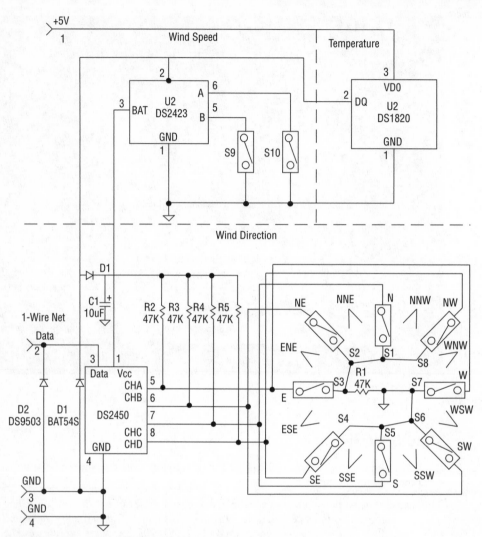

FIGURE 6-2: Schematic of the 1-Wire Weather Instrument.

The Wind Speed section uses a magnet and reed switch to determine wind speed. As you may recall from Chapter 3, the wind rotates a magnet that passes over a reed switch. This causes a momentary closure of the reed switch. If you count the number in a given time period, you can determine the wind speed. The WS-1 uses single DS2423 1-Wire Counter to count the number of reed switch closures. The wind speed is calculated in miles per hour (MPH) using this formula:

```
Wind Speed = counts / time / 2 * 2.453
```

where count is the number of reed switch closures in the time period. Because there are two reed switch closures for every revolution (the rotor has two magnets), you have to divide by two. The factor 2.453 is a constant provided by the manufacturer to convert rotation to MPH. Note that there are two reed switches in this circuit. One is connected to counter A, and the second is connected to counter B. You will only read counter A in your software.

The Wind Direction circuit uses a novel approach employing a single a DS2450 4-Channel A-to-D converter to decode one of 16 possible wind directions. Rather than use 16 reed switches, the WS-1 uses eight reed switches that are physically grouped so that two adjacent switches may be on at the same time. Therefore, if the magnet on the wind position rotor is in between two switches, both switches are activated.

The reed switches are electrically grouped into four pairs so that both switches in the pair can never be activated at one time. One side of the pair is tied to one of the DS2450 inputs and a 47K pull-up resistor to +5V (bus power). On the other side of the pair, one switch is tied directly to ground, while the other switch in the pair is tied to ground through a 47K resistor, creating a divide-by-two voltage divider. What this provides is a means of detecting which switch in the pair is closed with a single A-to-D input. If neither switch is closed, the voltage will be +5.0V. If one switch is closed, the voltage will be 2.5V. If the other switch is closed, the voltage will be 0V. The four A-to-D channel voltages are summarized in Table 6-2.

Table 6-2   A-to-D Input Voltages

A-to-D Channel	Switch Closed	Volts
A	None	5.0
A	S3	2.5
A	S7	0.0
B	None	5.0
B	S2	2.5
B	S6	0
C	None	5.0
C	S1	2.5
C	S5	0
D	None	5.0
D	S4	0
D	S8	2.5

When a wind direction reading is taken, the four A-to-D channels are sampled, and then compared to a lookup table as shown in Table 6-3. Depending on which switches are closed, the software can now decode the wind direction. In practice, the actual bus voltage varies, and is usually lower than the 5.0V used in this example.

Table 6-3   Decoding the A-to-D Voltages

CH A	CH B	CH C	CH D	Direction	Value
5.0	5.0	2.5	5.0	N	0
5.0	2.5	2.5	5.0	NNE	1
5.0	2.5	5.0	5.0	NE	2
2.5	2.5	5.0	5.0	ENE	3
2.5	5.0	5.0	5.0	E	4
2.5	5.0	5.0	0.0	ESE	5
5.0	5.0	5.0	0.0	SE	6
5.0	5.0	0.0	0.0	SSE	7
5.0	5.0	0.0	5.0	S	8
5.0	0.0	0.0	5.0	SSW	9
5.0	0.0	5.0	5.0	SW	10
0.0	0.0	5.0	5.0	WSW	11
0.0	5.0	5.0	5.0	W	12
0.0	5.0	5.0	2.5	WNW	13
5.0	5.0	5.0	2.5	NW	14
5.0	5.0	2.5	2.5	NNW	15

Building the Hardware

Before you put together your wind instrument, you're going to disassemble it! Here are few things that you should do before you install your unit:

- Oil the bearings
- Balance the rotor for longer life
- Align north on the wind vane for easier calibration
- Jumper pins 1 and 4 so you can use the WS-1 in a daisy-chain network

Preparing the Unit

Before you start disassembling your WS-1, pick a suitable work area and round up a Phillips screwdriver and the supplies listed in Table 6-1.

Disassemble the Unit

1. Remove the four screws holding the two halves together.

2. Lay the unit down on a flat surface.

3. Carefully lift off one side of the outer case as shown in Figure 6-3. Note the positions of the rotors and circuit board.

 The circuit board contains glass reed switches. Handle the board carefully.

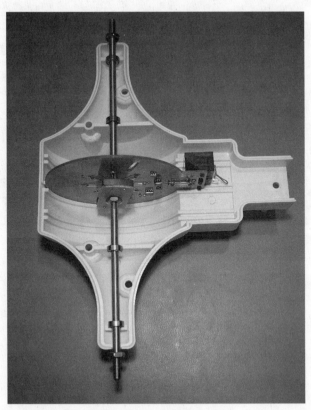

FIGURE 6-3: WS-1 with one case half removed.

Oil the Bearings

1. Remove rotor assemblies.

2. Working over a paper towel or rag, add one drop of household "3-in-1" or motor oil to each of the four bearings. Rotate the bearing to work the oil into the races. See Figure 6-4.

3. Once the oil has been absorbed, wipe off the excess oil.

FIGURE 6-4: Oiling the bearings.

Align the Wind Vane Rotor

1. To keep it balanced, only one of the two magnets on the wind vane rotor is really a magnet. The other is simply a plastic weight. Determine which one is the "real" magnet by using a screwdriver or iron nail and gently touching it to each magnet. Mark the magnet as "north" using a permanent marker.

2. Make sure one of the 10/32 nuts is fully screwed onto the wind vane shaft. Do not tighten the nut.

3. Insert the wind vane on the shaft, aligning the pointer with the north magnet and the 10/32 nut. You may have to slightly loosen the inner nut to get a good alignment.

4. Once the two are aligned, install the outer 10/32 nut. Tighten the nut snugly, but not too tight so you don't break the plastic. The completed assembly should look like Figure 6-5.

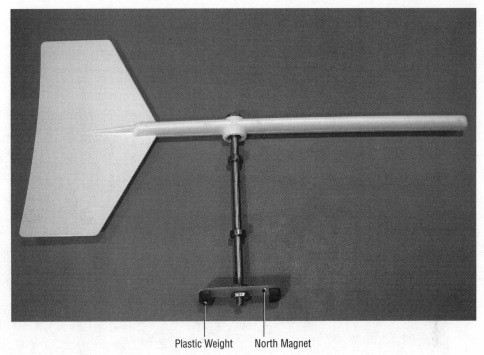

Plastic Weight   North Magnet

FIGURE 6-5: Align the pointer with the magnet.

Add a Jumper (Optional)

The temperature sensor in the WS-1 is wired for external power from the factory. For parasitic power, a jumper is plugged into one of the two RJ-14 connectors, connecting pins 1 and 4. Unfortunately, this ties up one of the RJ-14s preventing you from daisy-chaining to other units. If you desire this capability, you can add an internal jumper freeing up both RJ-14s.

1. Remove the circuit card from the remaining half of the housing.

2. Make a jumper wire by stripping ⅛ inch of insulation off of a 1-inch piece of 24 gauge stranded wire and tin both ends.

3. Position the circuit board component-side up with the RJ-14 connectors facing away from you.

4. Locate the two unused solder pads to the left and below the RJ-14 pins.

5. Clean the area thoroughly with isopropyl alcohol.

6. Carefully solder the jumper wire across the two pads shown in Figure 6-6. You may have to get the solder a bit hotter than normal to burn through the circuit board's coating.

7. Make sure the wire lays flat against the board because the wind speed rotor passes directly over this area. The wire must not be higher than the reed switches.

8. Clean the area thoroughly with alcohol.

9. You may want to waterproof the pads by applying a small coating of silicon RTV over the connections.

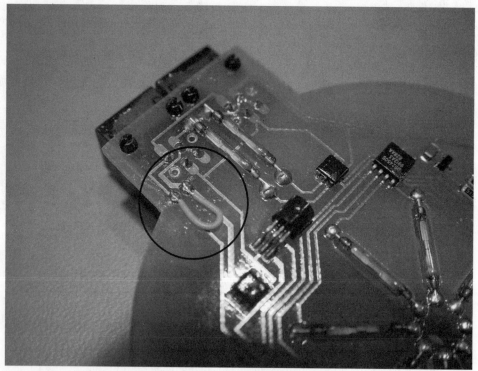

FIGURE 6-6: Jumper the temperature sensor for parasitic power.

Warning After installing this modification, *do not* plug this unit into a device that supplies power using a 4-conductor data cable. Because pins 1 and 4 are shorted, this could damage the device supplying power. Instead, use a 2-conductor (1-pair) data cable connected to pins 2 and 3 only.

Reassemble the Unit

1. Place the circuit card back in the housing. The circuit board components should face down as shown earlier in Figure 6-3.

2. Place the wind speed rotor in the top section, making sure the bearings are on the inside of the mounting tabs as shown in Figure 6-7. If you installed the jumper, spin the rotor and make sure it does not touch the jumper.

3. Place the wind vane rotor on the lower section making sure the bearings are on the inside of the mounting tabs.

4. Carefully set the top half of the housing over the bottom half. Make sure the bearings and circuit board slide into the correct grooves.

5. Install the four 6/32 screws and lock nuts. The shorter screws go toward the ends of the unit. Tighten the screws securely.

6. Place the wind cup assembly on the wind speed rotor shaft, aligning it with the hex nut.

7. Install the 1-inch washer and 10/32 nut from the parts bag. Center the washer and tighten the nut snugly.

8. You may want to add a drop of Silicon RTV to where the nut contacts the threaded shaft to prevent the nut from loosening in high winds.

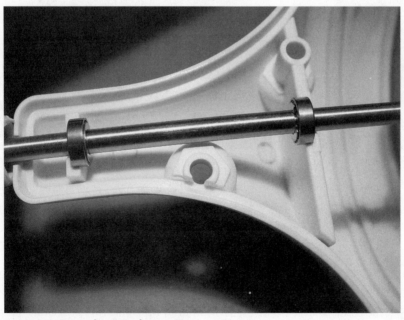

FIGURE 6-7: Proper bearing placement.

Balance the Wind Speed Rotor and the Wind Vane

1. Locate the 1-inch square mounting tube. Temporarily insert the arm into the wind unit.

2. Orient the wind unit horizontally and clamp the mounting arm to a flat surface as shown in Figure 6-8. Make sure both the wind speed rotor and weather vane can rotate a full 360 degrees.

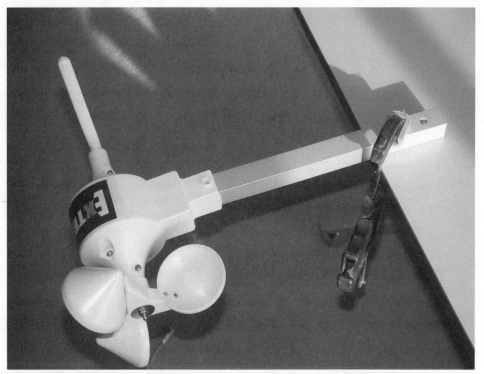

FIGURE 6-8: Preparing to balance the rotors.

3. Gently rotate the wind speed cups. Notice which cup is heavier and falls to the bottom. Apply a small amount of metal tape to the back side of the topmost cup.

4. Gently rotate the cups again. Continue adding tape to the top cup(s) until the wind speed rotor seems balanced. You don't have to balance it perfectly. The rotor should spin slowly and stop smoothly without swinging back and forth. Figure 6-9 shows the placement of the metal tape.

5. Note the position of the weather vane. Most likely the pointer will be pointing up.

6. Wrap the metal tape around the pointer end until it balances and rotates smoothly through 360 degrees. It doesn't have to be perfectly balanced, but from the factory, it's way off! Check out Figure 6-10.

FIGURE 6-9: Balancing the wind speed rotor.

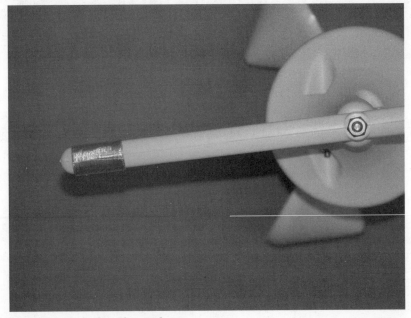

FIGURE 6-10: Balancing the wind vane.

Prepare the Mounting

1. Open the mast mount bracket hardware you purchased.

2. Locate the mounting plate and place approximately 1 inch from the end of the mounting arm. Make sure that you're using the opposite end from the one with the holes pre-drilled to secure the WS-1, as shown in Figure 6-11.

3. Mark and center-punch the holes on the mounting plate.

4. Drill two ⁵⁄₁₆-inch holes where marked. De-burr both the inside and outside with a file.

5. Insert the U-Bracket through the mounting plate and into the mast.

6. Install the lock washers and nuts provided in the kit. You may want to add a couple of ³⁄₈-inch washers under the lock washers. There's no need to tighten them yet.

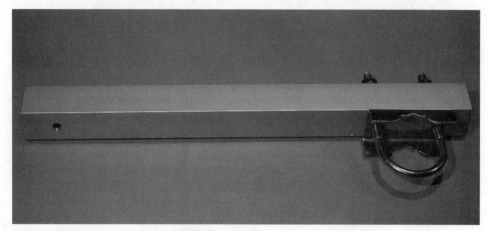

FIGURE 6-11: Mast mount bracket installed.

Note

There are many options for mounting the wind instrument. For mast or tower mounting, using a mast mount bracket is required. If you're not sure how you are going to mount your unit, you may wish to postpone this section.

Assemble the Instrument

1. You will need a data cable to connect the WS-1 to your 1-Wire network. The WS-1 uses RJ-14 connectors, and Table 6-4 lists the connector pin-outs for reference. If you've completed the first project, you have already built the data cable to the WS-1, built an in-line temp sensor with an RJ-14, or are using the Hobby-Boards temp sensor with an RJ-45. If you're using the in-line sensor (or want to connect directly to the DS9097U),

you'll need an RJ-14 to RJ-14 data cable. If you're using the Hobby-Boards module, you'll need an RJ-14 to RJ-45 data cable.

2. Slide the RJ-14 end of the data cable down the mounting arm.

3. Connect the cable to the WS-1.

4. If you decided not to install the soldered jumper, install the supplied RJ-14 jumper in the second RJ-14 jack on the WS-1.

5. Carefully slide the WS-1 on to the mounting arm, lining up the holes on the body with the holes in the mounting tube. Because you're not mounting the unit yet, you don't need to install the mounting screw to attach the WS-1.

6. Connect the WS-1 to the module you built in the last project or directly to the DS9097U.

7. You are now ready to add wind speed and direction to your SimpleWeather software project.

Table 6-4   WS-1 Connector Pin-outs

Function	RJ-14
DS18S20 External Power	1
Data	2
Return	3
Return	4

Add Wind Capability to Your Software

Before you start adding new capability to SimpleWeather, quickly review what you did in the last project:

- First, you added the TempSensor class.

- Next, you added a user constant for the device serial number, a class variable to point to the TempSensor class, and a float variable to hold the temperature.

- Then you added a call to instantiate the TempSensor class.

- Finally, you called the TempSensor class's `GetTemperature()` method to get the temperature, and called `System.out.println()` to print the temperature.

In this project, you are going to do nearly the same thing, except that there are two 1-Wire devices; one for wind speed and one wind direction. You will need two variables to hold the results from the two sensors.

Add the WindSensor Class

Locate your WeatherToys directory and then find the project source code you downloaded from the companion web site. Next, find the Chapter 6 directory. You will find a single file and a folder. The file `WindSensor.java` is the source file for this project. The folder named Chap 6 Project is a completed project for this chapter. If you have problems getting your project to work, you can refer to this project.

Copy the `WindSensor.java` file to your SimpleWeather Project directory ➪ src ➪ simple-weather folder.

Modify SimpleWeather

Launch NetBeans and open your SimpleWeather project. In the Projects window, double-click the SimpleWeather project, double-click Source Packages, and then double-click the SimpleWeather package to see a list of your source files. If you put `WindSensor.java` in the right spot, you will see three source files, one of them being `WindSensor.java`.

Open `SimpleWeather.java` in the NetBeans editor. Scroll down just past the `// 1-Wire Device` comment at line 30. Add the following two lines:

```
// 1-Wire Devices
private final String TEMP_SENSOR_ID = "3F00080008F42810";
private final String WIND_SPD_ID = "";
private final String WIND_DIR_ID = "";
```

Just like the last project, you'll get the serial number for these two constants in just a second.

Scroll down to `// class variables` near line 40 and add two variables for wind. One is a float for the wind speed, and the other is an int (integer) variable for the wind direction:

```
// class variables
public float temp;
public float windSpeed;
public int windDir;
```

Next, add a class variable for the WindSensor class after the `// sensors` comment at line 50:

```
// sensors
private TempSensor ts1;
private WindSensor ws1;
```

Now that you have the constants for the two serial numbers, and a variable to hold the class, you can instantiate the WindSensor class. Notice that you pass both serial numbers to this class. Scroll down to the `// initialize sensors` comment near line 94 and add the following:

```
// initialize sensors
ts1 = new TempSensor(adapter, TEMP_SENSOR_ID);
ws1 = new WindSensor(adapter, WIND_SPD_ID, WIND_DIR_ID);
```

Finally, in the main loop locate the comment `// get the weather`. Add the following lines to get the wind speed and direction and print it to the screen:

```
// get the weather
// get temperature
temp = ts1.getTemperature();
System.out.println("Temperature = " + temp + " degs F");

// get wind speed & direction
windSpeed = ws1.getWindSpeed();
windDir = ws1.getWindDirection();
System.out.println("Wind Speed = " + windSpeed + " MPH " +
                    "from the " + ws1.getWindDirString(windDir));
```

Notice in the preceding code that you get the wind speed and wind direction. Wind speed is returned in miles per hour (MPH), and the wind direction is returned as an integer value. Unless an error occurs during the measurement, the returned value will be 1 of 16 possible values (0 to 15) that represent the 16 compass points. This value is passed to the `getWindDirStr()` method of WindSensor to convert the value from an integer to a String (text) representation of the direction, such as "N," "SE," or "NNW."

Save your work by selecting File ➪ Save or clicking the Save All icon in the toolbar.

Find Your Device Serial Number

Make sure your WS-1 is connected to your 1-Wire bus and that the DWS9097U is connected to your serial port.

Without quitting NetBeans, find the `OneWireLister.jar` program you created in the last project. Launch the jar file and select your serial port. You will see at least three 1-Wire devices listed. Referring back to Table 3-1, you know that family code of 1D is the counter family, which, as you learned earlier, is used for the wind speed counter. The family code of 20 is used for the 4-channel A-to-D converter device. Copy and paste the serial number with the family code 1D into the `WIND_SPD_ID`, and the family code of 20 into the `WIND_DIR_ID`. Your code should look similar to this:

```
// 1-Wire Devices
private final String TEMP_SENSOR_ID = "3F00080008F42810";
private final String WIND_SPD_ID    = "87000000013E301D";
private final String WIND_DIR_ID    = "B200000000F3B620";
```

Notice that I've added a few extra spaces to line things up. This is strictly for readability only; Java doesn't mind if you add extra spaces in your code.

Once you've copied your serial numbers, you can quit OneWireLister and save your NetBeans project.

Testing the Project

At this point, you should be ready to compile and run SimpleWeather. Press F6 (or Run ⇨ Run Main Project).

After the start-up data, your output should start displaying temperature and wind data once a minute:

```
init:
deps-jar:
compile:
run:
Starting SimpleWeather 1.0
Found Adapter: DS9097U on Port /dev/tty.USA19QW4b44P1.1
Resetting 1-wire bus
Time = Fri Jan 13 10:28:52 MST 2006
Temperature = 75.2 degs F
Wind Speed = 0.0 MPH from the  N

Time = Fri Jan 13 10:29:00 MST 2006
Temperature = 75.2 degs F
Wind Speed = 0.0 MPH from the  N
```

Now spin the wind speed cups and give the wind vane a push. Your data should show the wind speed and the new wind direction:

```
Time = Fri Jan 13 10:30:00 MST 2006
Temperature = 75.2 degs F
Wind Speed = 1.8012383 MPH from the SE

Time = Fri Jan 13 10:32:00 MST 2006
Temperature = 75.2 degs F
Wind Speed = 3.4391294 MPH from the  E
```

Go ahead and play with your weather toy for a while. Check out each of the wind directions to make sure that your WS-1 is working properly. If you spin the wind vane quickly, you may occasionally get a wind direction value of 16 and an "ERR" message. This is normal. What's happening is that each of the 4 DS2450 A-to-D channels gets read in sequence. If the wind vane is moving very quickly, the reed switches may change state during the read process, resulting in a value that is not listed in the lookup table. If this occurs, the getWindDirection() method returns a value of 16. If this value is passed to the getWindDirStr() method, it returns the text "ERR," which is displayed.

If your code didn't compile, check your typing. NetBeans will underline errors in red. Look through your code to see if there are any underlines. If so, fix the code. If you get really stuck, there is a complete project in the source code directory for this chapter. If you use this file, don't forget to change the serial port and 1-Wire devices to your values.

Walk Through the Code

You already have learned how the WS-1 works electrically, and you covered the changes to SimpleWeather.java as you coded it. Now, walk through the WindSensor class code. There's quite a bit of code here because you're using two sensors.

Starting with the class declaration, you define the class variables. You will notice that there is a constant called NORTH_OFFSET. This is where you can adjust the software to compensate for mounting the WS-1 in an orientation that is not pointed north.

To measure wind speed, you have to know two things: how many counts did you get and how long has it been since you last checked the count. You use two variables to keep track of the last measurement count and time: lastCount and lastTicks. You also have two OneWireContainers; one is used for the wind speed and the other is for wind direction:

```
public class WindSensor
{
  // calibration constants
  private final int NORTH_OFFSET = 0;

  // class variables
  private DSPortAdapter adapter;
  private long lastCount = 0;
  private long lastTicks = 0;
  private OneWireContainer1D windSpdDevice = null;
  private OneWireContainer20 windDirDevice = null;
  private static boolean debugFlag = SimpleWeather.debugFlag;
```

Next is the class constructor. Just like you read in the last chapter, constructors have the same name as the class and get called automatically when the class is created. Here you create instances of the two 1-Wire device containers:

```
public WindSensor(DSPortAdapter adapter, String windSpdDeviceID,
                                 String windDirDeviceID)
  {
    // get instances of the 1-wire devices
    windSpdDevice = new OneWireContainer1D(adapter, windSpdDeviceID);
    windDirDevice = new OneWireContainer20(adapter, windDirDeviceID);
  }
```

Here is the wind speed method. First you check to make sure that you have a valid wind speed device (is not equal to null):

```
public float getWindSpeed ()
  {
    float windSpeed = 0f;

    if (windSpdDevice != null)
    {
      try
      {
```

```
if (debugFlag)
{
  System.out.print("Wind Speed: Device = " + windSpdDevice.getName ());
  System.out.print("  ID = " + windSpdDevice.getAddressAsString () +
                        "\n");
}
```

Then you read the wind speed counter and the current time in milliseconds. Once again, here's where the beauty of a good Java library comes in. Notice that it only takes a single call to readCounter() to get the wind speed counter value:

```
// read wind counter & system time
long currentCount = windSpdDevice.readCounter (15);
long currentTicks = System.currentTimeMillis ();
```

If this isn't the first time you've used this routine (last ticks does not equal 0), you calculate the wind speed by determining the elapsed time (currentCount-lastCount), the number of counts since the last measurement (currentTicks-lastTicks), and convert the time to seconds (/ 1000). These values are plugged into the equation presented earlier in this chapter to calculate wind speed:

```
if (lastTicks != 0)
{
  // calculate the wind speed based on the revolutions per second
  windSpeed = ((currentCount-lastCount) / ((currentTicks-lastTicks)
              / 1000f)) / 2.0f * 2.453f;    // MPH
}

if (debugFlag)
  System.out.println ("Count = " + (currentCount-lastCount) +
  " during " + (currentTicks-lastTicks) + "ms calcs to " + windSpeed);
```

After calculating wind speed, you need to remember the current time and counter value so you can use it next time this routine is called:

```
// remember count & time
lastCount = currentCount;
lastTicks = currentTicks;
```

Once again, here is the "error catcher." If something bad happens in this routine (maybe the WS-1 got unplugged) the wind speed is set to an error value and the call returns:

```
catch (OneWireException e)
{
  System.out.println ("Error Reading Wind Speed: " + e);
  windSpeed = -999.9f;
}
}
return windSpeed;
}
```

Here's the method for getting the wind direction:

```
public int getWindDirection ()
{
  int windDir;

  try
  {
    if (debugFlag)
    {
      System.out.print ("Wind Dir: Device = " + windDirDevice.getName ());
      System.out.print ("  ID = " + windDirDevice.getAddressAsString () +
                        "\n");
    }
```

First, you have to set up the DS2450 A-to-D channels. In your application, you will set all four for an 8-bit reading at 5.12 V full-scale:

```
// set up A to D for 8 bit readings, 5.12 V full-scale
byte[] state = windDirDevice.readDevice ();

windDirDevice.setADResolution (OneWireContainer20.CHANNELA, 8, state);
windDirDevice.setADResolution (OneWireContainer20.CHANNELB, 8, state);
windDirDevice.setADResolution (OneWireContainer20.CHANNELC, 8, state);
windDirDevice.setADResolution (OneWireContainer20.CHANNELD, 8, state);

windDirDevice.setADRange (OneWireContainer20.CHANNELA, 5.12, state);
windDirDevice.setADRange (OneWireContainer20.CHANNELB, 5.12, state);
windDirDevice.setADRange (OneWireContainer20.CHANNELC, 5.12, state);
windDirDevice.setADRange (OneWireContainer20.CHANNELD, 5.12, state);
windDirDevice.writeDevice (state);
```

Then, each channel is commanded to measure the voltage at its input:

```
// command each channel to read voltage
windDirDevice.doADConvert (OneWireContainer20.CHANNELA, state);
windDirDevice.doADConvert (OneWireContainer20.CHANNELB, state);
windDirDevice.doADConvert (OneWireContainer20.CHANNELC, state);
windDirDevice.doADConvert (OneWireContainer20.CHANNELD, state);
```

Then you read it back:

```
// read results
float chAVolts = (float)windDirDevice.getADVoltage(
                        OneWireContainer20.CHANNELA, state);
float chBVolts = (float)windDirDevice.getADVoltage (
                        OneWireContainer20.CHANNELB, state);
float chCVolts = (float)windDirDevice.getADVoltage (
                        OneWireContainer20.CHANNELC, state);
float chDVolts = (float)windDirDevice.getADVoltage (
                        OneWireContainer20.CHANNELD, state);
```

Next, the four voltage readings are passed to the routine that implements the lookup table discussed in the "How It Works" section, which returns a value between 0 and 16. After that, a check is performed to see if value from the `lookUpWindDir()` method returned an error value (16). If so, some diagnostic messages are printed and then you return to the calling program:

```
// convert the 4 A to D voltages to a wind direction
windDir = lookupWindDir (chAVolts, chBVolts, chCVolts, chDVolts);

if (windDir == 16)
  System.out.println ("Wind Direction Error: ");

if (debugFlag || windDir == 16)
{
  System.out.println ("Wind Dir AtoD Ch A = " + chAVolts);
  System.out.println ("Wind Dir AtoD Ch B = " + chBVolts);
  System.out.println ("Wind Dir AtoD Ch C = " + chCVolts);
  System.out.println ("Wind Dir AtoD Ch D = " + chDVolts);
  System.out.println ("Wind Direction     = " + windDir + "\n");
}
}
catch (OneWireException e)
{
  System.out.println ("Error Reading Wind Direction: " + e);
  windDir = 16;
}


return windDir;
}
```

This is the lookup routine called from the `getWindDirection()` method. Notice the list of values near the end of the routine. Look familiar? This is almost the same as Table 6-3, except that you are using a nominal value 4.5V instead of 5V. This routine compares the four voltages passed into the table to determine direction. Because the voltage readings can fluctuate depending on the 1-Wire bus voltage, a tolerance of +/− 1.0V is applied in the lookup. Once a match is found, this method returns with the value of the direction. If you reach the bottom of the list without a match, this method returns 16, indicating an error:

```
// convert wind direction A to D results to direction value
private int lookupWindDir (float a, float b, float c, float d)
{
  int i;
  int direction = 16;


  for (i=0; i<16; i++)
  {
    if(((a <= lookupTable[i][0] +1.0) && (a >= lookupTable[i][0] -1.0)) &&
       ((b <= lookupTable[i][1] +1.0) && (b >= lookupTable[i][1] -1.0)) &&
       ((c <= lookupTable[i][2] +1.0) && (c >= lookupTable[i][2] -1.0)) &&
       ((d <= lookupTable[i][3] +1.0) && (d >= lookupTable[i][3] -1.0)) )
    {
```

```
        direction = i;
        break;
      }
    }
  }
  return direction;
}

static final float lookupTable[][] = {
            {4.5F, 4.5F, 2.5F, 4.5F}, // N    0
            {4.5F, 2.5F, 2.5F, 4.5F}, // NNE 1
            {4.5F, 2.5F, 4.5F, 4.5F}, // NE   2
            {2.5F, 2.5F, 4.5F, 4.5F}, // ENE 3
            {2.5F, 4.5F, 4.5F, 4.5F}, // E    4
            {2.5F, 4.5F, 4.5F, 0.0F}, // ESE 5
            {4.5F, 4.5F, 4.5F, 0.0F}, // SE   6
            {4.5F, 4.5F, 0.0F, 0.0F}, // SSE 7
            {4.5F, 4.5F, 0.0F, 4.5F}, // S    8
            {4.5F, 0.0F, 0.0F, 4.5F}, // SSW 9
            {4.5F, 0.0F, 4.5F, 4.5F}, // SW  10
            {0.0F, 0.0F, 4.5F, 4.5F}, // WSW 11
            {0.0F, 4.5F, 4.5F, 4.5F}, // W   12
            {0.0F, 4.5F, 4.5F, 2.5F}, // WNW 13
            {4.5F, 4.5F, 4.5F, 2.5F}, // NW  14
            {4.5F, 4.5F, 2.5F, 2.5F}, // NNW 15
                            };
```

Lastly, you have the method that converts the numeric direction value to a String (text) representation. To convert from an integer value to a String, you simply create an array containing the String values in order from 0 to 16. The routine simply returns the corresponding value in the array indexed by the input value. This routine also adds the NORTH_OFFSET calibration factor coded above:

```
// convert direction value into compass direction string
  public static String getWindDirStr(int input)
  {
    String[] direction = {" N ", "NNE", "NE ", "ENE",
                          " E ", "ESE", "SE ", "SSE",
                          " S ", "SSW", "SW ", "WSW",
                          " W ", "WNW", "NW ", "NNW",
                          " ERR"};

    if (debugFlag)
      System.out.println("GetWindDirectionString input = " +
            input + " and cal = " + NORTH_OFFSET);

    // valid inputs 0 thru 16
    if (input < 0 || input >= 16)
      input = 16;
    else
      input = (input + NORTH_OFFSET) % 16;
```

```
if (debugFlag)
  System.out.println("Wind Direction Decoded = " +
        input + " = " + direction[input]);

return direction[input];
}
```

Calibration

The WS-1 has been pre-calibrated for wind speed, and in spite of its low cost, is fairly accurate. Figure 6-12 shows the results of a wind tunnel test performed on a WS-1. At the slower speeds, the WS-1 is about 5% low. At about 10 MPH, the error increases to approximately 10%.

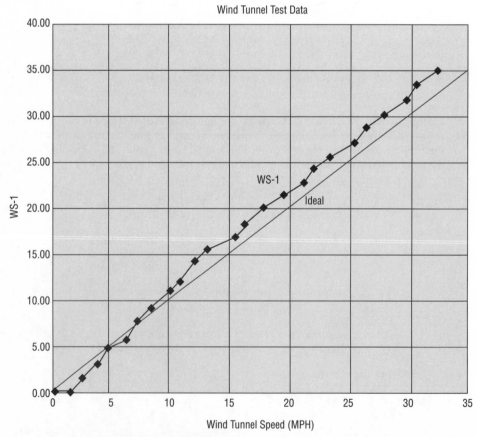

FIGURE 6-12: Wind tunnel test results for the WS-1.

When installing the WS-1, you will have to either orient the mounting tube north because the wind vane north was aligned to the tube, or change the NORTH_OFFSET constant. Referring to Table 6-3, you can determine the offset value based on the orientation of the wind vane. For example, if your WS-1 (the mounting tube) is pointed west, looking up west in the table shows the value of 12. Copy this value to your user constant NORTH_OFFSET.

Wrap Up

Congratulations, you've completed another weather module! You now have temperature, wind speed, and wind direction capability. With just a few more modules, you will have a complete weather station and be ready to install it outdoors.

This is your second project, and you may have noticed a pattern. With each project, you get to pick from the options, build the hardware, add the software, and then test the project. Hopefully, you're finding building your weather station fun.

You might have also noticed a pattern in the software part of the project: You add a constant for the device serial number, add a variable for the module or sensor output, add a variable for the class, instantiate the class, and finally call the class's get method(s) to get the weather data. If you're a seasoned programmer, you are probably already thinking of how to hack this software. If you're a novice, I hope you've been able to follow along and it's starting to make sense.

What's the Humidity?

The next step in building your weather station is adding humidity measurement capability. The humidity module described in this chapter uses the Honeywell HIH sensor you read about in Chapter 1. Similar to the past few chapters, you will build the hardware, add the software, and test your new module. Although the design presented here doesn't require it, I'll also show you how to calibrate your humidity module.

As an added bonus, once you get your humidity sensor working, you can add dewpoint to your list of weather data with just a few lines of code and no extra hardware!

This chapter requires that you have completed the software framework in Chapter 4 as well as the temperature sensor in Chapter 5. Also, your 1-Wire bus must be connected and functional to test the module.

Project Overview

Before you get started, you will need to decide which hardware option you want. If you're experienced at soldering, building from scratch or from a kit may appeal to you. As I get older, the parts keep getting smaller and I can't solder them without cooking them to a crisp or losing them!

Options

This project has several options:

- **Option 1: Build it from scratch.** If you choose this option, you can use the parts list and schematic shown later in the chapter. Because this project uses surface-mount components, using the PC board from Hobby-Boards is strongly recommended. If you used the Hobby-Boards PC board for the temperature module in Chapter 5, this project can be built on the same board.

- **Option 2: Build the Hobby-Boards kit.** This kit uses surface-mount components, which can be pretty tricky to solder. If you're not experienced at soldering, this option may not be for you.

- **Option 3: Buy an assembled module.** Hobby-Boards offers a humidity sensor module with an optional temp sensor and light level sensor. AAG offers its TAI-8540 Humidity Module. Both modules use a nearly identical circuit for humidity measurement.

How It Works

Humidity is measured using a Honeywell HIH-3610 humidity sensor. This device has very low current consumption, so it can be powered directly from the 1-Wire bus. Referring to the schematic in Figure 7-1, power for the humidity sensor is supplied from the 1-Wire bus through diode D1 and capacitor C2. The output of the HIH-3610 is filtered slightly by R1 and C1, and is then applied to the A-to-D input of the DS2438. Diode D2 protects the circuitry from reverse voltages on the 1-Wire bus.

The DS2438 is a 1-Wire Battery Monitor IC. Why are you using a Battery Monitor? It turns out that this device has three A-to-D inputs: one is connected directly to the chip's power supply pin (Vdd), one is a general-purpose input (Vad), and one is used in a differential mode to monitor current flow. For your humidity application, you aren't interested in current. But you do need two A-to-Ds. You need to measure the HIH-3610's output, and because the HIH-3610 is a ratiometric device, its output is proportional to its supply, so you also need to know the supply voltage value. Because both devices are connected to the same power source, you can use the DS2438's Vdd A-to-D to measure the supply voltage for the humidity sensor.

The HIH-3610 data sheet provides the necessary equation for converting the measured voltage output and supply voltage to percent relative humidity at 25 °C:

$$\text{Sensor RH} = (\text{Vad/Vdd} - 0.16) / 0.0062$$

Digging a bit deeper into the data sheet, the manufacturer also provides an equation for temperature compensating the HIH-3610's output:

$$\text{Compensated RH} = \text{sensor RH} / (1.0546 - 0.00216 * T)$$

where T is the sensor's temperature in degrees C. Seems like you also need the temperature to apply this equation. Well, guess what! The DS2438 also has its own built-in temperature sensor. With just a few more lines of code, you can read back the temperature and apply the compensation. You will get to see this in the code later in this chapter.

Parts List

If you're building the kit from scratch, Table 7-1 lists the parts in the schematic. It does not list the additional parts needed to use the external power that the Hobby-Boards module supports. For some of the common parts, I have not listed part numbers. Because building this from scratch is not for beginners, I leave the final parts selection up to you.

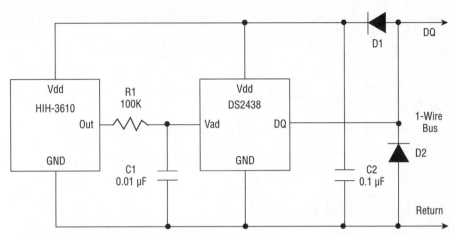

FIGURE 7-1: Output voltage at nominal station pressure.

Table 7-1 Basic Parts List

Qty	ID	Description	Vendor
1	U1	DS2438Z	Digikey, Maxim, Hobby-Boards
1	U2	HIH-3610-001	Hobby-Boards
1	D1, D2	IN5617 or BAT54S	Digikey
1	R1	100K 1/8W	Digikey
1	C1	0.01 MF 10V Ceramic	Digikey
1	C2	0.1 MF 10V Ceramic	Digikey

Note The HIH-3610-001 is being updated to an HIH-4000-001 but the new part wasn't available as this was being written. Both parts are interchangeable and should function identically in this application. Check the companion web site to see if there is any updated information on the HIH-4000-001.

Building the Hardware

Once you have chosen which option you want to use, you are ready to start building the hardware.

Your humidity module is ESD sensitive. Always use an anti-static wrist strap when handling or calibrating your module.

Option 1: Building from Scratch

If you're building from scratch, component placement is not critical. You can use a surface-mount prototyping board offered from several online electronics dealers. Try to keep the overall size as small as possible, because you will need to install the completed module outdoors in some sort of protective container.

Option 2: Building the Kit

1. Make sure you have all the parts as shown in Figure 7-2.

FIGURE 7-2: Hobby-Boards kit parts.

2. Clean the board. Wipe the solder-side of the PC board with alcohol to remove any residue or contamination.

3. Solder the parts. Insert the parts in the circuit board and solder them carefully.

4. Clean-up. Clean the solder connections with a soft cloth dampened with isopropyl alcohol to remove any flux residue. Over time, this residue will collect moisture and short out your 1-Wire bus. Use a pair of good quality diagonal cutters to trim excess lead length from the components.

5. Inspect. Double-check your work for correct component installation and make sure there are no shorted traces. Figure 7-3 shows the completed board.

6. Test. If you added the circuitry for external power, apply +12V to the screw terminals marked +12V and GND. Using a voltmeter, measure the voltage from GND to the jumper connector pin 1. You should read +5V. This project won't be using external power, so make sure to remove the jumper when you are done testing.

Note The HIH-3610's chip is exposed to allow air to contact the sensor. Be very careful and do not touch the exposed semiconductor material. Also, this part is sensitive to volatile chemicals. When cleaning the board, do not let the alcohol contact the sensor element.

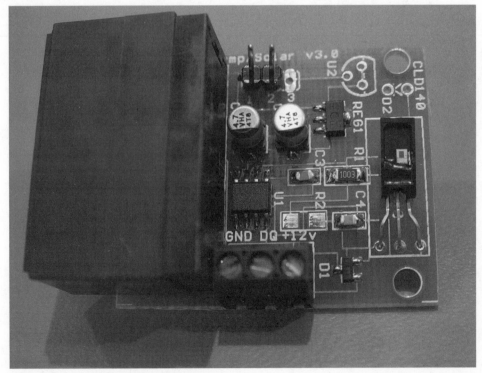

FIGURE 7-3: Completed Hobby-Boards module.

Option 3: Using the Completed Module

If you've purchased the completed Hobby-Boards or AAG humidity modules, or built your own, you need to build a data cable to connect the module to your 1-Wire bus. Decide where you want to connect the module in your test configuration and build the necessary data cable. Table 7-2 lists the pin connections for both the Hobby-Boards and AAG module. Remember, you only need Data and Return for these projects. Once your cable is complete and tested, go ahead and connect your humidity module to your 1-Wire bus.

Note

If you're using the Hobby-Boards module, it comes with a jumper that allows external power operation. Because you will be using the humidity module in a bus-powered mode, be sure to remove the jumper before proceeding.

Table 7-2   Humidity Module Pin-outs

Function	Hobby-Boards RJ-45	AAG RJ-14
Data	4	2
Return	3,5	3
External Power (+12V)	7	N/A
Unused	1,2,6,8	1,4

Add Humidity Capability to Your Software

Here's a quick review what you did in the last project:

- Added the WindSensor class.
- Added two user constants for the two device serial numbers: one for wind speed and one for wind direction.
- Added a class variable to point to the WindSensor class.
- Added a float and int variable to hold the wind speed and wind direction, respectively.
- Instantiated the WindSensor class.
- Called the method GetWindSpeed() to get the wind speed and called getWindDirection() to get the wind direction value.
- Finally, you used System.out() to print the wind speed and wind direction to the screen.

You are going to continue with this pattern for the humidity module in this project.

Add the HumiditySensor Class

Locate your WeatherToys directory and find the project source code you downloaded from the companion web site. Next, find the Chapter 7 directory. You will find a single file and a folder. The file HumiditySensor.java is the source file for this project. The folder named Chap 7 Project is a completed project. If you have problems getting your project to work, you can refer to this project.

Copy the HumiditySensor.java file to your SimpleWeather Project directory ⇨ src ⇨ simpleweather folder.

Modify SimpleWeather

Launch NetBeans and open your SimpleWeather project. Look in your Source Packages, and check your list of source files to make sure HumiditySensor.java is listed.

Open SimpleWeather.java in the NetBeans editor. Scroll down just past the // 1-Wire Device comment to line 35. Add the following line:

```
// 1-Wire Devices
private final String TEMP_SENSOR_ID    = "550008000881E510";
private final String WIND_SPD_ID       = "C20000000143E01D";
private final String WIND_DIR_ID       = "3A00000000CE6920";
private final String HUMIDITY_SENSOR_ID = "";
```

Scroll down to // class variables near line 41. Add a float variable for the humidity results at the end of the list:

```
// class variables
public float temp;
public float windSpeed;
public int windDir;
public float humidity;
```

Next, add a class variable for the HumiditySensor class after the // sensors comment at line 53:

```
// sensors
private TempSensor ts1;
private WindSensor ws1;
private HumiditySensor hs1;
```

Here's where you instantiate the HumiditySensor class. Scroll down to the // initialize sensors comment near line 98 and add the following:

```
// initialize sensors
ts1 = new TempSensor(adapter, TEMP_SENSOR_ID);
ws1 = new WindSensor(adapter, WIND_SPD_ID, WIND_DIR_ID);
hs1 = new HumiditySensor(adapter, HUMIDITY_SENSOR_ID);
```

Finally, in the main loop, locate the comment `// get the weather`. Add the following lines to get the humidity and print it to the screen:

```
// get the weather
...

// get wind speed & direction
windSpeed = ws1.getWindSpeed();
windDir = ws1.getWindDirection();
System.out.println("Wind Speed = " + windSpeed + " MPH " +
                   "from the " + ws1.getWindDirString(windDir));
// get humidity
humidity = hs1.getHumidity();
System.out.println("Humidity = " + humidity + "%");
```

By the time you finish the next project, you'll be able to do this in your sleep! Save your work by selecting File ⇨ Save or clicking the Save All icon in the toolbar.

Find Your Device Serial Number

Make sure your humidity module is connected to the 1-Wire bus.

Launch the `OneWireLister.jar` program. If the rest of your sensors are connected, you should see several 1-Wire devices listed. As you read earlier, the humidity sensor uses the DS2438 Battery Monitor device. Referring back to Table 3-1, you can see that this device belongs to family code 26. Copy and paste the serial number with family code 26 into the `HUMIDITY_SENSOR_ID` constant in the `HumiditySensor.java` file:

```
// 1-Wire Devices
private final String TEMP_SENSOR_ID     = "3F00080008F42810";
private final String WIND_SPD_ID        = "87000000013E301D";
private final String WIND_DIR_ID        = "B200000000F3B620";
private final String HUMIDITY_SENSOR_ID = "3A00000074AE1126";
```

Notice that I've again added a few extra spaces to line things up. Once you've copied your serial numbers, you can quit OneWireLister and save your NetBeans project.

Testing the Project

At this point, you are ready to compile and run SimpleWeather. Press F6 (or Run ⇨ Run Main Project).

After the start-up data, your output should start displaying temperature, wind, and humidity data once a minute:

```
Starting SimpleWeather 1.0
Found Adapter: DS9097U on Port /dev/tty.USA19QW4b44P1.1
Resetting 1-wire bus
Time = Sat Jan 14 18:55:41 MST 2006
Temperature = 75.2 degs F
```

```
Wind Speed = 0.0 MPH from the  N
Humidity = 16.082481%

Time = Sat Jan 14 18:56:00 MST 2006
Temperature = 75.2 degs F
Wind Speed = 0.0 MPH from the N
Humidity = 15.314216%
```

Spin the wind speed cups and give the wind vane a push to make sure wind speed still works:

```
Time = Sat Jan 14 18:57:00 MST 2006
Temperature = 75.2 degs F
Wind Speed = 1.5039222 MPH from the SE
Humidity = 15.742126%
```

To make sure your humidity module is working, gently blow on the humidity sensor. The moisture in your breath should cause the humidity to rise:

```
Time = Sat Jan 14 18:59:00 MST 2006
Temperature = 75.2 degs F
Wind Speed = 0.0 MPH from the E
Humidity = 43.736526%
```

Go ahead and play with your new weather toy for a while. When you're done, quit SimpleWeather.

If your code didn't compile, check your typing. NetBeans will underline errors in red. Look through your code to see if there are any red underlines. If so, fix the code. If not, look at the error messages in the output window. Is there anything that gives you a clue as to what went wrong? If you get really stuck, there is a complete project in the source code directory for this chapter. If you use this, don't forget to change the serial port and 1-Wire devices to your values.

Dewpoint

Before you look at the HumiditySensor class code, there's an addition to make to your code. If you have also built the temperature sensor, you now have both temperature and humidity. Using these two values, you can calculate dewpoint. This code is included as part of the HumiditySensor class, so all you have to do is call the routine.

First, add a new float variable to the class variable list to store dewpoint:

```
// class variables
public float temp;
public float windSpeed;
public int windDir;
public float humidity;
public float dewpoint;
```

Next, just after the call to getHumidity(), add a call to the calcDewpoint() method in the HumiditySensor class and print the results:

```
// get humidity
humidity = hs1.getHumidity();
System.out.println("Humidity = " + humidity + "%");

// calculate dewpoint
dewpoint = hs1.calcDewpoint(temp, humidity);
System.out.println("Dewpoint = " + dewpoint +  " degs F");
```

Run your program again. This time you will also see the dewpoint. Pretty cool, huh?

```
run:
Starting SimpleWeather 1.0
Found Adapter: DS9097U on Port /dev/tty.USA19QW4b44P1.1
Resetting 1-wire bus
Time = Sat Jan 14 19:33:45 MST 2006
Temperature = 75.2 degs F
Wind Speed = 0.0 MPH from the  E
Humidity = 16.681557%
Dewpoint = 27.356468 degs F
```

Debug Mode

As long as you're trying new stuff, let me show you one more thing. You have probably noticed all these if (debugFlag) statements in your code and may have been wondering what they do.

When programmers are developing code, they sometimes embed print statements at various spots throughout the code. That way, they can check the progress of the code and make sure it's doing what it is supposed to. I've done that too. If a sensor is not reading right, I can turn on the "debug mode" and get a more detailed printout. Here's how you can try it.

In the NetBeans Project window, right-click the SimpleWeather project and select Properties. In the Categories box, click Run, and then in the Arguments text field, type **-d** as shown in Figure 7-4. Then click OK. Back in the main NetBeans window, run your project again. This time, you'll see more info:

```
run:
Starting SimpleWeather 1.0
debug on
Found Adapter: DS9097U on Port /dev/tty.USA19QW4b44P1.1
Resetting 1-wire bus
Time = Sat Jan 14 19:48:27 MST 2006
Temperature: Device = DS1920  ID = 550008000881E510
Temperature = 75.2 degs F
Wind Speed: Device = DS2423  ID = C20000000143E01D
Count = 0 during 1137293308907ms calcs to 0.0
Wind Dir: Device = DS2450  ID = 3A00000000CE6920
Wind Dir AtoD Ch A = 2.3880832
Wind Dir AtoD Ch B = 4.665462
```

```
Wind Dir AtoD Ch C = 4.667493
Wind Dir AtoD Ch D = 4.6647587
Wind Direction     = 4

GetWindDirectionString input = 4 and cal = 0
Wind Direction Decoded = 4 =   E
Wind Speed = 0.0 MPH from the  E
Humidity: Device = DS2438   ID = 3A00000074AE1126
Supply Voltage = 4.69 Volts
Sensor Output  = 1.24 Volts
Temperature    = 23.46875 C / 74.24375 F
Uncomp RH      = 16.83747162803494%
Hum Gain       = 1.0
Hum Offset     = 0.0
Calibrated RH  = 16.771935%

Humidity = 16.771935%
Dewpoint = 27.48794 degs F
```

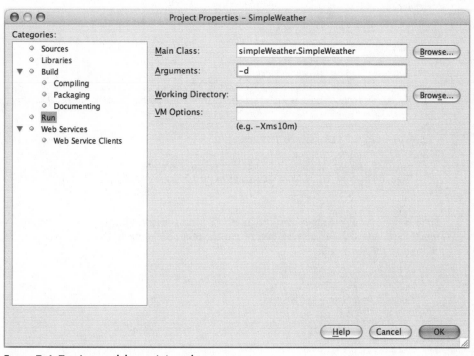

FIGURE 7-4: Turning on debug print mode.

With the -d option, you have enabled debug mode. Looking through the printed data, you can see intermediate steps, 1-Wire bus voltages, and raw sensor output. This is very helpful when things aren't working right. To turn it off, just go back to the Properties dialog and remove the -d option in the Run category Arguments field.

If you're curious, look at SimpleWeather's main() method. When SimpleWeather is run in NetBeans, -d gets passed to it as an argument. When the main() method detects the -d, it sets debugFlag to true.

Walk Through the Code

Take a look at the code for the HumiditySensor class. First, you will notice two constants: HUMIDITY_OFFSET and HUMIDITY_GAIN. This is how you adjust the software to calibrate your humidity output, which is covered in the next section. After that is the class variables and the constructor, which creates a OneWireContainer26:

```
public class HumiditySensor
{
  // calibration constants
  private final float HUMIDITY_OFFSET = 0.0f;
  private final float HUMIDITY_GAIN   = 1.0f;

  // class variables
  private DSPortAdapter adapter;
  private String deviceID;
  private OneWireContainer26 humidityDevice = null;
  private static boolean debugFlag = SimpleWeather.debugFlag;


  public HumiditySensor(DSPortAdapter adapter, String deviceID)
  {
    // get an instance of the 1-wire device
    humidityDevice = new OneWireContainer26(adapter, deviceID);
  }
```

Here is the start of the main humidity method. First you assign the output variable to a value to indicate an error, so if an error occurs and the program jumps to the catch() method, it will return the error value:

```
public float getHumidity()
{
  float humidity = -999.9f;

  if (humidityDevice != null)
  {
    if (debugFlag)
    {
```

```
System.out.print("Humidity: Device = " + humidityDevice.getName());
System.out.print("  ID = " + humidityDevice.getAddressAsString() +
                 "\n");
}
```

This is where you read the humidity and temperature from the device. First, you read the device's state, which is used in the next few methods. Next you command the device to perform a temperature conversion, and then read the temperature. After the temperature, you command the DS2438 to perform an A-to-D conversion on the Vad input and read the voltage back. This is repeated on the Vdd A-to-D:

```
try
{
  // read 1-wire device's internal temperature sensor
  byte[] state = humidityDevice.readDevice();
  humidityDevice.doTemperatureConvert(state);
  double temp = humidityDevice.getTemperature(state);

  // Read humidity sensor's output volatge
  humidityDevice.doADConvert(OneWireContainer26.CHANNEL_VAD, state);
  double Vad = humidityDevice.getADVoltage(OneWireContainer26.CHANNEL_VAD,
                                           state);

  // Read the humidity sensor's power supply voltage
  humidityDevice.doADConvert(OneWireContainer26.CHANNEL_VDD, state);
  double Vdd = humidityDevice.getADVoltage(OneWireContainer26.CHANNEL_VDD,
                                           state);
```

Now that you have the supply voltage (Vdd), the humidity sensor output (Vad), and the temperature (temp), these values are plugged into the equations discussed earlier, and ta-da, you have the current humidity:

```
  // calculate humidity
  double rh = (Vad/Vdd - 0.16) / 0.0062;
  humidity = (float)(rh / (1.0546 - 0.00216 * temp));
```

This is the line of code that applies the calibration constants to the measured humidity. Notice that the humidity is multiplied by HUMIDITY_GAIN and then HUMIDITY_OFFSET is added to the result:

```
  // apply calibration
  humidity = humidity * HUMIDITY_GAIN + HUMIDITY_OFFSET;
```

Here is where the debug mode prints the diagnostic data if debug mode is enabled:

```
      if (debugFlag)
      {
        System.out.println("Supply Voltage = " + Vdd + " Volts");
        System.out.println("Sensor Output  = " + Vad + " Volts");
        System.out.println("Temperature    = " + temp + " C / " +
                           ((temp * 9/5) + 32) + " F");
        System.out.println("Uncomp RH      = " + rh + "%");
        System.out.println("Hum Gain       = " + HUMIDITY_GAIN);
        System.out.println("Hum Offset     = " + HUMIDITY_OFFSET);
        System.out.println("Calibrated RH  = " + humidity + "%\n");
      }
    }
    catch (OneWireException e)
    {
      System.out.println("Error Reading Humidity: " + e);
    }
  }
  return humidity;
}
```

The dewpoint method is included as part of this class. It simply requires the temperature (in °F) and the percent humidity to be passed to the method, and it returns the dewpoint. This code has a comment for each line to explain what it is doing. If you modified your software to work in °C, comment out the lines that convert the temperature from °F to °C and back again.

```
public float calcDewpoint(float temp, float hum)
{
  // compute the dewpoint from relative humidity & temperature

  // if necessary, convert to degrees C
  temp = ((temp - 32.0f)/9.0f * 5.0f);

  // now convert to degrees K
  double tempK = temp + 273.15;

  // calc dewpoint
  double dp = tempK/((-0.0001846 * Math.log(hum/100.0) * tempK) + 1.0);

  // convert back to degrees C
  dp = dp - 273.15;

  // and if necessary, convert back to degrees F
  dp = (dp * 9/5) + 32;

  return (float)dp;
}
```

Calibration

By itself, the HIH-3610 humidity sensor boasts accuracy of +/− 2% . However, as you add additional electronics, the errors get compounded. Therefore, checking the calibration of the completed module is suggested, but not required. As you just read, the software provides constants for fine-tuning the calibration to within a few percent of error.

Why Is Calibrating Humidity a Challenge?

You must contend with two issues to calibrate a humidity sensor or module. The first is the wide dynamic range the sensor must operate. Consider a typical range of 10% to 90% from freezing to 100 °F. At the low end of 10%, you have about 10,000 parts per billion of moisture. At the other end of 90%, you have approximately 200,000,000 parts per billion of moisture. That's a 20,000:1 range. Think of it this way: 10% is a swimming pool with a single drop of water, whereas 90% is the swimming pool nearly full. This means that creating a calibration environment that is stable is a bit tricky.

The second issue is that humidity varies with temperature. When calibrating a humidity device, both the humidity and the temperature must be taken into consideration. For example, at room temperature and 50% RH, a 1 °C change results in approximately a 3% RH change. That said, here are two methods for calibrating your module.

 Warning The HIH-3610 Humidity Sensor IC is sensitive to light. When calibrating or installing your humidity module, keep the IC shielded from bright light.

The Calibration Options

You have several options for calibrating your humidity sensor. The option you choose depends on how much effort you want to expend. "Out of the box" accuracy should be better than 10%, which is suitable for most applications. You can check this with a simple voltmeter reading. To achieve accuracy around 5%, you will have to do some calibration, and I'll show you the 2-point method I use. To achieve better than this, you will probably have to visit a calibration lab and have your humidity sensor professionally calibrated.

The 10% Solution

This method will provide an accuracy of 10% or better, and for most users this should be accurate enough. It's easy and only requires a good digital voltmeter. After you've used your humidity sensor for a while and desire higher accuracy, you may want to try the method listed in the next section.

To verify your module is working correctly, do the following:

1. Place the module in a small box or container to protect it from drafts and light. Mount it component-side up and cut a hole in the top of the box so that you can probe the humidity IC's leads. The location should be at a constant temperature and humidity (away from heaters or air conditioner ducts, and so on)

2. Connect it to your 1-Wire bus.

3. Launch the SimpleWeather project and run it so that it is displaying the measured humidity.

4. Allow the humidity module to stabilize to at least 30 minutes.

5. Using a digital voltmeter, reach into the box and very carefully measure the voltage across the Vdd pin and the Ground pin on the humidity sensor. Record the voltage.

6. Repeat the measurement, this time measuring between Vout and Ground. Record the voltage.

7. Calculate the %RH using this equation: RH = (Vout/Vdd − 0.16) / 0.0062;

8. Compare your calculated value to the value displayed using SimpleWeather. You may want to try this a few times and average the results.

9. If the measured value is less than 5% different than the displayed value, you're done. Otherwise:

   ■ Divide the measured value by the displayed value. For example, you calculated 55% but SimpleWeather is displaying 45%, 55 / 45 = 1.22.

   ■ Stop SimpleWeather. Open `HumiditySensor.java` in the editor window.

   ■ Locate the constant `HUMIDITY_SCALE_FACTOR`. Multiply the current value of `HUMIDITY_SCALE_FACTOR` by the value calculated in step a and enter the new value back in `HUMIDITY_SCALE_FACTOR`. For example, initially, `HUMIDITY_SCALE_FACTOR` is 1.0. Multiplying by 1.22 from Step A results in 1.22 (I didn't need a calculator for that one). Remember, `HUMIDITY_SCALE_FACTOR` is only 1.0 initially and that Java requires an "f" on the end of your number so that it knows it is a floating-point value (you'll get an error otherwise).

   ■ Run SimpleWeather again. The new, corrected humidity value is displayed, and should be within 5%. Repeat this process to double-check your work.

The Greater Than 5% Solution

To achieve better than 5% calibration across the 10% to 90% range, you will have to do a 2-point calibration. This requires two reference environments, one at low humidity and one at high humidity.

To create a low-humidity environment, I use a high-quality desiccant to dry out the air in a small closed container. A good choice is Drierite anhydrous calcium sulfate. It is inexpensive and available for purchase online at www.drierite.com. I use the "indicating" version, Drierite part number 23001, which is colored light blue when dry as shown in Figure 7-5. As it absorbs moisture, it will turn a light pink color.

FIGURE 7-5: Drierite desiccating granules.

One way to create a high-humidity environment is with the saturated salt method. Certain salts, when saturated with distilled water, will maintain a constant humidity level in a sealed container. Table 7-3 lists some of the common salts and their humidity levels. For your calibration, you will use Sodium Chloride, commonly known as table salt, which can be purchased at most grocery stores. Make sure you purchase the non-iodized type, because this may impact the results.

Table 7-3   Saturated Salt Humidity Levels at 25°C

Salt	Humidity
Lithium Bromide	6.37%
Lithium Chloride	1.30%
Potassium Acetate	22.51%
Magnesium Chloride	32.80%
Potassium Carbonate	43.16%
Magnesium Nitrate	52.89%
Sodium Bromide	57.57%
Potassium Iodide	68.86%
Sodium Chloride	75.30%
Potassium Chloride	84.34%
Potassium Sulfate	97.30%

To perform the 2-point calibration, you need to make a humidity chamber using a small sealable container, such as a 24-oz. disposable plastic container used to store food in your refrigerator. Here are the steps:

1. Cut a small hole in one end of the plastic container just large enough to fit the data cable's RJ connector through.

2. Insert the data cable through the hole and connect your humidity module inside the container.

3. Seal the hole around the cable using plumber's putty, sticky-tack, or Silly-Putty. Don't use a water-based material, such as modeling clay, because the moisture it contains will affect the results. Make sure to form an airtight seal.

4. Locate the container in an area that has a constant temperature around 77°F/25°C. If the area is subjected to temperature variations, place your plastic container inside an ice chest to help minimize temperature fluctuations. The actual temperature in this section is not critical.

5. Open the Drierite container and pour enough desiccant to cover the bottom of your container about 1 inch deep. Close your Drierite bottle right away to keep it fresh. Your container should look like the one in Figure 7-6. I added a reference humidity probe so you can see the actual humidity levels.

6. Seal the container.

FIGURE 7-6: Plastic container with Drierite desiccant.

7. Launch SimpleWeather and start monitoring the humidity levels. Depending on how dry it was when you started, you should see the humidity level dropping.

8. Let your test chamber stabilize for at least 8 to 12 hours. Fresh Drierite will produce a humidity level around 2%.

9. After the humidity measurements stabilize, record the humidity level SimpleWeather reports. This is one point of your 2-point calibration. In my test, I measured 0.7% RH.

10. Open your humidity chamber and carefully pour out the Drierite.

11. Place a small bowl that will hold about a ¼ cup of salt inside your container.

12. Fill the bowl about ¾ full of the non-iodized (plain) table salt.

13. Carefully add distilled water to the salt until the salt is completely moistened, but not covered with water. If there is standing water in the bowl, add a little more salt to cover the water. Do not allow the salt solution to contact the humidity module. Your chamber should look like the one shown in Figure 7-7.

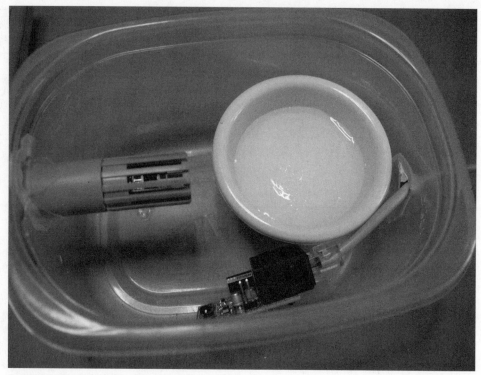

FIGURE 7-7: Humidity chamber with salt solution.

14. Seal the container again.

15. Check the humidity level using SimpleWeather. You should see the humidity slowly rising. Periodically check the humidity levels and allow the chamber to stabilize for 10 to 12 hours at a constant 77°F/25°C. Don't leave your humidity module in the chamber for longer than 24 hours because the salt vapors may cause corrosion of your module's electronics.

16. Check the humidity level. It should be somewhere near 75%. Record the value. In my calibration, the measured value was 70%. In Figure 7-8 the reference probe measured 73.6% RH, pretty close to the expected 75%.

17. Open the container and remove the salt solution. You may want to gently wipe off your humidity module's circuit board with a soft cloth moistened with distilled water to remove any salt residue. Don't wipe the sensor element.

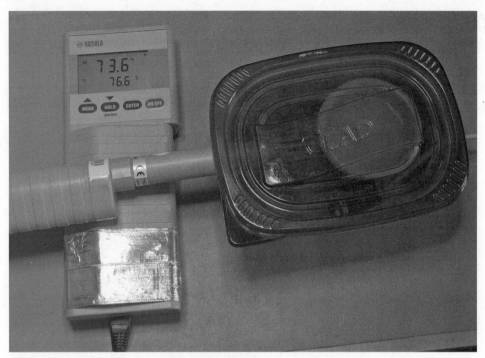

FIGURE 7-8: High humidity setup after 12-hour soak.

18. Calculate the calibration constants as follows:

    HUMIDITY_OFFSET = 2% – your low reading

    HUMIDITY_GAIN = 75% ÷ your high reading

    For example, in my case:

    HUMIDITY_OFFSET = 2% – 0.7% = –1.3

    HUMIDITY_GAIN = 75% ÷ 70% = 1.07

19. Enter your new calibration values in the HumiditySensor.java file.

That's it. You now have a calibrated humidity sensor. Keep your supplies because you may want to recalibrate your module about once a year to make sure it is working properly.

Wrap Up

If you are going to build a second humidity module to monitor indoor humidity, go ahead and build it now while you have your calibration equipment out.

If you didn't buy or build the humidity module with the temperature option, it is easy to add. If you're using the Hobby-Boards module or PC board, all you have to do is purchase a DS18S20 and solder it onto the board in the area indicated.

Congratulations, you've completed another weather module! You now have temperature, wind speed, wind direction, and humidity capability. With just a few more modules, you will have a complete weather station and be ready to install it outdoors.

Getting the Pressure

In this project, you are going to add barometric pressure measurement to your list of measured weather parameters. Barometric pressure is one of the few weather parameters that you can use to predict weather changes. Rising barometric pressure generally indicates clear skies and falling pressure indicates possible stormy weather.

Project Overview

The barometric pressure module, like the humidity module in the last chapter, can be assembled in one of three ways: from scratch, from a kit, or by purchasing the completed module. This module is also the only module in the basic weather station that requires external power.

Options

This project has several options:

- **Option 1: Build it from Scratch.** If you have been building your modules from scratch, you have most likely developed a good method for building your circuits by now. Once again, because this project uses surface-mount components, using the Hobby-Boards PC board is highly recommended.

- **Option 2: Build the Hobby-Boards Kit.** This kit uses just a few surface-mount components. If you're reasonably good at soldering, you may want to try the kit.

- **Option 3: Buy an Assembled Module.** Hobby-Boards offers a pressure sensor fully assembled, along with an optional enclosure.

Note The AAG Barometric Pressure Module uses a different sensor and 1-Wire interface than this project and will not work with this software.

in this chapter

- ☑ Build a Barometric Pressure Module

- ☑ Add Barometric Pressure Capability to Your Software Project

- ☑ Test the Project

- ☑ Calibrate Your New Module

How It Works

Referring to the schematic in Figure 8-1, external power is applied to the +14 to +18V input. It is then regulated to +12V by U4 and +5V by U5.

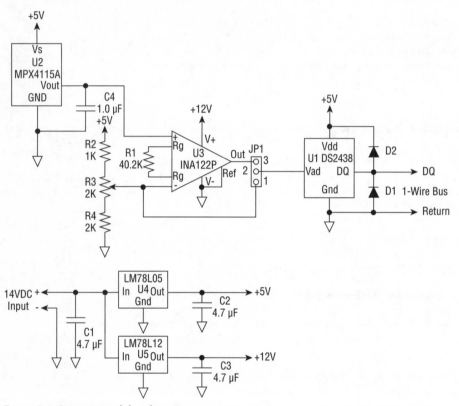

FIGURE 8-1: Pressure module schematic.

The Motorola MPX4115A provides a voltage output proportional to pressure, defined by

$$Vout = Vs * (0.009 * (P - 0.095))$$

where Vs is the supply voltage (+5.0 in your case) and

P is the barometric pressure.

The output of the pressure sensor is applied to an INA122 Instrumentation Amplifier. This device was chosen primarily for its excellent temperature stability and single-supply operation. This device also operates from "rail-to-rail," which, in this application, can output voltages from 0 to the 12V supply. The INA122 is set for a fixed gain of 10 via resistor R1. The output is filtered slightly by C4 and is connected to the DS2438's Vad input.

At normal pressures, the MPX4115A is close to its upper operating range; therefore it has a large DC output as shown in Figure 8-2. When multiplied by the instrumentation amp's gain of 10, you would exceed the 10.23V max input of the DS2438. U3 conveniently has an offset correction input, which is set by resistor R3. Using the offset calculation tool presented later in this chapter, you will set the offset for your altitude so that the instrumentation amp's output is optimized for your altitude. Properly calibrated, this circuit is capable of better than 0.01 inHg resolution from sea level to 10,000 feet.

Jumper JP1 allows the use of the DS2438 to set the offset value without the need for an external voltmeter. The jumper is placed across pins 1 and 2, which allows offset voltage readings in software via the DS2348. Once the offset is set, the jumper is placed back across pins 2 and 3 to read the pressure.

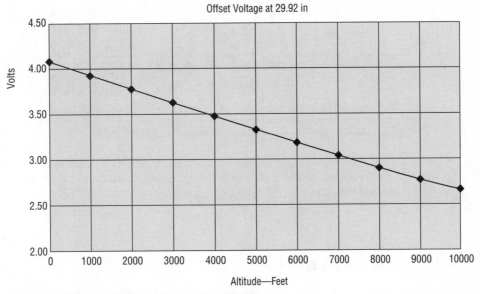

FIGURE 8-2: MPX4115A offset voltage versus altitude.

Parts List

If you're building the kit from scratch, Table 9-1 lists the parts in the schematic. I have not listed part numbers for some of the common parts. Because building this from scratch is not for beginners, I will leave the final parts selection up to you. You'll also need a data cable to connect your humidity module to your other 1-Wire modules and sensors.

Table 8-1   Basic Parts List

Qty	ID	Description	Vendor
1	U1	DS2438Z 1-Wire Battery Monitor	Digikey, Maxim-IC.com, Hobby-Boards
1	U2	MPXA4115A Pressure Transducer	Digikey, Hobby-Boards
1	U3	INA122 Instrumentation Amp	Digikey
1	U4	LM78L05 +5V Regulator	Digikey
1	U5	LM78L12 +12V Regulator	Digikey
1	D1,D2	IN5617 or BAT54S Diode	Digikey
1	R1	40.2K Resistor 1/8W	Digikey
1	R2	1K Resistor 1/8W	Digikey
1	R3	2K Potentiometer	Digikey
1	R4	2K Resistor 1/8W	Digikey
1	C1,C2,C3	4.7MFD Electrolytic Capacitor, 25V	Digikey
1	C4	1 MFD 10V Tantalum Capacitor	Digikey
1	---	14-18 V 200 ma AC Adapter	Radio Shack

Note Even though the minimum voltage requirement for this project is 14V, most 12V unregulated wall transformers supply 14–18V unloaded. Because this design uses very low current, many 12V wall transformers will work.

Building the Hardware

After you have selected the option, gather up the necessary parts and supplies. As noted earlier, this project requires a separate power source. If you are going to be mounting the barometer indoors near the DS9097U, check out the Hobby-Boards power injector module. It provides a regulated 12VDC output that you can use to power this module through the 1-Wire bus cable.

Warning Your barometer module is ESD sensitive. Always use an anti-static wrist strap when handling or calibrating your module.

Option 1: Building from Scratch

If you're building from scratch, component placement is not critical. You can use a surface mount prototyping board offered from several online electronics dealers. Because the barometric pressure module should be installed indoors, overall size is not critical. I suggest that you add screw terminals or a coaxial connector to make it easy to connect the power source.

Option 2: Building the Kit

1. Make sure you have all the parts.

2. Clean the board. Wipe the solder-side of the PC board with alcohol to remove any residue or contamination.

3. Solder the parts. Insert the parts in the circuit board and solder them carefully.

4. Clean-up. Clean the solder connections with a soft cloth dampened with isopropyl alcohol to remove any flux residue. Trim any excess lead length from the components with diagonal cutters.

5. Inspect. Inspect each connection to make sure there are no shorted parts. Figure 8-3 shows the completed board.

6. Test. Connect the +14V power source to the screw terminals observing proper polarity. Plug in the wall transformer to an AC outlet and check the following voltages:

   - +5V — From U4 (REG1) pin 3 to GND. It should be +5V +/− 0.25V

   - +12V — From U5 (REG2) pin 3 to GND. It should be +12V +/− 0.25V

7. Preliminary calibration. Remove Jumper JP1. Connect your voltmeter (−) lead to GND. Connect your voltmeter (+) lead to JP1 pin 1. Adjust R3 for a reading of 3.6V +/− 0.1V.

Option 3: Buy the Completed Module

You can buy the completed module from Hobby-Boards, which also offers a case for the completed module. If you've purchased the completed Hobby-Boards Pressure Module or built your own, you will also need the wall transformer in the parts list. If you want, Hobby-Boards will perform initial calibration of your module if you supply your altitude.

Connecting

Decide where you want to connect the module in your test configuration and build the necessary data cable. If your wall transformer has a connector on the end, cut it off and split the two wires back approximately 1 inch. Strip approximately ¼ inch of insulation off the ends of wire and tin the ends with solder.

FIGURE 8-3: Completed Hobby-Boards module.

Table 8-2 lists the pin connections for the Hobby-Boards module. To keep things simple, you will attach power to the screw terminals shown in Figure 8-3. Use a voltmeter to determine which wire is positive and connect it to the screw terminal marked +14V. Connect the negative wire to the GND terminal. Double-check your work! If you get the wires connected backwards, you'll fry your barometer module!

Table 8-2   Humidity Module Pin-outs

Function	Hobby-Boards RJ-45
Data	4
Return	3,5
External Power (+14V)	7
Unused	1,2,6,8

Add Pressure Capability to Your Software

As usual, quickly review what you did in the last project:

- Added the HumiditySensor class.
- Added a user constant for the device serial number.
- Added a class variable to point to the HumiditySensor class.
- Added two float variables to hold the humidity and dewpoint.
- Instantiated the HumiditySensor class.
- Called the method `GetHumidity()` to get the humidity and called `calcDewpoint()` to get the dewpoint value.
- Used `System.out()` to print the humidity and dewpoint.
- Tried out "debug print mode" to enable printing diagnostic data to the screen.

You are going to continue with this pattern for the pressure module in this project.

Add the PressureSensor Class

Navigate to your WeatherToys directory and find the project source code you downloaded from the companion web site. Next, find the Chapter 8 directory. You will find a single file and a folder. The file `BaroSensor.java` is the source file for this project. The folder named Chap 8 Project is a completed project. If you have problems getting your project to work, you can refer to this project.

Copy the `BaroSensor.java` file to your SimpleWeather Project directory ⮑ src ⮑ simpleweather folder.

Modify SimpleWeather

Launch NetBeans and open your SimpleWeather project. Look in your Source Packages, and check your list of source files to make sure `BaroSensor.java` is listed.

Open `SimpleWeather.java` in the NetBeans editor. Scroll to the `// 1-Wire Device` comment around line 30. Add the following line to the end of the list:

```
// 1-Wire Devices
private final String TEMP_SENSOR_ID     = "5D00080005236F10";
private final String WIND_SPD_ID        = "8800000000CC1E1D";
private final String WIND_DIR_ID        = "3A00000000CE6920";
private final String HUMIDITY_SENSOR_ID = "3C00000073B6F426";
private final String BARO_SENSOR_ID     = "";
```

Scroll down to `// class variables` near line 42. Add a float variable for the pressure results at the end of the list:

```
// class variables
public float temp;
public float windSpeed;
public int windDir;
public float humidity;
public float dewpoint;
public float pressure;
```

Next, add a class variable for the PressureSensor class after the `// sensors` comment at line 55:

```
// sensors
private TempSensor ts1;
private WindSensor ws1;
private HumiditySensor hs1;
private BaroSensor bs1
```

Here's where the PressureSensor class is instantiated. Scroll down to the `// initialize sensors` comment near line 101 and add the following:

```
// initialize sensors
ts1 = new TempSensor (adapter, TEMP_SENSOR_ID);
ws1 = new WindSensor (adapter, WIND_SPD_ID, WIND_DIR_ID);
hs1 = new HumiditySensor(adapter, HUMIDITY_SENSOR_ID);
bs1 = new BaroSensor(adapter, BARO_SENSOR_ID);
```

Finally, in the main loop, locate the comment `// get the weather` around line 130. Add the following lines to get the pressure and print it to the screen:

```
// get the weather
...
// calculate dewpoint
dewpoint = hs1.calcDewpoint(temp, humidity);
System.out.println("Dewpoint = " + dewpoint +  " degs F");

// get barometric pressure
pressure= bs1.getPressure();
System.out.println("Pressure = " + pressure + " inHg");
```

That completes the code modifications. Save your work buy selecting File ⇨ Save or clicking the Save All icon in the toolbar.

Find Your Device Serial Number

Make sure your barometric pressure module is connected to the 1-Wire network and to the 14V power supply.

Launch the `OneWireLister.jar` program. If the rest of your sensors are connected, you should see several 1-Wire devices listed. As you read earlier, the pressure sensor uses the same DS2438 Battery Monitor device as used in the humidity module. Referring back to Table 3-1, you can see that this device belongs to family code 26. Looking through the list of 1-Wire devices, you should see two devices that have a family code of 26. One is for the humidity sensor and one is for the new barometer. How do you know which is which? Well, you could disconnect one of the modules and re-run OneWireLister to see which one disappears. However, if you look at your code, you have the serial number of the humidity module already listed. Copy the serial number that is not the same as the one listed for the humidity device and paste it in the code for the pressure sensor:

```
// 1-Wire Devices
private final String TEMP_SENSOR_ID     = "550008000881E510";
private final String WIND_SPD_ID        = "C20000000143E01D";
private final String WIND_DIR_ID        = "3A00000000CE6920";
private final String HUMIDITY_SENSOR_ID = "3A00000074AE1126";
private final String BARO_SENSOR_ID     = "6D00000073A34F26";
```

Testing the Project

At this point, you should be ready to compile and run SimpleWeather. Press F6 (or Run ➪ Run Main Project).

After the start-up data, your output should start displaying temperature, wind, humidity, and pressure data once a minute:

```
Starting SimpleWeather 1.0
Found Adapter: DS9097U on Port /dev/tty.USA19QW4b44P1.1
Resetting 1-wire bus
Time = Sat Jan 21 17:35:07 MST 2006
Temperature = 73.4 degs F
Wind Speed = 0.0 MPH from the SSW
Humidity = 20.29591 %
Dewpoint = 30.651836 degs F
Pressure = 30.583584000000002 inHg
```

Because you have not calibrated your module yet, don't worry if the value displayed isn't correct.

Walk Through the Code

Take a look at the code for the BaroSensor class. Because the pressure sensor uses the same DS2438 1-Wire device as the humidity module, the code is very similar.

First is the class definition. Notice there are two constants: PRESSURE_GAIN and PRESSURE_OFFSET. These are used to calibrate the pressure sensor output. There is also the declaration of the OneWireContainer for family code 26 that is used to hold the instance of your 1-Wire device:

```
public class BaroSensor
```

```
{
  // calibration constants
  private final double PRESSURE_GAIN    = 0.7171;
  private final double PRESSURE_OFFSET = 26.2523;

  // class variables
  private DSPortAdapter adapter;
  private OneWireContainer26    baroDevice = null;
  private static boolean debugFlag = SimpleWeather.debugFlag;
```

Next is the class constructor. Just like in previous projects, the constructor instantiates a OneWireContainer26 using the device serial number passed in:

```
public BaroSensor(DSPortAdapter adapter, String deviceID)
{
  // get instances of the 1-wire devices
  baroDevice = new OneWireContainer26(adapter, deviceID);
}
```

This is the method that gets the pressure. First, you command the Vad A-to-D to make a reading, and then read it back. This is the output of the pressure sensor after being amplified by the instrumentation amplifier. Then you command an A-to-D conversion on the Vdd channel, and read back the supply voltage. This is very similar to the humidity sensor in Chapter 7.

```
public float getPressure()
{
  double pressure;

  if (debugFlag)
  {
    System.out.print("Pressure: Device = " + baroDevice.getName());
    System.out.print("  ID = " + baroDevice.getAddressAsString() + "\n");
  }

  try
  {
    byte[] state = baroDevice.readDevice();


    // Read pressure A to D output
    baroDevice.doADConvert(OneWireContainer26.CHANNEL_VAD, state);
    double Vad = baroDevice.getADVoltage(OneWireContainer26.CHANNEL_VAD,
                                  state);


    // Read Supply Voltage (for reference only)
    baroDevice.doADConvert(OneWireContainer26.CHANNEL_VDD, state);
    double Vdd = baroDevice.getADVoltage(OneWireContainer26.CHANNEL_VDD,
                                  state);
```

After you have Vad and Vdd, you simply apply the calibration factor and offset to convert to inches of mercury or millibars:

```
        // apply calibration
    pressure = Vad * PRESSURE_GAIN + PRESSURE_OFFSET;

    // scale to mb if required
    //pressure *= 33.8640;

    if (debugFlag)
    {
      System.out.println("Sensor Output  = " + Vad + " Volts");
      System.out.println("Supply Voltage = " + Vdd + " Volts");
      System.out.println("Scale Factor   = " + PRESSURE_GAIN);
      System.out.println("Offset         = " + PRESSURE_OFFSET);
      System.out.println("Baro Pressure  = " + pressure + "\n");
    }
  }
  catch (OneWireException e)
  {
    System.out.println("Error Reading Baro Sensor: " + e);
    pressure = -99.99;
  }

  return (float)pressure;
}
```

Calibration

One of the topics discussed in Chapter 1 was station pressure. It was defined as altitude-compensated pressure measurement. As you read in the "How It Works" section, the circuit is optimized for a specific altitude by adjusting R3.

To calibrate your barometer, you need to know your (or actually your barometer's) altitude. Included in the Tools folder on the companion web site is an Excel spreadsheet that provides you with the calibration information needed by simply entering your altitude. Look inside the Tools folder for calibration.xls. Open it and click the Baro Cal tab. The good folks at Hobby-Boards have also provided a calibration calculator online at www.hobby-boards.com/catalog/baro_calc.php.

If you don't know your altitude, most hand-held GPS devices provide an altitude feature that displays altitude within 20 to 40 feet, or you may be able to find it online. If you're going to install your barometer module at another location, use that location's altitude for the following steps.

Calibrating Your Barometer

1. Open the Excel calibration file for the barometer or use the Hobby-Boards online version.

2. Enter the altitude that you want to calibrate your barometer for.

3. Note the offset voltage, gain (slope), and offset (intercept).

4. To use a digital voltmeter to set the offset (faster), follow these steps:

    a. Turn on your digital voltmeter and set it for Volts DC.

    b. Remove the jumper from the Barometer Module pins.

    c. Connect your voltmeter positive lead to the jumper block pin 1 and negative lead to power ground to measure offset as shown in Figure 8-4.

    d. Adjust R3 to the offset voltage obtained in step 3.

FIGURE 8-4: Using a voltmeter to set the offset voltage.

5. To set the voltage using the on-board DS2438 (more accurate), follow these steps:

   a. Stop SimpleWeather if it is running.

   b. Locate and "comment-out" the line `if (minute != lastMinute)` at line number 125. It should read `//if (minute != lastMinute)` when you're done.

   c. Enable debug mode by following the steps presented in the previous chapter.

   d. Run SimpleWeather again. SimpleWeather should take constant readings without waiting for 1 minute. You should see the debug data for the barometer module, showing the Vad voltage.

   e. Move the jumper on your barometer module to the 2-3 location to measure offset.

   f. While monitoring the Vad Voltage on the screen, adjust R3 to the offset value obtained in step 3.

   g. Un-comment the line you commented-out in step b (Remove the "//").

6. Stop SimpleWeather if it is running.

7. Open the `BaroSensor.java` file in the NetBeans editor window.

8. Locate the BARO_SENSOR_SLOPE constant and set it to the value obtained in step 3.

9. Locate the BARO_SENSOR_OFFSET constant and set it to the value obtained in step 3.

10. Reinstall the jumper across pins 1-2 to measure the pressure.

11. Disable debug mode.

12. Run SimpleWeather. The barometric module is now adjusted for your location.

Fine-Tuning

Fine-tuning is accomplished by monitoring your pressure and comparing it with a known calibrated reference source. You can use another calibrated barometer or a nearby online weather station such as the National Weather Service at www.nws.noaa.gov/.

Start by adjusting the intercept. When the reference station indicates a pressure near mid-scale (30.00 inHg), adjust the software offset until your weather station matches. Then, over time, monitor the pressure extremes to determine if the gain needs adjustment. When the reference station indicates high pressure (at least 30.30 inHg), check your barometer. If your pressure reads lower, you will need to increase the gain. Conversely, if the reference station is reporting a high pressure, and yours reads higher, you will need to reduce the gain.

In either case, divide the reference pressure by your reading to get a correction factor. Then multiply the correction factor by the present value of PRESSURE_GAIN to get your new PRESSURE_GAIN. If you tweak the gain, you'll also have to re-adjust the offset. Repeat this process until you are satisfied with the calibration.

Wrap Up

You've now completed another weather station sensor and are more than halfway to completing your basic weather station. Because the barometer requires external power, and is somewhat sensitive to temperature, you will want to install it indoors. You can start thinking about where you are going to mount the indoor portion of your weather station. It will need power and should be located near your computer because you need the serial port. Also, start thinking about how you will want to get the 1-Wire cable outside. There's a whole chapter dedicated to installing your weather station, but picking the right location can take some time.

Count the Rain

R ain, I find, is one of the more interesting weather measurements. Maybe it's because the rain occurs during a storm and sometimes there's lightning and thunder. Or maybe it's just because rain is a big deal living here in the southwest.

When measuring rain, you want to know several things. How much has it rained this year? How much has it rained since midnight? How hard is it raining now? The rain counter described in this chapter will provide this information and more.

Project Overview

Similar to the WS-1 wind instrument, the rain counter is usually purchased as an assembly. Trying to design and build a tipping bucket mechanism is just too time consuming and expensive. But you do have several options.

Options

This project has several options:

- **Option 1:** Buy a complete 1-Wire rain gauge. Both Hobby-Boards and AAG offer 1-Wire rain gauges.

- **Option 2:** Buy a regular rain gauge and convert it to 1-Wire using a pre-built module.

- **Option 3:** Buy a regular rain gauge and convert it to 1-Wire using a scratch-built module.

Both the Hobby-Boards and the AAG units are really Option 2 gauges that have been converted to 1-Wire for you. The Hobby-Boards rain gauge is a modified RainWise RAINEW model 111 rain counter, and the AAG model is a modified LaCrosse WS-7074U rain counter. Both are shown in Figure 9-1.

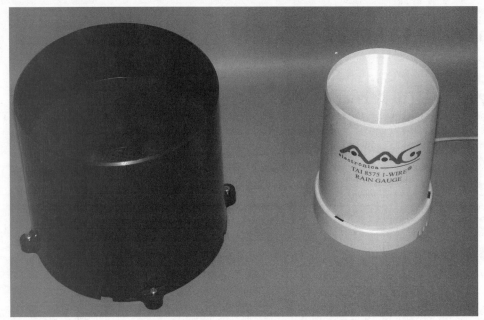

FIGURE 9-1: Hobby-Boards rain counter (left) and AAG rain counter (right).

If you chose Option 2 or 3, there are many online sites offering rain gauges. Here are just a few I've found:

- Davis Instruments Rain Collector II — www.davisnet.com
- RainWise RAINEW 111 — www.rainwise.com
- NovaLynx 260-2501 Professional Rain Gauge — www.novalynx.com

Make sure the rain gauge you choose uses a tipping bucket mechanism that employs a reed switch that is triggered by each tip, and it measures 0.01 inches per tip, because that is how you're going to set up your software. Also, to meet the NOAA rain collection standard for statistical accuracy, your rain gauge should have at least a 6-inch diameter collector, or 25 square inches of collecting area. The RainWise and Davis collectors both exceed this requirement. Both Hobby-Boards and AAG offer a retrofit module to convert the reed switch closures to 1-Wire counts. Both use nearly identical circuitry.

If Option 3 is your choice, then I suggest you use the Hobby-Boards PC board, because this project also uses surface-mount components.

How It Works

Tipping bucket rain gauges were discussed back in Chapter 1. Taking another look, Figure 9-2 shows the mechanical view of a typical unit. Rain is gathered in the top collector, and drips through a small hole in the center. This hole directs the water to drip into the "top" bucket and also acts as a regulator to keep the water from dripping too fast. As the bucket fills, it gets heavier. At a certain point, the weight causes the bucket to fall. As the bucket assembly moves, the magnet mounted near the bottom swings past the reed switch, which momentarily closes, triggering the counter circuit.

After the bucket falls, the water spills out and the other bucket is now in the "top" position to catch the dripping water. When it gets full, it falls, triggering the reed switch as the tipping bucket assembly swings back. This process repeats as long as water is dripping from the collector. Many of the tipping bucket rain gauges are designed to tip at 0.01 inches of rain. The first tip usually takes a little more than this, because some of the first few raindrops "stick" to the collector surface.

A pair of adjustment screws usually provides calibration. By shortening the screws so that the bucket assembly is tipped further, it takes more weight to tip it. This has the effect of reducing the counts for the same amount of water. Conversely, by lengthening the screws, there is less travel and buckets tip more often.

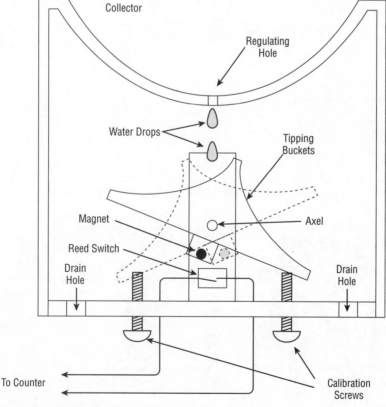

FIGURE 9-2: Tipping bucket mechanics.

The reed switch is connected to a DS2423 1-Wire counter. For each switch closure the counter increments by one. Looking at the schematic in Figure 9-3, the counter input is pulled low via resistor R1. When the reed switch closes, the counter input is pulled high momentarily, producing a count. The DS2423 incorporates internal filtering on the counter inputs to minimize contact "bounce."

In order to maintain the rain count during 1-Wire bus power outages, battery BAT1 provides power to the counter via D3. Using a CR2035, the battery should last at least 5 years. With the counter always powered, there is no way to reset the counts. In the section "Set the Rain Counter to Zero" you will zero the counter in your software.

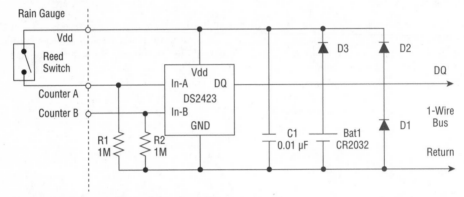

FIGURE 9-3: Rain counter module schematic.

Parts List

If you're building the kit from scratch, Table 9-1 lists the parts in the schematic. You'll also need a data cable to connect your humidity module to your other 1-Wire modules and sensors.

Table 9-1	Basic Parts List		
Qty	ID	Description	Vendor
1	U1	DS2423 1-Wire Counter	Digikey, Maxim, Hobby-Boards
1	BAT1	CR2035 Solder-Tab Battery	Digikey
1	R1, R2	1M 1/8W Resistor	Digikey
1	C1	0.01uF Tantalum Capacitor	Digikey
1	D1-D3	IN5617 or BAT54S Diode	Digikey

Building the Hardware

Gather the supplies needed for your chosen option. Remember that you'll also need a cable to connect this unit to your 1-Wire bus. Both the Hobby-Boards and AAG units do not have connectors like you are used to, so they can't be daisy-chained. For your indoor test bus, it will have to be the last device in the chain.

 Warning The rain counter module is ESD sensitive. Always use an anti-static wrist strap when handling your module.

Option 1: Building from Scratch

If you're building from scratch, component placement is not critical. You can use a surface mount prototyping board offered from several online electronics dealers. Because the rain module will most likely be installed in the rain collector, make the board as small as possible. I suggest that you use screw terminals for the 1-Wire and reed switch connections.

Option 2: Building the Kit

1. Make sure you have all the parts.

2. Clean the board. Wipe the solder-side of the PC board with alcohol to remove any residue or contamination.

3. Solder the parts. Insert the parts in the circuit board and solder them carefully.

4. Clean-up. Clean the solder connections with a soft cloth dampened with isopropyl alcohol to remove any flux residue. Trim excess lead lengths from the components.

5. Inspect. Inspect each connection to make sure there are no shorted parts. Figure 9-4 shows the completed board.

6. Test. Check the battery circuitry using a voltmeter. Measure the voltage from the DS2423 Vdd pin to the GND pin. It should read 2.5 volts or higher. You will test the rest of the circuit once the software is complete.

Option 3: Buy the Completed Module

You can buy the completed counter module from Hobby-Boards or AAG. The Hobby-Boards module uses screw terminals for connections as shown in Figure 9-4. The AAG module uses attached wires and supplies crimp-on connectors to attach it to the reed switch and 1-Wire bus.

FIGURE 9-4: Completed Hobby-Boards module.

Weatherproofing the Module

The rain counter board typically gets mounted in the rain collector, so it is a good idea to protect the rain counter from moisture. Chapter 11 looks at ways of waterproofing circuit boards, but because this board could get splashed by water as the buckets tip, the entire board will be put in a water-tight enclosure:

1. Find what size PVC pipe your rain counter board will fit in. If you are using the Hobby-Boards rain counter, it fits nicely in 1-inch PVC pipe about 2 inches long.

2. Buy two end caps that fit your pipe.

3. Glue one PVC cap on the pipe.

4. On the other cap, drill a hole just large enough to fit the 1-Wire data cable and the rain counter reed switch wire through. See Figure 9-5 for an example.

5. Run the wires through the hole and connect to the rain counter board.

6. Slide on the cap firmly, but do not glue it in case you need to open it later.

7. Seal the hole where the wires are routed with RTV sealer.

8. Find a location in your rain collector to install the sealed unit. Secure with a wire tie or RTV adhesive.

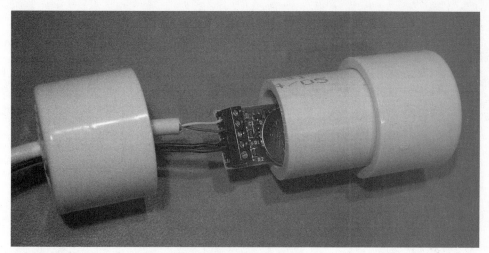

FIGURE 9-5: Waterproof container for the rain counter board.

If you purchased the RainWise rain gauge, the manufacturer already provides a water-tight housing inside the collector to install the rain counter, as shown in Figure 9-6.

FIGURE 9-6: RainWise built-in circuitry housing.

Modifying the Hobby-Boards Rain Collector

The Hobby-Boards rain collector comes with about 60 feet of cable to connect to your 1-Wire bus. Unfortunately, it is standard two-conductor cable. Not shielded, not twisted. Although this may work initially, this could lead to trouble. It's a simple operation to replace the cable with something more reliable. Here's how:

1. Decide how long your cable needs to be and cut an appropriate length of CAT5 cable. Because you haven't reached the installation chapter yet, you may not be sure. You can take a guess (guess on the long side) or you can skip this section and come back when you know.

2. Remove the rain collector top and then remove the blue cap sealing the rain counter housing.

3. Loosen the screws on the circuit board for the 1-Wire bus. They're labeled GND and DQ.

4. Turn the unit over and carefully remove the sealer covering the access hole. Remove the old cable.

5. Insert the CAT5 cable into the access hole.

6. Turn the unit back over and pull the CAT5 cable out far enough so that you can strip back about 2 inches of the outer insulation.

7. Cut off all but the blue/blue-white pair. Strip the ends of this pair about ¼ inch.

8. Insert the blue wire into the rain counter DQ terminal and tighten the screw. Insert the blue-white wire into the terminal marked GND and tighten the screw.

9. Leave a couple of inches of service loop in the housing and replace the housing cover.

10. Turn the unit over and apply RTV sealer or hot glue in the hole to completely seal it as shown in Figure 9-7.

11. Attach an RJ-45 connecter to the other end of the cable.

Connecting

You are now ready to connect the rain counter to your test 1-Wire bus. Table 9-2 lists the connections for the Hobby-Boards and AAG units.

If you haven't already done so, attach the blue/blue-white pair of a length of CAT5 cable to the rain counter board. Connect the blue wire to the DQ terminal and tighten the screw. Attach the blue-white wire to the terminal marked GND and tighten the screw. Cut off the remaining wires. Crimp an RJ-45 connector to the other end of the cable.

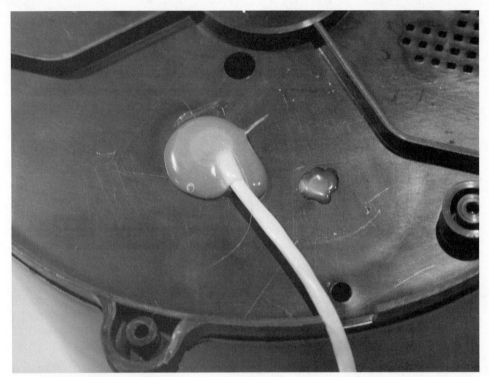

FIGURE 9-7: Sealing the cable holes.

Next, connect the counter to your reed switch. Referring to Table 9-2, connect one side of the reed switch to Counter A and the other side of the reed switch to +5V/Vdd.

If you are using the AAG rain gauge, the 1-Wire connection is under the battery cover on the bottom of the unit. You will need to drill a small hole in the cover to access the wires. Follow the previous instructions for building the data cable, but instead of attaching the rain gauge end to screw terminals, use the waterproof telephone splice connectors you read about in Chapter 3. Run CAT5 cable through the hole in the battery door and connect the blue wire of the CAT5 cable to the red wire of the rain gauge. Connect the blue-white wire to the rain gauge's black wire as shown in Figure 9-8. Tuck the splices in the battery compartment and replace the door. Seal the hole in the door with RTV or hot glue.

You're now ready to plug in the rain counter to your 1-Wire bus and add the software. Because the rain gauge has a cable with a connector on the end, it will have to plug into the last device on your 1-Wire bus.

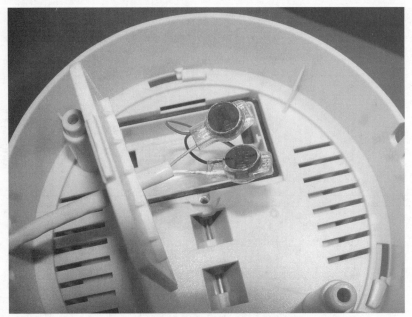

FIGURE 9-8: Connecting the 1-Wire cable to the AAG rain gauge.

Table 9-2	Rain Counter Module Pin-outs	
Function	*Hobby-Boards Terminal*	*AAG Rain Gauge*
GND/Return	1	Black
DQ	2	Red
+5V / Vdd (Switch Power)	3	N/A
Counter A	4	N/A
Counter B	5	N/A

Add the Rain Gauge to Your Software

Once again, quickly review what you did in the last software project:

- Added the BaroSensor class.
- Added a user constant for the device serial number.

- Added a class variable to point to the BaroSensor class.

- Added a float variable to hold the pressure.

- Instantiated the BaroSensor class.

- Called the method `GetPressure()` to get the pressure.

- Used `System.out()` to print the humidity and dewpoint.

- Entered a gain and offset value to calibrate your barometric pressure module.

By now, you most likely have these steps memorized. Good, that's the point. This is the process you'll follow when you start adding your own sensors.

Add the RainSensor Class

Locate your `WeatherToys` directory and find the folder for Chapter 9. Locate the file `RainSensor.java`. Copy this file to your SimpleWeather project source code folder src ⇨ simpleweather.

If you have trouble getting your code to run, there is also a completed project located in the same folder. If you use this project, make sure to change the serial port name and 1-Wire device serial numbers to yours.

Modify SimpleWeather

Launch NetBeans and open your SimpleWeather project. Look in your Source Packages, and check your list of source files to make sure `RainSensor.java` is listed.

Open `SimpleWeather.java` in the NetBeans editor. Scroll down just past the `// 1-Wire Device` comment at line 30. Add the following line:

```
// 1-Wire Devices
private final String TEMP_SENSOR_ID     = "5D00080005236F10";
private final String WIND_SPD_ID        = "8800000000CC1E1D";
private final String WIND_DIR_ID        = "3A00000000CE6920";
private final String HUMIDITY_SENSOR_ID = "3C00000073B6F426";
private final String BARO_SENSOR_ID     = "6D00000073A34F26";
private final String RAIN_COUNTER_ID    = "";
```

Scroll down to `// class variables` near line 43. Add a float variable for rain at the end of the list:

```
// class variables
public float temp;
public float windSpeed;
public int windDir;
public float humidity;
public float dewpoint;
public double pressure;
public float rain;
```

Next, add a class variable for the RainSensor class after the `// sensors` comment at line 57:

```
// sensors
private TempSensor ts1;
private WindSensor ws1;
private HumiditySensor hs1;
private BaroSensor bs1;
private RainSensor rs1;
```

Scroll down to the `// initialize sensors` comment near line 104 and add the rain sensor to the bottom of the list:

```
// initialize sensors
ts1 = new TempSensor (adapter, TEMP_SENSOR_ID);
ws1 = new WindSensor (adapter, WIND_SPD_ID, WIND_DIR_ID);
hs1 = new HumiditySensor(adapter, HUMIDITY_SENSOR_ID);
bs1 = new BaroSensor(adapter, BARO_SENSOR_ID);
rs1 = new RainSensor(adapter, RAIN_COUNTER_ID);
```

Finally, in the main loop, locate the comment `// get the weather` near line 134. Add the following lines to get the rain count and print it to the screen:

```
// get the weather
...

// get barometric pressure
pressure= bs1.getPressure();
System.out.println("Pressure = " + pressure + " inHg");

// get rain count
rain= rs1.getRainCount();
System.out.println("Rain = " + rain + " in");
```

Double-check your typing and then save your file.

Find Your Device Serial Number

Make sure your rain counter module is connected to the 1-Wire bus. Use the OneWireLister program to find the device serial number. If you built the wind instrument, you may remember that 1-Wire counters belong to family 1D. Use the technique you learned in the previous chapter to determine which 1-Wire counter is the rain gauge:

```
// 1-Wire Devices
private final String TEMP_SENSOR_ID     = "550008000881E510";
private final String WIND_SPD_ID        = "C20000000143E01D";
private final String WIND_DIR_ID        = "3A00000000CE6920";
private final String HUMIDITY_SENSOR_ID = "3A00000074AE1126";
private final String BARO_SENSOR_ID     = "6D00000073A34F26";
private final String RAIN_COUNTER_ID    = "3A000000065C741D";
```

Testing the Project

You're now ready to compile and run SimpleWeather. Press F6 (or Run ⇨ Run Main Project).
Your output should include the rain:

```
Starting SimpleWeather 1.0
Found Adapter: DS9097U on Port /dev/tty.USA19QW4b44P1.1
Resetting 1-wire bus
Time = Wed Feb 01 19:06:05 MST 2006
Temperature = 74.3 degs F
Wind Speed = 0.0 MPH from the SW
Humidity = 26.093096 %
Dewpoint = 37.706905 degs F
Pressure = 29.988391 inHg
Rain = 14.82 in
```

Unless you just connected the battery, your counter will most likely be some non-zero value.
That's OK. Open the rain gauge and manually tip the buckets. For every tip, the measured rain
should increase by 0.01 inches.

Walk Through the Code

In the declaration of the class there is one user constant. RAIN_OFFSET is the value subtracted
from the current measurement to allow you to zero the result. For example, if out-of-the-box
your rain counter already reads 1.87 inches, you can enter 187 in this user constant so the net
result is zero:

```
public class RainSensor
{
  // calibration constants
  private final float RAIN_OFFSET = 1482;


  // class variables
  private DSPortAdapter adapter;
  private OneWireContainer1D rainDevice = null;
  private static boolean debugFlag = SimpleWeather.debugFlag;
```

Next, an instance of the 1-Wire device is created using the device serial number and the
adapter object:

```
  public RainSensor(DSPortAdapter adapter, String deviceID)
  {
    // get instances of the 1-wire devices
    rainDevice = new OneWireContainer1D(adapter, deviceID);
  }
```

In the body of the method, the DS2423 counter A is read (page 15 in the DS2423 memory map) using the `readCounter()` method. The `RAIN_OFFSET` value is subtracted from the counter A value:

```java
public float getRainCount()
{
  float rain;

  try
  {
    if (debugFlag)
    {
      System.out.print("Temperature: Device = " + rainDevice.getName());
      System.out.print("  ID = " + rainDevice.getAddressAsString() + "\n");
    }

    // read rain count from counter 15 and subtract offset
    rain = rainDevice.readCounter(15) - RAIN_OFFSET;
```

The counter value is divided by 100 to convert it to a floating-point value representing inches. To convert to centimeters, un-comment out the line that multiplies the result by 2.54:

```java
    // convert to inches
    rain /= 100F;

    // convert to centimeters if required
    //rain *= 2.54f;

    if (debugFlag)
      System.out.println("Rain Count: " + rain + " inches\n");
  }
```

As in all the 1-Wire functions, there is a try-catch mechanism so that if an error occurs in the try block, the code will jump to the catch statement. In this case, an error is displayed on the screen and the rain value is set to indicate an error:

```java
  catch (OneWireException e)
  {
    System.out.println("Error Reading Rain Counter: " + e);
    rain = -99.99f;
  }
  return rain;
  }
}
```

Calibration

If you purchased your rain gauge, it may have already been calibrated. However, it never hurts to check. You will also want to periodically check the rain counter's calibration. To do this, you'll need to calculate the volume of water that will measure 1 inch of rain. If your rain gauge measures 0.01 inch for every tip, there are 100 tips (or counts) for every inch. If your rain gauge is circular, the equation for volume is

Volume = $\pi * (d/2) * h$

where

$\pi = 3.141$

d = the inside diameter of your rain collector

h = 1 (for 1.00 inch)

I've included the calculation in a spreadsheet in the same excel file you used to calculate station pressure.

 Note The rain counter cannot be reset to zero unless the battery is removed. The Hobby-Boards and AAG versions both have the battery soldered permanently in place.

Set the Rain Counter to Zero

Before you calibrate, here's how to set the rain counter to zero inches in your software using the tare method:

1. Run SimpleWeather and note the rain output.

2. Multiply the rain value by 100 to convert from inches to counts.

3. Find the current value for RAIN_OFFSET in RainSensor.java.

4. Add the existing offset to the new value from step 2, and replace RAIN_OFFSET with the new value.

5. Restart SimpleWeather. The rain value should now be zero.

Calibrating Your Rain Gauge

1. Open the calibration.xls Excel file and click the tab labeled Rain Cal.

2. If your rain gauge is circular, enter the inside radius in the Circular Gauge field. If your gauge is square or rectangular, enter the length and width in the Rectangular Gauge fields.

3. The spreadsheet will convert the required volume to both cups and ounces.

4. Fill a measuring cup to the amount indicated in the spreadsheet.

5. Start SimpleWeather and record the current rain count.

6. *Slowly* pour the water through the rain gauge. The buckets should tip no faster than once every 5 seconds.

7. Check the reading in SimpleWeather. The count should have increased by 1.00 +/– 0.05 inches (100 +/– 5 counts). If your value isn't close, then:

   ■ If the adjustment screws are located inside your rain gauge, remove the cover. Figure 9-9 shows the adjustment screws on the RainWise gauge.

   ■ If the count was higher, you need to *lengthen* the travel of the tipping bucket arm. On gauges with the screws inside like the RainWise gauge in Figure 9-9, this means turning the screws clockwise (in). If your adjustment screws are on the outside, like Figure 9-2, you need to turn the screws counterclockwise (out).

   ■ If the count was low, you need to *shorten* the travel of the tipping bucket arm. On gauges with the screws inside the gauge, this means turning the screws counterclockwise (out). If your adjustment screws are on the outside, you need to turn the screws clockwise (in).

   ■ Start by turning each screw ¼ turn to see what effect this has on the count.

8. Repeat these steps until you are satisfied with the calibration.

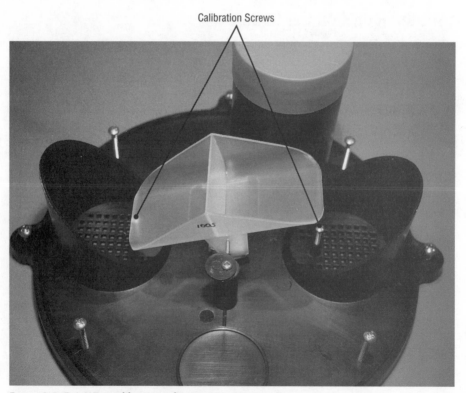

Calibration Screws

FIGURE 9-9: RainWise calibration adjustment screws.

Adding a Heater

If you live in a cold climate, a potential problem you face is the rain gauge collecting snow or freezing. One solution is to install a small heater in your rain collector. There are many ways to do this, but one of the safer ways is to install one or more low-voltage lights in the collector. These are the kind normally used for landscape lighting. A single 11-watt 12-volt bulb provides just enough warmth to melt light snow. If you live in a colder climate, you may have to use higher wattage or additional lamps. Mount a connector or screw terminals on the outside of the collector top and drill a small hole to get the wires inside the collector. I found some inexpensive tier lights at a local hardware store and removed the lamp socket and mounted it inside the collector as shown in Figure 9-10. Make sure you pick a spot that won't interfere with the tipping bucket mechanism.

If you already have low-voltage landscape lighting, you can use it to power the lamp. Otherwise, you will need to purchase a low-voltage transformer and cable. If you do install a separate transformer, Chapter 16 shows you how to turn it on only when the temperature is below freezing.

FIGURE 9-10: Low voltage landscape lighting used as a rain gauge heater.

Wrap Up

Congratulations, you have completed another weather sensor. At this point, your software only shows the total rainfall. To give you a preview of Part III of the book, there's a chapter on adding a web page to SimpleWeather that will show the rain since midnight and rain rate. There's also a chapter that shows you how to turn on or off appliances if it's raining. I don't want to give too much away, but there's also a chapter showing you how to build a smart sprinkler timer that checks to see if it's raining.

If you're like me, by now your workbench has several weather sensors and cables and I bet you are getting anxious to install them outdoors. Hang on, there's only one more project before for you get to the installation chapter.

Is There Lightning?

The lightning detector described in this chapter is, by far, my favorite weather sensor. Maybe it's because I live in the southwest and get some great summer thunderstorms or maybe it's because I designed this sensor and I enjoy sharing it. Most likely, it's both.

Before you start building the lightning detector, let's quickly review what lightning is. Most lightning occurs within a cloud. Only about 10% of lightning discharges go from cloud to ground. Most lightning occurs within the cloud itself. The cloud-to-ground variety is the one you're most interested in, so let's focus on that.

During the formation of a thunderstorm, moist warm air rises, cooling as it goes. As the air cools, the moisture it contains condenses, forming a thundercloud. Inside the cloud, the condensing moisture starts forming water and ice. When the water or ice particles become heavy enough, they begin to fall to earth. In a process scientists and meteorologists don't fully understand, the falling water and ice causes a separation of charges within the cloud. When these charges become sufficient, they begin to travel toward earth in short bursts, called *stepped leaders*. When the stepped leader gets close to the earth surface, a return charge starts forming, rising up to meet the stepped leader. These return charges are called *streamers*. When the streamer meets the stepped leader, an ionized discharge path is formed, allowing the intense discharge we are all familiar with. This discharge is called a *return stroke*, and radiates a large amount of Electro-Magnetic Energy (EME). The flash you typically see usually rises from the earth to the cloud, even though it appears to start at the cloud. Figure 10-1 shows a cloud-to-ground lightning stroke.

Once the discharge is complete, some charges may remain in other areas of the cloud. These can quickly "jump" to the area that just discharged (intracloud lightning). If this happens quickly enough, the path to earth may still be ionized, allowing another flash along the same path. That is why lightning sometimes appears to flicker.

There are many variations of lightning and some very unusual displays can be seen. For more information on lightning, you can find some great discussions and tutorials on the Internet.

FIGURE 10-1: A cloud-to-ground lightning stroke.

There are several ways to detect lightning. Most involve sophisticated electronics that detects the radiated energy, and most are quite expensive. Some provide a distance measurement to gauge how far away the lightning is, and some can even provide the direction. This project is quite the opposite. It is designed to be very inexpensive and it provides a simple lightning detection scheme that detects lightning activity in Strokes per Minute (SPM). By looking at the SPM, you can determine the storm's intensity. By looking at the trend, you can tell if the storm is approaching or leaving the area.

Project Overview

You have three basic options when building your lightning sensor. Like most of the projects, you can build it from scratch, build it from a kit, or buy the completed circuit. In all three cases, you will need to build some sort of enclosure for your lightning detector. I'll give you some guidelines and show you how I packaged my detector.

Options

You have several options for building the circuit:

- **Option 1: Build it from Scratch.** Because this circuit employs a surface-mount component, you may want to buy the bare PC board from Hobby-Boards. Overall, this circuit is fairly simple and easily built. If you built your own rain counter, then with some minor modifications, you could connect the output of the opto-coupler to the spare counter input on your rain counter board.

- **Option 2: Build from a Kit.** Hobby-Boards offers a kit of the parts, including the PC board.

- **Option 3: Buy the Completed Module.** Hobby-Boards offers a fully assembled circuit board with RJ-45 connectors for the 1-Wire bus and screw terminals for connecting the antenna and earth ground.

How It Works

Referring to Figure 10-2, the energy radiated from the lightning stroke is received from the antenna and is connected to a high-gain Darlington-pair amplifier via R1. If the input is above the 0.7 volt threshold, T1 conducts, turning on T2. When transistor T2 conducts, it provides power from the 9-volt battery through current limiting resistor R4 to the LED side of the opto-isolator U2.

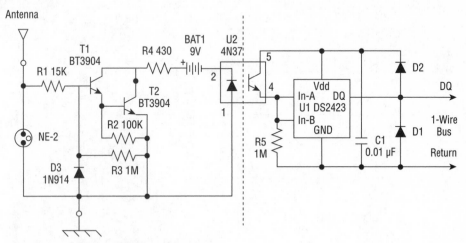

FIGURE **10-2: Lightning detector schematic.**

R5 pulls the counter inputs low. When the LED activates U2's phototransistor, the short pulse caused by the lightning triggers the DS2423's counter by pulling it to Vdd. Note that when the lightning has multiple flashes in a single stroke, this circuit is fast enough to count each of the individual flashes. Therefore, you may get several counts for every flash you see.

Diode D3 provides clamping protection for T1. Neon lamp NE-2 is connected from the antenna input to earth ground to provide input protection clamping from nearby lightning strikes.

The dashed line shows the isolation between sections of the circuit. The lightning detection circuitry is connected to earth ground while the counter circuitry is connected to the 1-Wire return to provide a small amount of lightning protection and prevent ground loops.

Parts List

If you're building the kit from scratch, Table 10-1 lists the parts in the schematic, but I have not listed part numbers for some of the common parts. Because building this from scratch is not for beginners, I will leave the final parts selection up to you. You'll also need a test data cable to connect your lightning module to your other 1-Wire modules and sensors.

Table 10-1   Parts List

Qty	ID	Description	Vendor
1	U1	DS2423 1-Wire Counter	DigiKey, Maxim, Hobby-Boards
1	U2	4N37 Type Opto-Isolator	DigiKey
2	T1, T2	BT3904 Transistor	DigiKey
1	R1	15K 1/8W Resistor	DigiKey
1	R2	100K 1/8W Resistor	DigiKey
2	R3, R5	1M 1/8W Resistor	DigiKey
1	R4	430 Ohm 1/8W Resistor	DigiKey
1	C1	0.01uF Tantalum Capacitor	DigiKey
1	D1- D2	IN5617 or BAT54S Diode	DigiKey
1	D3	1N914 Diode	DigiKey
1	NE-2	NE-2 Neon Lamp	DigiKey
1	BAT1	9V Alkaline Battery	Various
1	—	9V Battery Connector	Radio Shack, DigiKey

Building the Hardware

Round up the necessary hardware for the option you have chosen. Remember, you will also need the parts to build a data cable to connect this module to your 1-Wire test bus. You may also want to read the section "Packaging Your Lightning Sensor" later in this chapter now so you can keep the final assembly in mind as you buy your parts and build the circuit.

 Warning

The lightning module is ESD sensitive. Always use an anti-static wrist strap when handling your module.

Option 1: Building from Scratch

If you're building from scratch, component placement is not critical. You can use a surface-mount prototyping board offered from several online electronics dealers. Some do-it-your-selfers have divided the circuit into two parts: They have installed the detection circuit with the antenna at the top of the mast and installed the counter circuit and battery in a weatherproof enclosure at ground level.

Option 2: Building the Kit

1. Make sure you have all the parts.

2. Clean the board. Wipe the solder-side of the PC board with alcohol to remove residue or contamination.

3. Solder the parts. Insert the parts in the circuit board and solder them carefully.

4. Clean-up. Clean the solder connections with a soft cloth dampened with isopropyl alcohol to remove any flux residue.

5. Inspect. Inspect each connection to make sure there are no shorted parts. Figure 10-3 shows the completed board.

6. Test. Because there isn't much to test until you connect it to the 1-Wire network, wait until the software is working to test the circuit.

Option 3: Buy the Completed Module

You can buy the completed lightning module from Hobby-Boards. The Hobby-Boards module uses screw terminals for the antenna and earth ground connections and uses the standard RJ-45 for the 1-Wire network connection.

FIGURE 10-3: Completed Hobby-Boards module.

Connecting

Once your module is built, you are ready to connect the lightning counter to your test 1-Wire network. Table 10-2 lists the connections for the Hobby-Boards circuit. Make the appropriate data cable and connect the board. Cut approximately 12 inches of general-purpose hook-up wire and attach it to the antenna terminal. Connect another wire between the earth ground terminal and a good earth ground, such as a water pipe or power outlet ground wire. Install a fresh 9V alkaline battery to the battery connector. Before you can complete the test and checkout, you need to add lightning capability to your software.

Table 10-2   Rain Counter Module Pin-outs

Function	Hobby Boards Terminals	Hobby-Boards RJ-45
Antenna	1	N/A
Earth Ground	2	N/A
DQ	3	4
1-Wire Return (DGND)	4	5
Unused	N/A	1,2,3,6,7,8

Add Lightning Capability to Your Software

Here's a quick review of what you did in the last software project:

- Added the RainSensor class.
- Added a user constant for the device serial number.
- Added a class variable to point to the RainSensor class.
- Added a float variable to hold the rain count.
- Instantiated the RainSensor class.
- Called the `getRainCount()` method to get the rain count.
- Used `System.out.println()` to print the rain data to the screen.

Again, you'll follow the same pattern for this project.

Add the LightningSensor Class

Open your `WeatherToys` directory and find the project source code you downloaded from the companion web site. Locate the Chapter 10 directory and the file `LightningSensor.java`. Copy it to your SimpleWeather ⇨ src ⇨ SimpleWeather folder. You will add the other file, `DataLogger.java`, later in this chapter.

Modify SimpleWeather

Launch NetBeans and open your SimpleWeather project. Check your list of source files to make sure `LightningSensor.java` is listed. Open `SimpleWeather.java` in the NetBeans editor. Scroll down just past the `// 1-Wire Device` comment to about line 37. Add the following line:

```
// 1-Wire Devices
private final String TEMP_SENSOR_ID     = "550008000881E510";
private final String WIND_SPD_ID        = "C20000000143E01D";
private final String WIND_DIR_ID        = "3A00000000CE6920";
private final String HUMIDITY_SENSOR_ID = "3A00000074AE1126";
private final String BARO_SENSOR_ID     = "6D00000073A34F26";
private final String RAIN_SENSOR_ID     = "3A000000065C741D";
private final String LIGHTNING_SENSOR_ID = "";
```

Scroll down to `// class variables` near line 45. Add an integer variable for the lightning count at the end of the list:

```
// class variables
public float temp;
public float windSpeed;
public int windDir;
```

```
public float humidity;
public float dewpoint;
public float pressure;
public float rain;
public int lightning;
```

Next, add a class variable for the LightningSensor class after the `// sensors` comment near line 60:

```
// sensors
private TempSensor ts1;
private WindSensor ws1;
private HumiditySensor hs1;
private BaroSensor bs1;
private RainSensor rs1;
private LightningSensor ls1;
```

Here is where the LightningSensor class gets instantiated. Scroll down to the `// initialize sensors` comment near line 109 and add the lightning sensor to the bottom of the list:

```
// initialize sensors
ts1 = new TempSensor(adapter, TEMP_SENSOR_ID);
ws1 = new WindSensor(adapter, WIND_SPD_ID, WIND_DIR_ID);
hs1 = new HumiditySensor(adapter, HUMIDITY_SENSOR_ID);
bs1 = new BaroSensor(adapter, BARO_SENSOR_ID);
rs1 = new RainSensor(adapter, RAIN_SENSOR_ID);
ls1 = new LightningSensor(adapter, LIGHTNING_SENSOR_ID);
```

Finally, in the main loop, locate the comment `// get the weather`. Add the following lines to get the lightning count and print it to the screen:

```
// get the weather
...

// get rain count
rain= rs1.getRainCount();
System.out.println("Rain = " + rain + " in");

// get lightning count
lightning = ls1.getLightning();
System.out.println("Lightning = " + lightning + " SPM");
```

That completes the code modifications. Save your work buy selecting the menu item File ⇨ Save or clicking the Save All icon in the toolbar.

Find Your Device Serial Number

The lightning detector uses a 1-Wire counter. This is the third sensor that uses a counter to interface with the 1-Wire network; can you remember the family code? If not, Table 3-1 shows that the DS2423 belongs to family of 1D. Launch `OneWireLister.jar` and select your serial port. Compare the list of 1-Wire devices shown to the list of devices in your code. If your lightning detector is connected and working properly, you will see a new device with this family code. Copy the serial number (using the keyboard shortcut for copy) and then paste it in your code for the lightning sensor:

```java
// 1-Wire Devices
private final String TEMP_SENSOR_ID      = "550008000881E510";
private final String WIND_SPD_ID         = "C20000000143E01D";
private final String WIND_DIR_ID         = "3A00000000CE6920";
private final String HUMIDITY_SENSOR_ID  = "3A00000074AE1126";
private final String BARO_SENSOR_ID      = "6D00000073A34F26";
private final String RAIN_SENSOR_ID      = "3A000000065C741D";
private final String LIGHTNING_SENSOR_ID = "840000000771FE1D";
```

Testing the Project

At this point, you are ready to compile and run SimpleWeather. Press F6 or Run ➪ Run Main Project.

After the start-up data, your output should start displaying temperature, wind, humidity, pressure, and lightning data once a minute:

```
Starting SimpleWeather 1.0
Found Adapter: DS9097U on Port /dev/tty.USA19QW4b44P1.1
Resetting 1-wire bus
Time = Sat Feb 11 08:00:41 MST 2006
Temperature = 69.8 degs F
Wind Speed = 0.0 MPH from the SSW
Humidity = 23.729174 %
Dewpoint = 31.440002 degs F
Pressure = 30.268060000000002 inHg
Rain = 16.92 in
Lightning = 0 SPM
```

Reading the counter only tests about half the circuit. To test the rest of the circuit, you could wait for a lightning storm to hit or you could generate your own lightning. Well, sort of. An inexpensive "electronic" lighter as shown in Figure 10-4 generates a small electric spark that provides enough energy to trigger the lightning detector. Hold it about 4 inches away from the antenna while "sparking" the lighter. Every time you pull the trigger, the spark should cause the lightning detector to count. Try it. You should see some counts listed in SimpleWeather.

```
Time = Sat Feb 11 08:16:00 MST 2006
Temperature = 69.8 degs F
Wind Speed = 0.0 MPH from the SSE
Humidity = 23.296041 %
Dewpoint = 30.984756 degs F
Pressure = 30.282402 inHg
Rain = 16.92 in
Lightning = 28 SPM
```

FIGURE **10-4: Inexpensive lightning simulator.**

If you can't seem to get your detector to count, make sure that you have connected a fresh battery. Did you connect the detector to a good earth ground? Did you get the right serial number? Hold the lighter a little closer to the antenna and try again.

Depending on where you install your lightning detector, it could pick up extraneous noise and EME. Sitting on my test bench, the lightning detector triggers every time I turn on or off the room light. Electric motors, refrigerators, and electrical appliances can generate considerable EME that can be detected with this sensor. When installing your lightning detector outdoors, try to keep it as far away from electrical appliances as possible.

Logging Your Data

So far, all of the data you have collected has just been printed to the screen. Wouldn't it be nice if you could keep the data? It is a simple task to save the data to your hard drive, so add that capability now. If SimpleWeather is still running, go ahead and stop it. Copy the file `DataLogger.java` from the Chapter 10 source folder to your SimpleWeather source code folder. Open `SimpleWeather.java` in the NetBeans editor. In the `user constants` section near line 26, add a new constant for the path and file name:

```
// user constants
public static final String VERSION = "SimpleWeather 1.0";
public static final String ONE_WIRE_SERIAL_PORT = "COM1";
public static final String LOG_PATHNAME = "WeatherLog.txt";
```

Next, you need to add a new class variable. Locate the comment for the class variables near line 46 and add the following:

```
// class variables
public float temp;
...

private DSPortAdapter adapter;
private DataLogger logger;
```

Add the code to instantiate the class in the class Constructor:

```
public SimpleWeather()
{
  // get the 1-wire adapter
...
  }
  catch (OneWireException e)
  {
    System.out.println("Error Finding Adapter: "+ e);
    System.exit(1);
  }

  // initialize the data logger
  logger = new DataLogger(LOG_PATHNAME);
}
```

Next, in the main loop, add the code that calls the logger routine. It should go right after the call to read the lightning detector:

```
// get the weather
...

// get lightning count
lightning = ls1.getLightning();
System.out.println("Lightning = " + lightning + " SPM")

// log the data
logger.logData(date, this);
```

Run SimpleWeather again. This time you should see a few new lines indicating that the weather data is being logged:

```
Starting SimpleWeather 1.0
Found Adapter: DS9097U on Port /dev/tty.USA19QW4b44P1.1
Resetting 1-wire bus
Creating Log File WeatherLog.txt
Time = Sat Feb 11 08:54:25 MST 2006
Temperature = 71.6 degs F
Wind Speed = 0.0 MPH from the SSE
Humidity = 22.6576 %
Dewpoint = 31.84069 degs F
Pressure = 30.289573 inHg
Rain = 16.92 in
Lightning = 0 SPM
Updating Log WeatherLog.txt
```

By default, Java writes this file to the current project folder. If you open the SimpleWeather project folder, you will see a new file named `WeatherLog.txt`. Double-click to open it and examine the contents. You will see something like this:

```
Weather Data log Started Sat Feb 11 08:54:24 MST 2006
Time,Temp,WSpd,WDir,Hum,Baro,Rain,Ltng
08:54,71.6,0.0,7,22.7,30.29,16.92,0
08:55,71.6,0.0,7,22.3,30.29,16.92,0
08:56,72.5,0.0,7,22.4,30.28,16.92,0
```

The first line is a time stamp when the program was started, followed by a header showing what's in each column. Next is the actual weather data. By default, the data is separated by a comma. This delimiter can be easily changed in `DataLogger.java`:

```java
public class DataLogger
{
  // class constants
  public static final char DELIM = ',';   // use a comma as the delimiter

  // class variables
  private String pathFileName;
```

If you would like to change where SimpleWeather logs your data, all you have to do is change the simple file name `WeatherLog.txt` to a full path name. Decide where you want to save your data and get the full path name to the intended directory. Because Java uses the backslash symbol as the Escape character, you have to "escape the escape" in Windows. This simply means using two backslashes, such as: `C:\\Documents and Settings\\Tim\\My Documents\\Weather Toys\\WeatherLog.txt`, where `WeatherLog.txt` is the file name of the data file you will be creating.

On a Mac or Linux box, you use forward slashes, so you don't have to use the Escape character. Your path name will look like /Users/tbitson/Documents/Weather Toys/ WeatherLog.txt. Enter the new path and file name in SimpleWeather.java's user constant LOG_PATHNAME.

Walk Through the Code

I'll wrap up the software section with a quick look at the code for getting the lightning count. Because there are many similarities between this code and the wind speed code, I'm going to skip most of the class code and jump right to the measurement routine:

```java
public int getLightning()
{
  int lightning = 0;

  try
  {
    if (debugFlag)
    {
      System.out.print("Lightning: Device = " + lightningDevice.getName());
      System.out.print("  ID = " + lightningDevice.getAddressAsString() +
                                "\n");
    }
```

The method starts off by reading the current count from the 1-Wire Counter A (Counter A is page 15 in the DS2423's memory), followed by reading the current time in milliseconds:

```java
      // read lightning count & time
      long currentCount = lightningDevice.readCounter(15);
      long currentTicks = System.currentTimeMillis();
```

If this isn't the first time the software has read the counter (lastTicks != 0), then the method calculates the Strikes Per Minute by subtracting the previous counter value from the current counter value. This is divided by the elapsed time (currentTick - lastTicks), converted from milliseconds to seconds (/ TICKS_PER_SECOND) and then to minutes (* 60):

```java
      if (lastTicks != 0)
      {
        // calculate the lightning activity in strikes per minute since last time
          lightning = (int)((currentCount-lastCount)*(60L /
                          ((currentTicks-lastTicks)/ TICKS_PER_SECOND)));
      }

      if (debugFlag)
        System.out.println("Count = " + (currentCount-lastCount) + " during " +
                        (currentTicks-lastTicks) + "ms calcs to " + lightning +
                        "SPM\n");
```

Just like in the wind speed method, the current time and count value must be stored for the next time this method is called:

```
    // save the counts & time
    lastCount = currentCount;
    lastTicks = currentTicks;


}
catch (OneWireException e)
{
  System.out.println("Error Reading Lightning Counter: " + e);
  lightning = -9999;
}

  return lightning;
}
```

Packaging Your Lightning Sensor

So far, I have postponed packaging the sensors and modules you've built until the installation chapter. The lightning detector is a little different. Because it requires an antenna as part of its packaging, go ahead and complete the assembly by adding a weatherproof enclosure. There are many ways to package your lightning sensor. In this chapter you learn one way; if that doesn't fit your application, feel free to make modifications.

Here are some guidelines to installing the lightning detector:

- The higher the unit is mounted, the more detectable range it has. At about 10 feet up, it can detect lightning more than 50 miles away.

- The detector should have a clear 360-degree view of the ground. Mounting it on the side of a structure degrades its performance in that direction.

- Mounting on the top of a mast is best. Side mounting on a mast is acceptable, but does cause some attenuation of the signal in the direction of the mast.

- Try to mount the detector far away from electrical appliances and motors to minimize false detections.

- Proper earth grounding is important, both from a safety perspective and an operational standpoint. Make sure the mounting mast and the detector's earth ground are tied to known good earth ground.

 Warning If you think about it, mounting the sensor on a mast, sticking up in the air, and then running a wire to your computer to measure lightning could invite trouble (although it's not as bad as flying a kite during a thunderstorm with a metal key attached). Exercise caution during a lightning storm. You may want to disconnect your weather station during electrical storms to prevent lightning damage.

When packaging and installing the weather sensors, PVC pipe is your friend. It comes in many shapes and sizes, provides a good weather-tight enclosure, and except for being brittle at low temperatures, can handle severe environments. It is easy to work with, and very inexpensive.

Figure 10-5 shows you two options for mounting the lightning detector: a top-of-mast mount and a mast side mount. Figure 10-6 shows the various pieces of PVC pipe and fittings needed for the mast-top mount. The Hobby-Boards module just fits inside 1½ inch PVC pipe.

The two 1½-inch couplers connected with the 2-inch length of 1½-inch pipe form the main body for the lightning detector circuit board. A 1½- to ½-inch reducer on top allows a 16-inch length of ½-inch PVC pipe to hold the antenna vertically. A second reducer on the bottom adapts the main body to a 6-inch length of ¾-inch pipe. This pipe is inserted down the top of the mast.

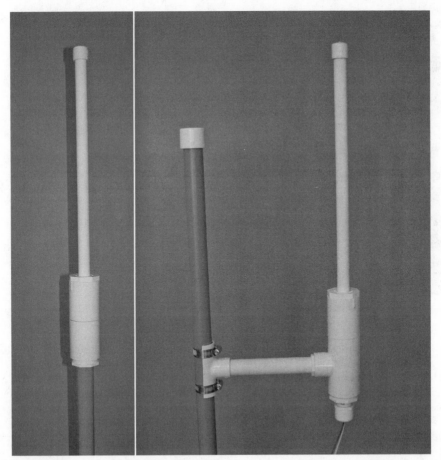

FIGURE 10-5: Completed top-of-mast mount (left) and mast side-mount (right).

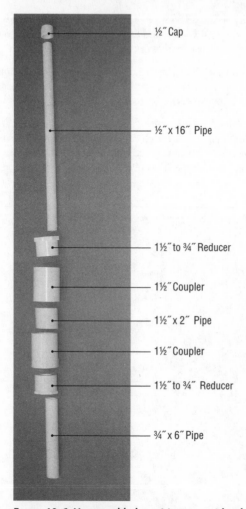

½″ Cap

½″ x 16″ Pipe

1½″ to ¾″ Reducer

1½″ Coupler

1½″ x 2″ Pipe

1½″ Coupler

1½″ to ¾″ Reducer

¾″ x 6″ Pipe

FIGURE 10-6: Unassembled mast top-mount hardware.

The side-mount hardware version details are shown in Figure 10-7. The bottom coupler has been replaced with a 1½ x 1 x 1½ inch tee and a 1-inch to ¾-inch reducer (you can use a 1½ x ¾ x 1½-inch tee if you can find one). A 7-inch length of ¾-inch PVC pipe connects it to a modified 1 x ¾ x 1-inch tee. To make a modified tee, one half of the long part of the tee has been cut off to allow it to be mounted to the mast. Two 2-inch hose clamps hold the tee to the mast. To get the wires out of the body, attach a short length of ¾-inch pipe to the bottom reducer. Drill a hole in a ¾-inch cap and slide it over the wires.

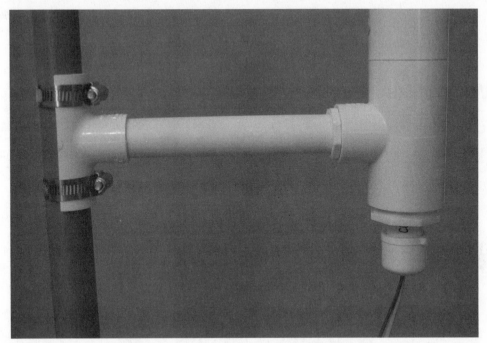

FIGURE 10-7: Completed side mount with modified tee mounted on mast.

When inserting the lightning detector circuit board into the PVC pipe, orient the board so that the screw terminals are toward the top (or antenna) end. Figure 10-8 shows how to pack everything in. A 16-inch length of 14-gauge solid wire is used as the antenna and is routed up through the ½-inch pipe. It's hard to see in the photo, but the 1-Wire network cable is connected behind the battery. The green earth ground wire is 14-gauge solid wire, and is routed down through the bottom along with the 1-Wire cable.

FIGURE 10-8: Close-up of module installation.

Once you have assembled the pieces and have tested the detector to make sure it still works, glue all fittings with PVC pipe glue except the antenna cap and the bottom reducer, because you will need to access the battery every couple of years to replace it.

Wrap Up

The lightning detector described in this project will give you the capability to detect lightning in your area and determine the intensity of the storm. By looking at the trend, you can determine whether a storm is approaching or receding. Figure 10-9 shows an actual plot of a summer thunderstorm captured with this design. You can clearly see the count values increase as the storm is getting closer, and decrease as the storm moves away.

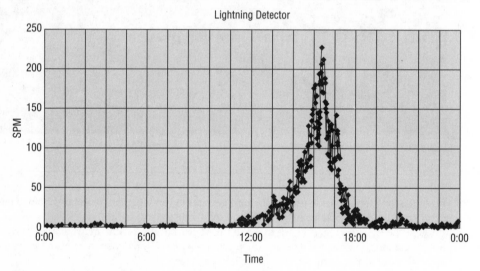

FIGURE 10-9: Lightning detector's response to a thunderstorm.

Every installation will differ in sensitivity. To get a feel for the detectable range, you can use the age-old method of counting the time between the flash and the sound to determine the approximate detection distance. Once you get your lightning detector installed, don't be surprised if you're able to detect lightning more than 50 miles away.

Warning This project is intended for entertainment purposes only and not as a safety-warning device! Lightning can literally strike "out of the blue" without warning. Do not use this project where incorrect operation could jeopardize someone's safety.

Your SimpleWeather software now has the ability to log data to your hard disk. Try letting your test setup run overnight and then plotting the data in Excel. Are you curious as to how often your heater (or air conditioner) runs at night? Mount the temperature sensor near a heater vent and collect the data. When you plot it, it will clearly show you how often and for how long your heater has been running.

Installing the Weather Station

I'm sure that by now you are getting anxious to install your weather station and start collecting real weather data. If you have been thinking about installation options as you've built each sensor, you may already have an idea where you want to install your sensors. If that's the case, use the first part of this chapter to make sure you have picked a good spot.

Every installation will be different. You may have a need to measure something specific or to install your weather station in a special way. This chapter is not a step-by-step guide; rather it is a collection of ideas, lessons learned, and tips gathered from various sources. The main focus is on three topics: location, mounting, and wiring. You will have to let your creative ideas flow for your particular installation. You can always experiment. For example, it took me three or four attempts before I was happy with the placement of my temperature sensor.

You may find it best to install most of your weather equipment on a metal mast, the kind commonly used to mount TV antennas as shown in Figure 11-1. Because this type of installation is the most common, this chapter is geared for using a mast mount.

As you begin to pick locations and start the installation, you may have a question or two. One place to get more information and to discuss 1-Wire installations is the 1-Wire mailing list. To subscribe, visit www.buoy.com/ mailman/listinfo/weather. There are currently more than 300 subscribers who have built and installed their own 1-Wire weather stations. If you have any questions, this is the place you can go for answers.

Installation Options

When installing your equipment, the proper location is paramount for getting good, accurate weather data. This section presents some guidelines for each of your sensors and modules. In general, you may find it easier to pick the location first, and then determine how to install it in that location.

FIGURE 11-1: Typical mast-mounted weather station.

Picking the Right Location

As you read though the following guidelines, keep in mind that you will need to run cable to that spot. In Chapter 3, there was a suggested goal of about 200 feet for the total 1-Wire bus length. Though this rule is not cast in stone, it is a good target. Some users have bus lengths as long as 500 feet that work reliably.

Temperature

Picking the right location for your temperature sensor can be challenging because there are so many factors that influence the temperature. The measured value should represent the ambient air temperature in the shade. For example, if you mount it too close to your house, radiated heat can cause it to read too high at night. Mount it too well-shielded on the north side of your house and it may stay too cold.

When choosing a potential location, keep these things in mind:

- Don't mount your sensor directly in the sun. Use some sort of a shield to keep the sun from influencing your measured temperature. You can mount your temperature sensor in a tree, use an existing structure for shade, or build the weather pagoda described later in this chapter.

- Mount your sensor at least 3 feet above the ground to minimize ground-heating effects. In general, higher is better.

- Choose a location over a "cool" surface such as grass or bushes. Mounting your temp sensor over a "hot" surface such as an asphalt driveway will cause high readings.

- Avoid mounting near a structure that will radiate heat. A block wall that gets direct sun exposure during the day will radiate heat for many hours after the sun sets.

- Temperatures above your roof could easily reach 20 degrees above the ambient temperature.

- The module needs to be protected from direct water contact and contamination caused by insects, dirt, birds, and debris.

- Consider weatherproofing the temperature module's circuit board as described later in this chapter.

- If you're mounting a temperature sensor indoors, keep it away from heater and air conditioning vents as well as heat producing appliances like televisions.

Once you have identified a potential location, try testing your module in that spot. Build a temporary data cable and install the temperature sensor. Run SimpleWeather for a day or two logging the temperature, then plot it using your favorite tool like Excel. Does the data look OK? Is it reasonably close to what the National Weather Service recorded for your area?

Humidity

Choosing the location for your humidity module is very similar to picking the location for your temperature sensor. In fact, most weather hobbyists mount them in the same location.

- Choose a location that provides an accurate temperature. Remember, humidity changes with temperature.

- Trees, grass, and other foliage give off moisture that can cause high humidity readings.

- Sprinklers and irrigation equipment add moisture to the air. Try to keep your humidity module some distance away.

- The humidity module's sensor element is exposed to air. Protect it from harsh elements such as ocean spray, insects, and direct contact with water.

- If you're going to use SimpleWeather's dewpoint feature, the humidity module needs to be at the same temperature as the temperature sensor.

- Consider weatherproofing the circuit board as shown later in this chapter.

Wind

Determining the location for the 1-Wire Wind Instrument is relatively easy:

- The anemometer and wind vane need a clear path in all directions. Trees, houses, block walls, and fences can disturb or block the wind flow.

- If you're mounting your wind instrument on the roof, it should be at least several feet above the highest point. See Figure 11-2 for an example.

- If you are mounting your wind instrument in an open area, try to mount it at least 10 feet above the ground, the higher the better.

FIGURE 11-2: Mount the wind instrument at least 3 feet above the roof line.

Pressure

The barometric pressure sensor is slightly sensitive to temperature, so it is best to install the module indoors or in a location that does not have a wide temperature swing.

- The pressure sensor requires 14 to 18V DC. It is desirable not to run power in the same cable as the 1-Wire data bus for more than a few feet. Try choosing a location near an AC power outlet to install your wall transformer.

- Avoid locations that are subject to temperature fluctuations such as near a heater vent, near hot appliances, or exposed to direct sunlight.

- Don't mount the circuit board in an airtight case. Believe it or not, I've seen a couple of installations where the user weatherproofed his barometer by putting it in a sealed case.

- If you do mount it outdoors, consider weatherproofing the board like the other sensors.

Rain

Selecting the location for your rain collector doesn't have a lot of constraints. Just keep the following in mind:

- Install it in a location where rain won't be blocked by trees or structures. A good rule of thumb is that the rain gauge should have a clear view of the overhead sky plus 45 degrees in all directions, as shown in Figure 11-3.

- If you live in an area with lots of falling leaves, you may want to install your rain gauge in a location where you can easily clean it out.

- If you installed a rain gauge heater, you will need access to power for the heater.

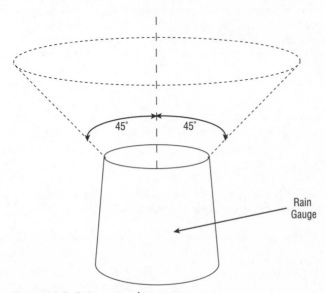

FIGURE 11-3: Rain gauge clear area.

Lightning

You read quite a bit about installing your lightning detector when you built it in Chapter 10. Here's a brief summary of the guidelines:

■ Your lightning detector needs a good earth ground, both from an operational point of view and for safety. If possible, install a ground rod near your lightning detector. Other options include connecting it to a metal water pipe or a building ground. Don't connect your lightning detector's ground wire to natural gas or propane pipes.

■ Like the wind instrument, the lightning detector needs a full unobstructed 360-degree view for best performance.

■ The higher you mount the detector, the longer the detectable range, but beware. The higher you mount it, the more susceptible you are to a direct lightning strike!

■ Make sure you ground the mounting mast.

Now that you have a good understanding of the issues with picking a location for your weather equipment, take a look at some of the mounting options.

Mounting

After you have determined a potential location for your weather modules and sensors, you need to determine a mounting method. Some of the modules, such as the wind instrument, require a mast mount. Here's a look at each of the modules from a mounting perspective.

Temperature and Humidity

The temperature and humidity modules are small enough that you can mount them in many ways. Hobby-Boards offers a small plastic case to protect the modules as shown in Figure 11-4. The AAG modules come pre-built in small plastic cases. Once you have determined a location, the modules can be mounted with wire ties, Velcro, double-stick foam tape, or screws. See the section "Building a Pagoda for Temperature and Humidity" later in this chapter for a mast-mount option.

Wind

The 1-Wire Wind Instrument is designed to be mast mounted. In Chapter 6, you read about installing a mast-mount bracket. If you haven't already done so, now is a good time to install the bracket.

Pressure

The pressure module, like the temperature and humidity modules, can be installed in a small case. Hobby Boards also provides cases for the barometric pressure module.

FIGURE 11-4: Hobby Boards temperature and humidity module case.

Rain

Depending on the type and size of your rain gauge, you have several options. The two most popular are to mount it on a mast with the rest of the weather equipment, or to mount it on a flat area of your roof. The rain gauge should be mounted level. If you're mounting it on your roof, make sure to attach it to the roof to keep the wind from blowing it over.

A third option is to set it in your yard on a level surface. You may want to attach it to a heavy object like a stepping-stone or a fence post to keep the wind or your pet from knocking it over. In Figure 11-5, the rain gauge has been attached to the top of a brick wall using masonry-adhesive Velcro.

Lightning

The lightning detector gets mounted on a mast as described in Chapter 10. This can be the same mast used for the wind instrument and temperature pagoda.

FIGURE 11-5: A block wall provides a good location for mounting the rain gauge.

Installation

Now that you have read about the various mounting options, it's time start assembling the hardware. In this section, you'll read about building a temperature pagoda, followed by mast-mount options for your rain gauge. You will also read about weatherproofing your sensors and doing the final assembly on your anemometer.

When installing your weather station, remember:

- Watch out for power lines when working with your mast.

- Be careful when working on the roof and ladders.

- Follow the grounding guidelines presented in this chapter.

Building a Pagoda for Temperature and Humidity

If the location you have chosen for your temperature and humidity modules doesn't have naturally occurring shade, then one option is to create your own. A search online of "solar radiation shield" will provide many sources. Unfortunately, most cost more than $100, with some being as high as $400. So why not build your own?

A solar radiation shield performs two functions: it provides protection from environmental conditions such as rain and snow, and creates "artificial" shade so your sensors read the ambient air temperature. They usually consist of round plates stacked closely together to block the sun but allow air to freely flow through. The sensor is placed in the center of the plates with as little contact with the plates as possible. Figure 11-6 shows the solar shield for the Davis weather station.

FIGURE 11-6: Davis Vantage Pro2 solar radiation shield.

One way to build a solar shield is to modify a garden "pagoda" or tier light. These are available at most home and garden stores in the landscape lighting section. They are available as either the low-voltage lighting style or AC powered. To convert the tier light to a solar pagoda, you will have to make a couple of modifications. Out of the box, the tiers are too widely spaced to offer much solar protection and are painted black or some other dark color, which absorbs heat.

1. Choose a tier light that is constructed of metal and has at least four tiers. If you desire more than four tiers, you can combine two or more lights to make one tall one. Make sure the tier light has a standard ½-inch threaded base to make mounting easier.

2. Disassemble the tier light, discarding all but the top, bottom, and tiers.

3. Cut the legs of each tier shorter. You want about a ½- to ¾-inch gap between tiers. Measure the existing gap between the tiers and subtract ½-inch to determine how much of the leg to cut off. Figure 11-7 shows a cutting guide constructed of metal stock that allows the unwanted portion of the leg to protrude. If you can't find metal stock the correct thickness, try stacking several washers to get the correct height.

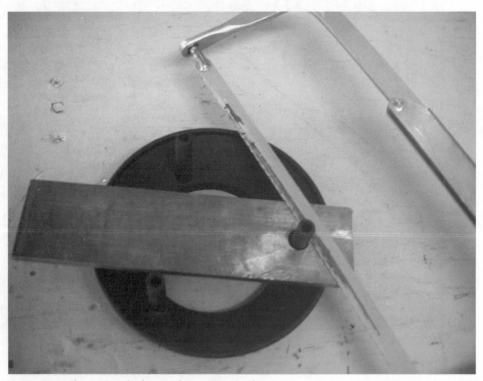

FIGURE 11-7: Shortening the legs with a cutting guide.

4. Temporarily assemble the pagoda and make sure the pieces fit. Chances are the screws that come with the tier light won't be the right length and you'll have to purchase new ones. The pagoda light I used needed 10/32 x 3-inch all-thread screws.

5. To minimize the absorption of heat, paint your pagoda white. Use a good quality, glossy white epoxy enamel.

6. Once the paint is dry, you will need to decide how to mount the temperature module inside the pagoda. Because blowing rain can enter your pagoda, you should provide some protection for your module. Figure 11-8 shows a 1½-inch PVC cap glued to the top of the pagoda to block rain and snow from reaching the board. Use Velcro to attach the module inside the cap. This scheme provides good air circulation around the sensors, but prevents access to the RJ-45 connectors, so the screw terminals must be used instead. For added protection, coat the board with a waterproof coating and install a sock over the humidity sensor element as described later in this chapter.

FIGURE 11-8: Mounting the temperature and humidity module.

7. Figure 11-9 shows two mounting options for the pagoda. The photo on the left shows a close-up of the mast mount and the photo on the right shows it on top of a 4-foot length of standard ½-inch EMT conduit that was sunk into the ground. Run the 1-Wire cable through the bottom of the pagoda and connect it to your temperature module.

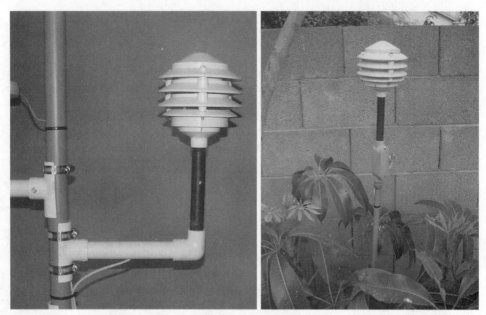

FIGURE 11-9: Completed pagoda mounted on a mast (left) and using a conduit pole (right).

Building a Mast Mount for Your Rain Gauge

All of the weather modules presented so far can be mast mounted, so why not add the rain gauge? It makes installation easier by having all the equipment on one pole. From an electrical point of view, it makes wiring a lot easier.

To mast-mount your rain gauge, attach it to a piece of wood that fits your rain gauge. The Hobby-Boards, Davis, and RainWise gauges fit nicely on a 12-inch diameter round wooden disk, available at many of the larger hardware stores and lumberyards. Figure 11-10 shows one way to mount the rain gauge to a mast. The wood was painted to protect it from water damage. Use a ½-inch plumbing floor flange to attach the wood to the PVC pipe frame constructed of ¾-inch PVC pipe and two of the ¾-inch modified tees presented in Chapter 10.

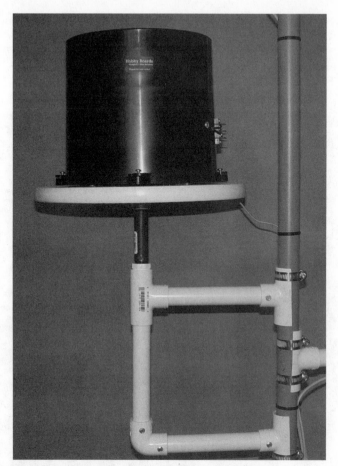

FIGURE 11-10: Mast-mounting the rain gauge.

Protecting the Circuit Boards

When installing your weather modules outdoors, they get subjected to harsh environments. Moisture, dust, insects, and salt spray can lead to early failure. Because you are trying to measure the weather, installing the circuit board in a weatherproof enclosure is not a good option. Instead, to protect your modules, coat the circuit board with a waterproof coating.

Weatherproof the Board

Several coatings on the market are designed specifically for PC boards. The two most popular used commercially are urethane and acrylic conformal coating. These coatings provide excellent protection against contamination and corrosion.

Another option is a product called liquid electrical tape. It can be found at most major hardware stores and comes in several colors. Its primary advantage is that it is inexpensive and easily removed if you need to work on the circuit board. It dries to a rubber-like coating as shown in Figure 11-11. You should coat all the circuit boards you are installing outdoors. Make sure to mask off the sensor and connectors when applying the coating.

FIGURE 11-11: Temperature and humidity board coated with liquid electrical tape.

Make a Humidity Sock

Some 1-Wire weather hobbyists have experienced problems with the humidity module when exposed to very high humidity (> 98%) for extended periods. The cause is unknown, but the problem seems to be in the sensor element itself. Some hobbyists have reported that installing a cloth "sock" over the humidity sensor eliminated the problem. Construct the sock out of a synthetic material, such as polyester or nylon because natural fabrics like cotton tend to absorb and hold moisture. Figure 11-12 shows a sock constructed by hot-gluing the sides and one end of a strip of polyester fabric together. The sock is then slipped over the humidity sensor. This also provides extra protection from contamination.

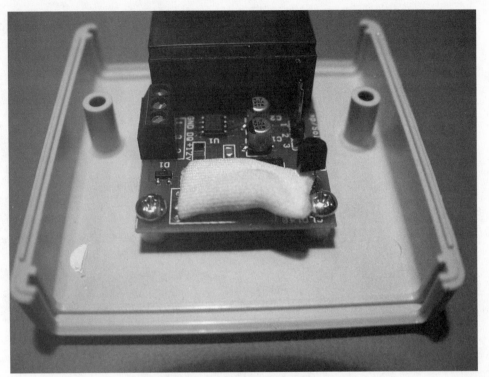

FIGURE 11-12: Polyester sock installed over the humidity sensor.

Finishing the Wind Instrument

When you built the Wind Instrument in Chapter 6, there were a few items that were not completed. Here's a list of things to do when mounting your wind instrument:

- To protect water from entering the wind instrument, slide it off the 1-inch tube and stuff some foam rubber in the tube around the cable to prevent water from entering the wind instrument from the tube.

- Install the end cap. File a slot in the cap that came with the 1-Wire Wind Instrument to allow the 1-Wire cable to pass through. See Figure 11-13 for an example. Make sure to leave some extra cable in case you need to slide the wind instrument off for repair.

- Install the screw that attaches to the wind instrument to the 1-inch square tube.

- Use wire ties to attach the cable to the mast to keep the wind from whipping it around.

FIGURE 11-13: Installing the end cap.

Wiring the Station

In Chapter 3 you read about what I call short length and long length cables. As a refresher, plain twisted-pair or CAT3 cable is used for the short distances to interconnect the various 1-Wire modules and the more expensive CAT5 twisted-pair cable is used for the longer distances. You also read about the various types of network topologies. Here is where you get to apply what you read about.

Connecting Your Weather Station

Once you have determined where you want to install your modules, you will have to run the 1-Wire network cable. Figure 11-14 shows an example 1-Wire network. In this example, the DS9097U 1-Wire serial adapter and the barometer module are mounted indoors and the remaining modules are installed outdoors with the rain counter located at the end of the chain and a considerable distance away. The majority of the cable length in this configuration is assumed to be between the indoor barometer and the first outdoor module (in this case it's the temperature module), and between the lightning detector and the rain counter. CAT5 cable is used for these long lengths.

Using the daisy-chain method is not always feasible. For example, the connections to the temperature and humidity module in the weather pagoda were made using the screw terminals. In these situations, a stubbed network may be an acceptable alternative. Stubbed networks are not as reliable as a straight daisy-chain, but if the stubs are kept short and originate from a single point, you should get satisfactory operation. Figure 11-15 shows the example network using a junction box to connect the various outdoor modules.

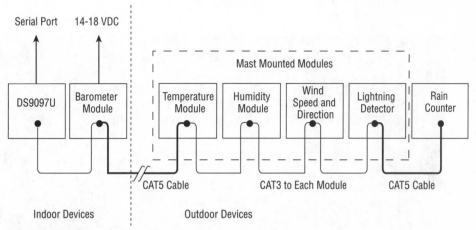

FIGURE 11-14: Example daisy-chaining the modules.

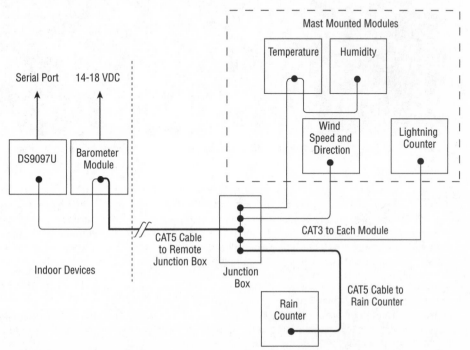

FIGURE 11-15: Example network using a junction box to connect modules.

A good option for your junction box is a multi-outlet modular phone connector. These can be purchased at many hardware stores and Radio Shack. You should cut off the existing modular cable, or open the case and remove the cable entirely. Using a modular junction box allows you to easily disconnect modules for troubleshooting wiring problems. Plus, it fits nicely in a dual-gang plastic outdoor waterproof electrical box, as shown with the cover removed in Figure 11-16. Other options for connecting the modules include barrier screw terminals and phone splices.

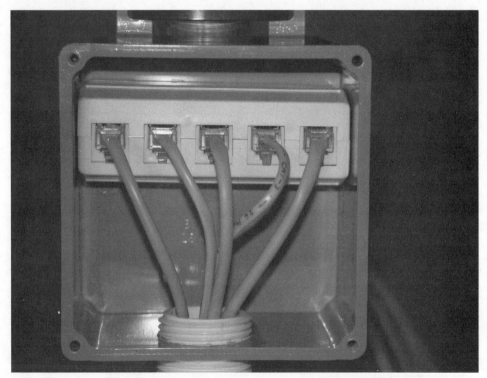

FIGURE 11-16: Five-outlet modular phone connector in a dual-gang outdoor electrical box.

Mounting the indoor section of your weather station is fairly straightforward. The DS9097U 1-Wire adapter connects to your serial port or USB-to-Serial adapter. A length of CAT5 or CAT3 twisted-pair cabling connects the 1-Wire adapter to the indoor barometer, which is then daisy-chained outdoors to the rest of your 1-Wire equipment. You should mount the circuit board inside a protective case. The case can be wall mounted, attached to your computer with Velcro, or simply set on a shelf.

Another option for your indoor equipment is to install it in a separate enclosure, such as a sprinkler timer box available at many hardware stores. As shown in Figure 11-17 with the front cover removed, these boxes provide plenty of room for additional modules, and many have an electrical outlet that provides power to the wall-mount transformer for the barometer module.

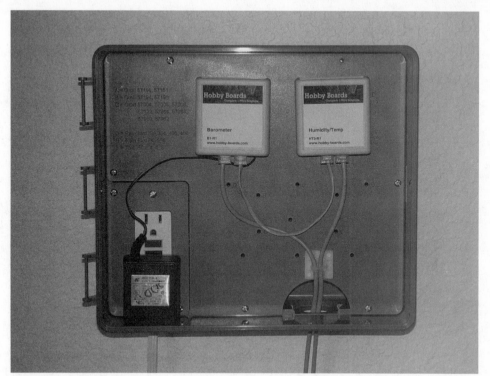

FIGURE 11-17: Sprinkler timer box used to house the indoor equipment.

Routing the Cables

When running your cables from your computer to the weather station, it is important to route the cables correctly. Here are few pointers:

- Don't run the cables through metal conduit. This adds capacitance to your 1-Wire bus, limiting the 1-Wire cable length and possibly causing data corruption.

- When running the cable through a wall to get outside, consider using an RJ-45 wall jack on the inside wall as shown in Figure 11-18. On the outside wall, use a cable bushing to protect the wires. Both the RJ-45 wall jack and cable bushings are available at most major hardware stores.

- When running the cable along the ground or under an object that critters could get under, you may want to install the cable in PVC pipe. For some reason, rodents love to chew through CAT5 cable.

- If you are running your cable underground, consider running it in the flexible ½-inch irrigation pipe. It is inexpensive and provides good protection for the CAT5 cable.

- If you need to make splices outdoors, make sure to waterproof the connections. Waterproof phone splices work well.

- Make sure you use outdoor UV-protected CAT5 cable anywhere the cable is exposed to the elements, especially the sun.

FIGURE 11-18: RJ-45 wall-mount plate (left) and outdoor cable bushing (right).

Grounding

Proper grounding is important, not just from an electrical perspective, but from a safety viewpoint as well. If you live in an area that gets thunderstorms, proper grounding is paramount.

In general, any metal mast or pole that's higher that 5 feet above ground level should be connected to earth ground. This provides a path to earth should lightning strike your mast. Finding a good earth ground is not always easy. In some cases, you will have to install a ground rod. These are 8-foot copper rods that are driven into the ground. Depending on your soil type, you may need more than one. For example, in Southwest Arizona, the soil is extremely dry, which limits its conductivity. I have three ground rods placed 3 feet apart and have added an irrigation "dripper" near each one to help keep the soil moist to improve conductivity.

From a safety perspective, it is best not to connect the mast to the house electrical ground. Should lightning strike your mast, channeling the current flow though your house wiring could lead to serious electrical damage or fire. Figure 11-19 shows a typical ground scheme for a weather station installation. Notice there is a separate ground for the mast.

If you are using the lightning detector, you also need to connect it to earth ground. This can be the same ground for the mast, but I suggest running a separate wire directly to the ground rod.

One more point to keep in mind is that one side of the 1-Wire bus is connected to your computer's case, which should already be connected to power ground. To prevent ground loops, don't connect the ground-side of the 1-Wire bus to a second earth ground. If you are concerned about the integrity of the house ground, install a separate ground rod and connect it directly to your computer's chassis. If you live in an area that is susceptible to lightning, Chapter 14 discusses additional measures for lightning protection and grounding.

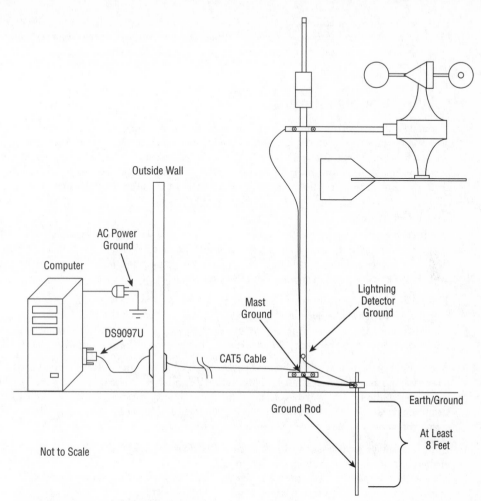

Outside Wall

AC Power
Ground

Computer

DS9097U

Mast
Ground

Lightning
Detector
Ground

CAT5 Cable

Earth/Ground

Ground Rod

At Least
8 Feet

Not to Scale

FIGURE 11-19: **Typical grounding scheme.**

Testing

Once you have your station installed and connected, you're ready to test your installation. If you have followed along with the book and built a "mini 1-Wire network" as you built each module, then you should have high confidence in each module and your software. So start by testing to see if you can detect the 1-Wire devices. If you haven't already done so, connect the weather equipment to your DS9097U 1-Wire serial port adapter and launch OneWireLister. If you have built all of the modules and sensors in this part of the book, you should see all seven 1-Wire devices. If all of your devices don't show up, refer to the 1-Wire troubleshooting in the next section.

If all your 1-Wire devices show up in OneWireLister, you are ready to launch SimpleWeather. Once SimpleWeather is running, watch the output window. You will see the weather data updated once every minute. Let it run for a while and check for any errors and whether the weather data looks correct.

1-Wire Troubleshooting

If your 1-Wire network doesn't work at first, don't worry. It is fairly simple to troubleshoot the network, because there are only two wires. If only one module is not working, start at the connection for that module and work backwards. If you have used the daisy-chain method to connect your modules, and one or more modules aren't working, start at the first module in the chain that isn't responding, because it may be the cause of the other modules not working.

Check the Voltage

The first thing to check when one or more devices aren't showing up in OneWireLister is the bus voltage. To make it easy to connect your voltmeter to the 1-Wire bus, build an RJ-45 or RJ-11 test aid by building a test data cable with a RJ plugs on one end and bare wires on the other (you can cut up one of the test cables you made earlier).

The 1-Wire bus is normally not powered unless it is enabled by software. The OneWireLister program only enables the bus for a few seconds, so you want to run SimpleWeather. Launch SimpleWeather and, ignoring the 1-Wire bus error messages, connect your voltmeter to the bus observing the proper polarity. In its idle state, the bus should read about 4.5 to 5.0 volts.

Another option is to use the 1-Wire test LED described in Chapter 3. Run SimpleWeather to power the bus and plug in the test LED to the 1-Wire adapter. It should light up green. (If it lights up red, the LED is wired backwards and you will need to swap the leads of the LED). Continue down the chain of devices, plugging the test LED into each device. If it lights up green, you know you have power and the polarity is correct.

If the bus voltage reads less than 1 volt or the LED does not light up, the bus is either shorted or there is a break in the cabling. Try disconnecting the first outdoor module and checking the bus coming from inside. Does the LED light or does the voltage read above 4.5 volts? If it does, then you know the problem is with the outside connections.

Check the Polarity

Another common mistake is to swap the wires when crimping on an RJ connector. If the wires are swapped, that module will cause a short on the bus (remember the diode across the 1-Wire bus in the schematics?). Again, use the test LED or a voltmeter to check for this. Another test aid you can build is to attach wires to an RJ-11 or RJ-45 jack. This way, you can connect your voltmeter to the end of a cable. If you're using the test LED, build a jack-to-jack adapter to connect to the end of the cable.

Occasional Errors

When collecting weather data, it is normal to experience a few 1-Wire network errors. But what is normal? In general you shouldn't see more than a dozen errors in a 24-hour period.

Because SimpleWeather reads the weather sensors once a minute, there are 1,440 readings in a 24-hour period. Multiply this times the number of sensors you have, and a dozen errors in a 24-hour period equates to an error rate of about 0.1%, which is acceptable.

There are two types of errors to expect: 1-Wire communication errors and sensor reading errors. Communication errors occur when the data sent or received from a device gets corrupted. This usually shows up as a CRC error in SimpleWeather or less commonly, a "device not found" error. These errors are usually caused by noise induced in the bus. For example, I experienced several CRC errors during the day. For a completely unrelated reason, I moved the indoor cabling to another side of my room and most of the CRC errors stopped. Upon investigation, I discovered that my air conditioner compressor was just on the other side of my wall. Every time it turned on it caused a large transient on the bus. If you are experiencing a large number of errors, start by checking the cable path, keeping your cable away from high-power devices, such as motors and fluorescent lights.

The only common sensor error you should see is an error reading the wind direction. This is caused by the small delay it takes to read each of the four A-to-Ds in the wind vane. If the wind vane is moving quickly, the remaining A-to-Ds could change value after the first or second A-to-D is read. During the look-up conversion discussed in Chapter 6, an incorrect value will return an error. This is a basic limitation of the wind vane, and should not be a cause of concern.

If you are getting errors higher than discussed, or the sensor reads the error value of −999.9 the problem may be with your 1-Wire network. Check each of your connections and re-read the guidelines for 1-Wire networks in Chapter 3. If you still can't pinpoint the problem, try installing a 100-ohm ¼-watt resistor in series with the 1-Wire data line just after the 1-Wire serial adapter as shown in Figure 11-20. This helps minimize cable reflections and lowers the slewing of the waveform, however it does increase the susceptibility of the 1-Wire network to interference from electrical noise because the impedance of the bus is increased.

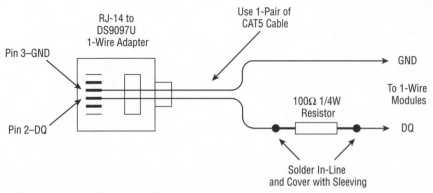

FIGURE 11-20: Installing a 100-ohm resistor in the bus.

If you are still experiencing higher than expected errors, Dallas/Maxim has an application note with several suggestions for minimizing 1-Wire network problems. Download Application Note 148 "Guidelines for Reliable 1-Wire Network" at www.maxim-ic.com/appnotes.cfm/ appnote_number/148.

Wrap Up

Now that you've got your weather station installed, you can start collecting real weather data. Let your weather station run for a day or two, then plot the logged data using your favorite graphing tool. Looking at the plots, you can see the weather changes throughout the day. Do they look as expected? You may be surprised at some of the variations in the data, whether they are real or caused by local conditions. Looking at my humidity plots, I can see a spike every morning at 6 A.M. when my sprinklers turn on, even though my humidity sensor is more than 20 feet away. Before jumping into Part III of this book, let your weather station run for a few days to make sure it works reliably.

This chapter completes Part II of this book, and you now have assembled all of the basic weather station sensors and have your weather station installed. I trust that you have enjoyed the projects so far, but don't stop yet. There's more to come in Part III.

If you have been following along with the software project, you have learned a great deal about how to communicate with 1-Wire devices and setting up the software. If you are new to programming, you should now be familiar with NetBeans and the basics of Java. You may already be thinking about ways you can hack the software to make custom modifications.

Before starting Part III, now would be a good time to make a backup copy of your SimpleWeather project. As you modify the code in the next chapters, you may want to come back to this code for reference.

Expanding Your Weather Station

In the last section, you completed and installed the 1-Wire weather station. Now that it's up and running, it's time to look at the many options you have to expand it.

This part of the book provides several projects to help you get started customizing your weather station to make it do what you want. You'll learn how to build a weather web server on your own computer, or post your collected information to an online weather data service so others can check the weather in your area. Other projects show you how to turn on or off appliances based on weather, add an LED sign to display the temperature, and how to protect your 1-Wire bus from surges and spikes. You can also add other environmental sensors and build your own weather-controlled sprinkler timer and home thermostat. There's also a chapter that shows you how to add the finishing touches to the SimpleWeather program and build it as a stand-alone application.

In this section, there's a chapter that shows you how to get rid of your PC and build a stand-alone microcomputer to run your weather station and post your data on the Internet.

Although there are many exciting projects in this section, they're a little more advanced than Part II and won't have as much detail. They are meant to provide you with ideas to expand and hack your weather station to meet your own personal needs. Since the application of each project will vary, each person's project will be a little different. I can show you the pieces of the puzzle, but you'll have to put it together yourself.

Add a Weather Web Server

chapter

12

in this chapter

☑ Add a Weather Data Cruncher

☑ Code the Web Page

☑ Add a Web Server to Your Software

☑ Configure SimpleWeather

☑ Share Your Weather Data on the Internet

Right now you should be feeling pretty proud of your accomplishments. You have built one or more weather sensors and coded the software that allows you to read the data. The weather sensors have been installed and are collecting data, saving it to your hard drive. But watching those little numbers scroll past in SimpleWeather is a little boring. Well, it's time to change that.

This chapter shows you one of the many ways you can share your data. It can be published locally to your machine, shared with everybody on your local network, or posted worldwide on the Internet. Sharing on your local network is easy and all you need is a networked computer. Setting up a web server so others on the Internet can see your weather data is a just a little more difficult and requires an always-on Internet connection like DSL or cable.

To web-enable SimpleWeather, you will do two things. First, you will add code that builds a simple web page from your weather sensor's data. To accomplish this, a class is added that takes the weather data and "crunches" it into text data that is used to build your web page. The second part is to add a mini web server in software that allows others to access the web page directly from your computer.

To allow Internet users to access the weather web page, you will need to know your Internet IP address and I'll show you how to discover this. Once you know your IP address, then anyone can type in that address and access your web page.

Project Overview and Design

Start by designing the weather web page. To keep things easy, the page will be a simple text-based table listing the weather data. It will need a header for station-specific data such as the location, current time, and a spot for a logo or picture. At the bottom of the page, a hit counter would be useful so you can track how many people have accessed your web page.

When you see a weather report on the news, it usually provides more than just the current conditions. Often, you get the daily highs and lows, the daily rain amount, and trend data for barometric pressure. You may also see the peak wind speed and the year-to-date rain amount. There's no reason why SimpleWeather can't display the same data. Table 12-1 shows the results of converting the desired parameters into a table.

Table 12-1   Weather Data Parameters for SimpleWeather's Web Page

Temperature	Current	High	Low	Trend
Wind	Current	High	Peak	Trend
Humidity	Current	High	Low	Trend
Dewpoint	Current	High	Low	Trend
Pressure	Current	High	Low	Trend
Rain	Year-to-Date	Daily	–	Rate
Lightning	Current	High	–	Trend

To complete the design, the web page update rate needs to be determined. There are several different thoughts on this. Some users prefer to update the page as fast as possible. Because the main loop in SimpleWeather runs once a minute, you could update your page that often. On the other hand, the National Weather Service updates its data once an hour. I believe the right answer is somewhere in between the two. For this project, I chose to update the page once every 10 minutes. If you desire, with just a few simple modifications, you can change this value to your preference.

Add a Weather Cruncher

Looking at Table 12-1, you may be wondering how to get the statistical data because all SimpleWeather has are the current values. That's where the Weather Cruncher class comes in. To get statistical data, code is needed to collect, process, and store data. To perform that function, a new class, WeatherCruncher.java, is added to SimpleWeather.

Add the WeatherCruncher Class

Open your WeatherToys directory and find the project source code you downloaded from the companion web site. Next, locate the Chapter 12 directory. Inside, you will find six files:

- WeatherCruncher.java
- WebPage.java

- `WebWorker.java`

- `HTTPServer.java`

- `HTTPServerException.java`

- `logo.jpg`

Copy only the `WeatherCruncher.java` file to your SimpleWeather ⇨ src ⇨ SimpleWeather folder. The remaining files will be added later. There is also a complete project folder for reference.

Modify SimpleWeather

Launch NetBeans and open your SimpleWeather project. Check your list of source files to make sure `WeatherCruncher.java` is listed. Open `SimpleWeather.java` in the NetBeans editor. Scroll down to the `// class variables` comment near line 46. Add a class variable for the WeatherCruncher to the list:

```
// class variables
public float temp;
public float windSpeed;
 ...
private DSPortAdapter adapter;
private DataLogger logger;
private WeatherCruncher wc;
```

In SimpleWeather's constructor, add the following code to instantiate the weather cruncher class after the code to instantiate the data logger:

```
// initialize the data logger
logger = new DataLogger(LOG_PATHNAME);

// initialize the weather data cruncher
wc = new WeatherCruncher(this);
```

In the `mainLoop()` near line 120, add an `int` variable for the current hour:

```
public void mainLoop()
{
    Date date = new Date();
    int minute, lastMinute = -99;
    int hour;
    boolean quit = false;
    String command;
    InputStreamReader in = new InputStreamReader(System.in);
```

In the `mainLoop()` after each of the sensors are initialized near line 132, add the code to clear the daily highs and lows, and reset the averages:

```
// initialize sensors
ts1 = new TempSensor(adapter, TEMP_SENSOR_ID);
```

```
...
ls1 = new LightningSensor(adapter, LIGHTNING_SENSOR_ID);

// reset the weather highs, lows, and averages
wc.resetHighsAndLows();
wc.resetAverages();
```

Add the code to read the hour just after the code to read the minutes in the main loop:

```
// main program loop
while(!quit)
{
  // sleep for 1 second
  ...
  // check current time
  date.setTime(System.currentTimeMillis());
  minute = date.getMinutes();
  hour = date.getHours();
```

Toward the end of the `mainLoop()` method near line 195, between the call to print the lightning count and where the data logger is called, add the following code:

```
// get lightning count
lightning= ls1.getLightning();
System.out.println("Lightning = " + lightning + " SPM");

// if it's midnight, reset the highs & lows
if (hour == 0 && minute == 0)
  wc.resetHighsAndLows();

// update the weather data
wc.update();

// log the data
logger.logData(date, this);
```

Let's walk through the proceeding code. First, there is a check to see if the hours and minutes are both zero. If that's the case, it must be midnight and WeatherCruncher's `resetHighsandLows()` method is called, which as you may have guessed, resets the highs and lows for various weather parameters.

Next is a call to the WeatherCruncher's `update()` routine, which updates the WeatherCruncher's data. This gets called once a minute, each time a weather measurement is taken. You learn more about this in the next section.

A Quick Look at WeatherCruncher.java

`WeatherCruncher.java` has more than 500 lines of code, so I can't go through the entire class in detail or you'd never get to the next project. However, it is important to understand a little bit about how the data is "crunched" so that if you later make changes to the web page, you know how to get the data you need. Not all of the code is listed here to save space, so if you prefer, you can follow along with the code in the NetBeans editor.

Starting at the class definition, there is only one user constant, TREND_SIZE. It defines the number of previous measurements to be used when calculating the trend data. By default it is set at 20, and because the data is updated once a minute, this provides the trend over the preceding 20-minute period.

The class variables list is quite extensive, because there are three or more values for each sensor. The variables have descriptive names, so you should be able to figure out what most of them are for. I've just listed the first few lines:

```
public class WeatherCruncher
{
  // user constants
  private static final int TREND_SIZE = 20;

  // class variables
  private float temp, tempHi, tempLo;
  private float hum, humHi, humLo;
  private float dp, dpHi, dpLo;
```

One of the more complicated parts of this class is the trending routine. Six arrays are used to keep the data for the last 20 measurements of the six sensors. In WeatherCruncher's constructor, the arrays are allocated and each element is initialized to a –1, which is checked later to indicate if real data has been stored. There is also an array x[], which is set to the values from 0 to the array size minus 1, which in this case is 19:

```
public WeatherCruncher(SimpleWeather sw)
{
  debugFlag = SimpleWeather.debugFlag;
  this.sw = sw;

  // arrays for trend data
  x = new int[TREND_SIZE];
  tempTrend = new float[TREND_SIZE];
...

  // initialize trend values
  for (int i = 0; i < TREND_SIZE; i++)
  {
    tempTrend[i] = -1;
    humTrend[i]  = -1;
    dpTrend[i]   = -1;
    windTrend[i] = -1;
    baroTrend[i] = -1;
    ltngTrend[i] = -1;
    x[i]         = i;
  }
}
```

Every evening at midnight, the highs and lows need to be reset. In this routine, the highs are set to an unrealistic low value and the lows to an unrealistic high. On the first measurement, both the low and the high will get set to the sensor's current measured value.

Rain does not have a high or low value. Instead, the rainfall since midnight is displayed. This is calculated by storing the current value of the rain counter at midnight. Then, each update subtracts this value to calculate the value since midnight:

```
public void resetHighsAndLows()
{
  // reset temperature
  tempHi = -999;
  tempLo = 999;
  ...

  // reset rain
  rainOffset = sw.rain;
  rainRate = 0;

  // reset lightning
  lightHi = 0;
}
```

The data between web page updates is averaged so that the web page is displaying a 10-minute average. Once the web page is built, the averages need to be reset for the next 10-minute period. The following code resets the average accumulator and number of samples to zero:

```
public void resetAverages()
{
  samples = 0;
  temp = 0;
  hum = 0;
  wind = 0;
  baro = 0;
  light = 0;
  avgSin = 0;
  avgCos = 0;

  if (debugFlag)
    System.out.println("Averages Reset");
}
```

Once a minute, the weather cruncher's `update()` method collects the data from each of the sensors and adds the data to the sensor's accumulator variable, and `samples` is incremented by one. When each sensor's data is requested later, the average value is calculated by dividing the sensor's accumulator's value by the number of samples.

After each sensor is updated, a call is made to the `UpdateTrendData()` method, which adds the current measurement to the sensor's trend data array.

The rain and lightning devices are a slightly different case. Because the count is persistently stored in the 1-Wire device, the device sums the counts for you and therefore a separate accumulator is not needed.

The web page displays the wind's high and peak values. The high value is the highest of the 10-minute averages, and the peak value is the highest of the 1-minute readings:

```java
public void update()
{
  // increment the sample counter
  samples++;

  if (debugFlag)
    System.out.println("Sample #" + samples);

  // update temperature
  temp += sw.temp;
  updateTrendData(sw.temp, tempTrend);

  ...

  if (sw.windSpeed > windPk)
    windPk = sw.windSpeed;

  updateTrendData(sw.windSpeed, windTrend);

  // update wind direction
  updateWindAvg(sw.windDir);

   ...

  // update rain
  rain = sw.rain;
  rain24 = sw.rain - rainOffset;

...
}
```

The `UpdateTrendData()` method adds the current measurement to the end of each sensor's trend array. To do this, each element in the array must be "shifted" one to the left, discarding the oldest data:

```java
private void updateTrendData(float value, float[] yArray)
{
  // shift trend data down one
  for (int i = 0; i < TREND_SIZE-1; i++)
    yArray[i] = yArray[i+1];

  // add new data to the end of the list
  yArray[TREND_SIZE-1] = value;
}
```

To get the trend for the sensor, a calculation is used to determine the rate of change over a series of values. If you remember your high school algebra, the line y = mx + b has slope m, which represents the linear rate of change. The x[] array was initialized earlier to contain the value from 0 to 19. This is the x values, and represents time in minutes. The six trend data arrays are the y values. To calculate the slope of the line that best fits the series of x[] and y[] values, the Least Squares Fit equation is used. It returns the rate of change per minute. Because the web page displays the trend in hours, the value is multiplied by 60, and the results are sent to the formatValue() routine. See the code in NetBeans for a listing of this method.

Averaging the direction is considerably more difficult from the other weather parameters, because the data is discontinuous. Referring back to Table 6-3, the wind vane values range from 0 to 15, and then start back at 0. If the weather vane is pointed north, for example, and is bouncing around a bit, its value could be jumping back and forth between 1 and 15. If you average these values, the result is 8, which according to Table 6-3 is south. To work around this problem, the wind direction is converted to its trigonometric sine and cosine values, which are continuous. The sine and cosine values are summed in the updateWindAvg() routine:

```
private void updateWindAvg(int windDir)
{

  // convert wind direction to radians
  double angle =  Math.toRadians(windDir * 22.5);

  avgSin += Math.sin(angle);
  avgCos += Math.cos(angle);
}
```

When the web page needs the wind direction, it calls the getWindDirAvg() method, which calculates the sine and cosine averages and then uses the arcsine and arccosine functions to determine the average angle. The resulting angle is converted back to a value between 0 and 15, then passed to the getWindDirStr() method, which returns the text string representing the average wind direction:

```
private String getWindDirAvg()
{
  // divide by the number of samples to get average
  avgSin /= samples;
  avgCos /= samples;

  // convert average to wind dir int value (0 thru 15)
  double angle = Math.toDegrees(Math.asin(avgSin));

  if (avgCos < 0)
    angle = 180 - angle;

  else if (avgSin < 0)
    angle = 360 + angle;

  int dir = (int)((angle + 0.5)/22.5);
```

```
   if (debugFlag)
     System.out.println("Avg Wind Angle = " + angle + " = " +
             WindSensor.getWindDirStr(dir));

   return WindSensor.getWindDirStr(dir);
 }
```

The `formatValue()` method takes a floating-point value and rounds it to the specified number of digits. The result is then converted into a String value:

```
// short routine format a float to 'digit' decimals and convert to
string
private String formatValue(float input, int digits)
{
  try
  {

  String arg1;
  String arg2;
  StringTokenizer t;

  if (digits < 1)
    digits = 1;

  if (digits > 3)
    digits = 3;

  float roundVal;

  if (digits == 1)
    roundVal = .05f;
  else if (digits == 2)
    roundVal = .005f;
  else // digits = 3
    roundVal = .0005f;

  t = new StringTokenizer(Double.toString(input + roundVal), ".");

  arg1 = t.nextToken();
  arg2 = t.nextToken();

  if (Math.abs(input + roundVal) < .001 && digits == 3)
    return ("0.000");
  else
    return arg1 + "." + arg2.substring(0, digits);
  }
  catch (Exception e)
  {
    return "error";
  }
}
```

When the SimpleWeather project was started, I made the sensor's result variables public, so that any method in the SimpleWeather program could access them. This is actually a bad programming practice. Java (and most Object-Oriented Programming languages) encourages *encapsulation*, where direct variable access is not allowed. Access to the variables is only achieved through the use of "getter" and "setter" methods, which get the value or set the value of a variable. This prevents inadvertently changing a variable's value.

WeatherCruncher.java uses this method. The remaining code is the "getters" that will allow the web page to get the data. Take a look at the temperature getters only, because the other sensors are similar. To retrieve the temperature, getTemp() is called. It calculates the current average and checks to see if the new average is a high or low for the day. The temperature is then formatted as a string using the FormatValue() method, which is then passed back to the calling function.

There are similar methods to get the high temperature, low temperature, and the temperature trend. You can look through the remaining 24 getters in WeatherCruncher.java.

```java
// temperature --------------------------------
  public String getTemp()
  {
    temp = temp/samples;

    if (temp > tempHi)
      tempHi = temp;

    if (temp < tempLo)
      tempLo = sw.temp;

    return formatValue(temp, 1);
  }

  public String getTempHi()
  {
    return formatValue(tempHi, 1);
  }

  public String getTempLo()
  {
    return formatValue(tempLo, 1);
  }

  public String getTempTrend()
  {
    return leastSquaresSlope(TREND_SIZE, x, tempTrend);
  }
  ...
}
```

Before continuing, compile and run your project to make sure there aren't any errors or typos. Although you won't see any differences in the output printed to the screen, this will catch any problems before you add the next class.

Add the Web Page Builder

Once the weather data has been crunched into text values, it needs to be converted into the format web browsers need, called HyperText Markup Language, or HTML. If you've never worked with HTML before, don't worry. I've already built the web page builder class for you. However, when you want to make changes or expand the web page, you'll need to know the basics of HTML programming. Many good books on HTML are available at your favorite bookstore or online.

Add the WebPage Class

Copy the following four files from your downloaded Chapter 12 source code to your SimpleWeather ➪ src ➪ SimpleWeather folder:

- `WebPage.java`
- `WebWorker.java`
- `HTTPServer.java`
- `HTTPServerException.java`

Modify SimpleWeather

In your SimpleWeather Project select File ➪ Refresh All Files and check the list of source files to make sure the four new files are listed.

Open `SimpleWeather.java` in the NetBeans editor and scroll down to the `// class variables` comment. After the declaration of the `debugFlag` near line 56, add a new object WebPageLock. A few lines down, right after the WeatherCruncher class, add the WebPage class:

```
// class variables
...
public static boolean debugFlag = false;
public static Object webPageLock = new Object();

private DSPortAdapter adapter;
private DataLogger logger;
private WeatherCruncher wc;
private WebPage wp;
```

In the SimpleWeather constructor, add the code to instantiate the new web page class, passing it a reference to the SimpleWeather:

```
// initialize the weather data cruncher
wc = new WeatherCruncher(this);

// initialize the web page
wp = new WebPage(wc);
```

In the main loop, after the code to call the weather cruncher, add the following code to update the web page once every 10 minutes and reset the averages:

```
// update the weather data
wc.update();

// and update the web page on the 10 minute mark
if (minute % 10 == 0)
{
  wp.updatePage(date);

  // reset the averages for the next set of data
  wc.resetAverages();
}
```

Save all your files before continuing.

Check the Web Page

Go ahead and run your project. You won't see any noticeable differences in the output window until the time hits an even 10-minute mark (7:10, 7:20, 7:30, and so on). At that time, you will notice a new line in the output window displaying Updating index.html. Leave the SimpleWeather project running and navigate to your SimpleWeather project folder. Open the folder and look for a new file named index.html. Open the file with your web browser. If all went well, you should be staring at a nicely formatted web page showing your weather data, similar to that shown in Figure 12-1.

Notice the trend data only shows dashes. Because the trend is calculated over a 20-minute period, the data isn't shown yet. Let SimpleWeather run for at least 20 minutes, and then on the *next* web page update, you should see trend data.

Looking Through the Web Page Code

A web page is nothing more than a text file containing the HTML code to display the page. The main task of the WebPage class is to create a file on your hard drive, get the weather data, and format it into the appropriate HTML text data.

My Weather Station
03/12/06 12:00 PM MST

Located in: My Town, My State

Temperature			
Current	Today's High	Today's Low	Trend
74.3 °F	74.3 °F	74.3 °F	- °F/Hr
Wind			
Current	Today's High	Today's Peak	Trend
0.0 MPH NNE	0.0 MPH	0.0 MPH	- MPH/Hr
Relative Humidity			
Current	Today's High	Today's Low	Trend
27.7 %	27.7 %	27.6 %	- %/Hr
Dewpoint			
Current	Today's High	Today's Low	Trend
43.1 °F	43.1 °F	39.1 °F	- °F/Hr
Barometric Pressure			
Current	Today's High	Today's Low	Trend
30.10 inHg	30.10 inHg	30.10 inHg	- inHg/Hr
Rainfall			
Year to Date	Since Midnight	-	Rate
0.00 in	0.00 in	-	0.00 in/Hr
Lightning Activity			
Current	Today's High	-	Trend
0.0 SPM	0.0 SPM	-	- SPM

Weather data collected using a 1-Wire Weather Station and SimpleWeather 1.0
JVM Free Mem = 1223 KB   Server Uptime: 0 Days, 0 Hours, 15 Mins   0 Hits since 06/01/06

Don't Use for Personal Safety

FIGURE 12-1: SimpleWeather's weather web page.

Looking at the code, you have many user constants that customize your web page. The first is the web page file path and name. It is constructed from the full path defined in the WebWorker class (which hasn't been covered yet) plus the file name in the user constant FILENAME. The path points to the root folder that is served by the web server, and FILENAME defines the weather page file name.

The next group of user constants defines the header and footer of the web page. Referring back to Figure 12-1, notice where the PAGE_HEADER and PAGE_SUBHEADER are located on the web page. These values should be changed to reflect your weather page name and location. The IMAGE_URL is the file name of a picture of your weather station or a logo. The maximum width of the data table is 640 pixels, which means the image should not be more than 320 pixels wide so the header data has room to be displayed. For now, leave the IMAGE_URL as is.

The START_DATE is displayed after the hit counter, and is used as an indicator of the web page creation date. The constant BANNER isn't used yet, and is presented in the next chapter. TITLE_COLOR and TABLE_COLOR define the color used for the sensor titles in the table and the general table color, respectively. These colors are defined in standard HTML hex color format (#RRGGBB).

The class constructor performs three functions: it assigns a copy of the WeatherCruncher class to a local variable, gets a copy of the web page lock, and gets the start time. The start time is used to display the server uptime in the table footer:

```java
public WebPage(WeatherCruncher wc)
{
  this.wc = wc;
  lock = SimpleWeather.webPageLock;

  // get the start time of this session
  Date d = new Date();
  startTime = d.getTime();
}
```

Escape... Escape...

Looking at some the Strings and color values in the code, you may be wondering why the code has multiple quotes and backslashes, such as "\"#bbccff\"".

In Java, enclosing the text within quotes denotes a String. If you need to include a quote in your String, it must to be preceded by the Java "escape" character, a "\" (backslash). If you want to include a backslash in your String, you must "escape the escape" and use two backslashes.

```java
public class WebPage
{
  private static Object lock;
  private WeatherCruncher wc;
  private static long startTime;
  private static boolean debugFlag = SimpleWeather.debugFlag;

  // user constants
  private static final String FILENAME       = WebWorker.ROOT_DIRECTORY +
                                                "index.html";
  private static final String PAGE_HEADER    = "My Weather Station";
  private static final String IMAGE_URL      = "logo.jpg";
  private static final String PAGE_SUBHEADER = "Located in: My Town, My State";
  private static final String PAGE_FOOTER    = "Don't Use for Personal Safety";
  private static final String START_DATE     = "06/01/06";
  private static final String BANNER         = "";
  private static final String TITLE_COLOR    = "\"#bbccff\"";
  private static final String TABLE_COLOR    = "\"#ddeeff\"";
```

The `updatePage()` method is where the web page is generated. The main body of this method is inside a `synchronized()` block. Synchronized sections of code can only be accessed by one portion or *thread* of the program at a time. Although the web server hasn't been started yet, it runs concurrently in its own separate thread of execution. A potential problem could occur if the `updatePage()` method was called just as the web server was serving the page to a requestor.

```
public void updatePage(Date dateTime)
{
   final char SPACE = ' ';

   //if (debugFlag)
   System.out.println("Updating " + FILENAME);

   try
   {
     synchronized(lock)
     {
```

In Java, a file is accessed using *streams*. Data is *input* from a file for reading or *output* to a file for writing. In the next block of code, a File Output Stream is created to write your HTML-formatted text to disk using the user constant `FILENAME`.

The code to build the web page header comes next, using the user constants mentioned earlier. As part of the header, there is a META tag `<meta http-equiv="refresh" content="600">`. This tells the web browser requesting the page to automatically refresh the page every 600 seconds, which is 10 minutes.

```
FileOutputStream file = new FileOutputStream(new File(FILENAME));

// write html page header
file.write(("<html>\n" +
           " <head>\n" +
           "   <title> 1-Wire Weather </title>\n" +
           "   <meta http-equiv=\"refresh\" content=\"600\">\n" +
           " </head>\n").getBytes());

file.write((" <body> <center>\n" +
           "   <table width=\"640\" align=\"center\"> \n" +
           "     <tr align=\"center\">\n" +
           "       <td> <img src=" + IMAGE_URL + ">" +  + "</td>\n" +
           "       <td> <h2>" + PAGE_HEADER + "<br>" +
                      getDateTimeString(dateTime) + "</h2></td>\n" +
           "     </tr>\n" +
           "     <tr align=\"center\">\n" +
           "       <td colspan=\"2\"> <H2>" + PAGE_SUBHEADER +
                                  "</h2></td>\n" +
           "     </tr>\n" +
           "   </table>\n").getBytes());
```

This is the code that formats the weather data into an HTML table. The table is set for a width of 640 pixels and has four columns. The border is set to 5 pixels and the entire table is aligned to the center of the browser window:

```
// weather data table
file.write(("    <font size=\"2\">\n" +
           "    <table border=\"5\" width=\"640\" align=\"center\"
                   bgcolor=" + TABLE_COLOR + "> \n").getBytes());
```

The weather data for each sensor is displayed in three rows. The first row spans the width of all four columns and displays the sensor's name, in this case "Temperature." The next row displays the name of the four parameters for the sensor. For temperature, "Current," "Today's High," "Today's Low," and "Trend" are displayed. The third row displays the measured and calculated values. Notice the use of the "getter" methods discussed earlier.

I've only included the code for the temperature display here. You can view the entire file in your NetBeans editor. If you have not built all the sensors in Part II of this book, you can delete or comment out the code for the missing sensors to make your web page look better.

```
// temperature
file.write(("    <tr>\n" +
           "      <th colspan=\"4\" bgcolor=" + TITLE_COLOR +
                  "><font size=\"4\">Temperature</th>\n" +
           "    </tr>\n" +
           "    <tr align=\"center\">\n" +
           "     <th> Current </th>\n" +
           "     <th> Today's High </th>\n" +
           "     <th> Today's Low </th>\n" +
           "     <th> Trend </th>\n" +
           "    </tr>\n").getBytes());

file.write(("    <tr align=\"center\">\n" +
           "        <td>" + wc.getTemp()      + " &#176F</td>\n" +
           "        <td>" + wc.getTempHi()    + " &#176F</td>\n" +
           "        <td>" + wc.getTempLo()    + " &#176F</td>\n" +
           "        <td>" + wc.getTempTrend() + " &#176F/Hr</td>\n" +
           "    </tr>\n").getBytes());
```

At the bottom of the table is a final row that displays a little information about the weather station including a hit counter, how long the server has been running, and Java's free memory. After the table, there is the web PAGE_FOOTER, which can be customized to display whatever you wish:

```
// print footer info
file.write(("        </td> \n" +
           "      </tr> \n" +
           "      <tr> \n" +
           "        <td align=\"center\" colspan=\"5\"><font
                        size=\"1\">\n" +
```

```
         "                 Weather data collected using a 1-Wire Weather
                           Station and SimpleWeather 1.0 <br>\n" +
         "                 JVM Free Mem = " + getFreeRam() + "    " +
                           getServerUpTime()).getBytes());

      file.write(("                 " + WebWorker.getHits() + " Hits since " +
                              START_DATE + "\n" +
         "           </td>\n" +
         "         </tr>\n" +
         "       </table>\n" +
         "     " + PAGE_FOOTER + "\n" + BANNER +
         "   </body>\n" +
         "</html>\n").getBytes());
    file.close();
    }
  }
  catch(Exception e)
  {
    System.out.println("Error Updating Web Page: " + e.toString());
  }
}
```

The getDateTimeString() method takes a Date object and converts it to a date string
that can be displayed in the web page. To modify the way the date is displayed, look up
SimpleDateFormat() in the Java Docs to read more about the formatting string.

```
// returns a date & time  string defined by the preferences date format string
public String getDateTimeString(Date d)
{
  SimpleDateFormat formatter = new SimpleDateFormat("MM/dd/yy h:mm a zzz");
  //System.out.println("Date = " + formatter.format(d));

  return formatter.format(d);
}
```

The getFreeRam() method simply returns the amount of free memory in kilobytes that is
available to the Java Virtual Machine (JVM):

```
// get JVM free ram
public String getFreeRam()
{
  return (Long.toString((Runtime.getRuntime().freeMemory())/1024) + " KB");
}
```

The last method determines how long your web server has been running. In the Web Page
constructor, a Date object was created called start time. This method subtracts the start time
from the current time and formats the data into a String for display in the web page:

```
// get how long this programs been running?
public static String getServerUpTime()
{
```

```
Date date = new Date();

long upTime = date.getTime() - startTime;   // calc uptime in millisecs
upTime /= 1000;                      // convert to seconds

int h = (int)(upTime/3600);
int m = (int)((upTime % 3600)/60);
int d = h / 24;
h = h % 24;

return ("Server Uptime: " + d +  " Days,  " + h + " Hours,  " + m + "
Mins");
  }
}
```

Looking through the preceding code, you may find it hard to visualize how the page is format-ted in HTML. If you notice, I did not follow the convention for formatting the code I have used so far. Instead, I lined up the open quote in each line, which helps to see the HTML format.

Another way to see what this class is doing is to look at the web page output in text form instead of web page form. An easy way to do this is to open the web page file in a text editor rather than a web browser. You can do this using the NetBeans editor. From the NetBeans menu, chose File ➪ Open File and navigate to the index.html file in your project folder. This will open index.html in a NetBeans editor window. You can now see how the text is formatted.

Add the Web Server

The final stage of this project is to add the web server. If your computer is not connected to a network or you are just content to view the web page locally, you can stop here. Even though the web server files have been added to the project, they are not running at this point.

First, create the folder you will designate as your web site folder and get the full path name. For now, name the folder "Web." For Windows users, your path will look similar to C:\\Documents and Settings\\Tim\\My Documents\\Web\\. Remember to use double backslash characters in Java file names. If you're a Mac user, your path would be similar to /Users/tbitson/Web/.

Next, open WebWorker in the NetBeans editor and add the full path name to the user constant ROOT_DIRECTORY. Make sure the end of the path has a forward slash (Mac users) or a double backslash (Windows users). The code should look like this:

Mac Users:

```
ROOT_DIRECTORY = "/Users/tbitson/Web/";
```

Or Windows Users:

```
ROOT_DIRECTORY = "C:\\Documents and Settings\\tbitson\\My Documents\\Web\\";
```

Copy the remaining file, logo.jpg, from your downloaded Chapter 12 source code to your new Web folder.

Modify SimpleWeather.java

The web server runs concurrently as a separate thread allowing SimpleWeather to essentially do two things at once: collect and crunch weather data and serve web pages as it receives requests. To accomplish this you need to start the WebWorker class in its own separate thread.

In SimpleWeather.java's main() method, add the following lines of code. This creates a new WebWorker, passing it the lock object previously discussed. Then you create a new webWorker thread called WebServer and then call webServer.start() method to start the web server:

```
public static void main(String[] args)
{
  ...
  try
  {
    // get instances to the primary object
    SimpleWeather weatherServer = new SimpleWeather();

    // start the web server
    WebWorker webWorker = new WebWorker(webPageLock);
    Thread webServer = new Thread(webWorker);
    webServer.setName("Web Server");
    webServer.start();

    // call the main program loop
    weatherServer.mainLoop();
  }
```

Threads are a very useful programming concept and you can read more about them in the Java Tutorial at http://java.sun.com/docs/books/tutorial/.

Save your SimpleWeather.java file, but don't run it yet. There are a few more topics to cover so you'll understand what's happening behind the scenes in case things don't work right.

Look at the WebWorker Code

The last bit of code to look at is the WebWorker class. There are four user constants used: PORT_NUMBER is the port that the webs server listens to. The ROOT_DIRECTORY was covered earlier, and is the folder that the web server uses as its main directory to serve pages from. This should be the path to the Web folder created earlier.

The DEFAULT_PAGE_NAME is the name of the file that the web server uses if the user (web page requestor) does not specify a file. The web server logs all requests to the file specified by WEB_SERVER_LOG.

```java
public class WebWorker implements Runnable
{
    private  Object      lock;
    private  String      threadName;
    private  HTTPServer  httpServer;
    private  boolean     debugFlag;
    private  static int  requestCount = 0;

    // user constants
    public static final int PORT_NUMBER = 8080;
    public static final String ROOT_DIRECTORY = "/Users/tbitson/Web/";
    public static final String DEFAULT_PAGE_NAME = "index.html";
    public static final String WEB_SERVER_LOG = "webLog.txt";
```

The constructor for WebWorker creates a new instance of the HTTPServer class, using the PORT_NUMBER defined in the user constants. Next, the ROOT_DIRECTORY and DEFAULT_PAGE_NAME are set.

The WebWorker also gets a copy of the web page lock object discussed earlier. It gets passed to the HTTPServer so the web server can lock out web access whenever the web page is being updated:

```java
public WebWorker(Object lock)
{
    try
    {
        this.lock = lock;
        debugFlag = SimpleWeather.debugFlag;

        // create an instance of HTTPServer on port httpPort
        httpServer  = new HTTPServer(PORT_NUMBER);

        // override the default HTTP root directory
        httpServer.setHTTPRoot(ROOT_DIRECTORY);

        // override the default index page
        httpServer.setIndexPage(DEFAULT_PAGE_NAME);

        if (debugFlag)
        {
            System.out.println("Web Server Path = " + ROOT_DIRECTORY);
            System.out.println("Web Server Port = " + PORT_NUMBER);
            System.out.println("Web Se4ver Page = " + DEFAULT_PAGE_NAME);
        }
    }
```

If an error occurred trying to configure the HTTPServer, it is caught here:

```
catch (HTTPServerException h)
{
  System.out.println("WebWorker Error: " + h);
}
```

Next the HTTPServer logging is enabled. If you want to disable logging, set the setLogging() value to FALSE.

```
try
{
  // enable or disable logging
  httpServer.setLogging(true);
}
catch (HTTPServerException h)
{
  // problem with log file
  System.out.println("WebWorker: Error Enabling Logging: " + h);
}
}
```

Do you recall in the WebPage class the update() method that gets the number of web page hits? This is the method that you were calling. It simply returns the value of the hit counter:

```
// get the number of server requests
public static int getCount()
{
  return (requestCount);
}
```

Here is the core of the method. When WebServer.start() is called in main(), it invokes the WebServer.run() method, which starts this method running in its own thread. The method is just a loop that calls the httpServer.serviceRequests() method, which sits idle waiting for a web request. When someone requests your weather page, it starts yet another thread (in the HTTPServer class) to handle the web request and immediately returns. The run() method then loops again to the httpServer.serviceRequests() call, waiting for the next web request. This scheme allows multiple web requests to be handled simultaneously. Don't worry if this is a little over your head, it gets pretty complicated inside the HTTPServer class and you really don't need to understand this to use SimpleWeather.

```
// Run the server thread
public void run()
{
  threadName  = Thread.currentThread().getName();

  System.out.println("Starting " + threadName);
```

```
      while(true)
      {
        try
        {
          // Threaded web server blocks on accept
          httpServer.serviceRequests(lock);

          requestCount++;

          // free up memory used
          java.lang.System.gc();
        }
        catch(HTTPServerException h)
        {
          System.out.println("WebWorker: HTTPServerException in Run(): " + h);
        }
        catch(Throwable t)
        {
          System.out.println("WebWorker: Exception in Run(): " + t);
        }
      }
    }
  }
```

I'm not going to walk through the code for the HTTPServer class. The code is a little more advanced and is beyond the scope of this book. However, feel free to take a look at the code on your own.

Note When working on this part of the project, I discovered a potential bug in NetBeans. Although I changed the full path name for the user constant ROOT_DIRECTORY and re-ran SimpleWeather, the value didn't change in the running code. To force NetBeans to recompile the WebWorker class, I had to do a Build ➪ Clean and Build Main Project command from the NetBeans menu.

Using the Web Server

Before running the web server, you need to understand a little bit about IP addresses and port numbers. If you are already familiar with these concepts, then this is just a quick review.

Getting Your IP Address

To test the web server, will you need to know the IP address of your computer. If you already know it, great, if not, here's how to find it.

Mac Users

Open your System Preferences and select Network. On the Show pull-down menu select Network Status. In the status dialog, look through the currently enabled network interfaces. You will see some text showing which interface is currently active and the IP address it is currently using. Write this number down.

Windows Users

Open a command window and type **ipconfig /all** and press return. You will see some network information. Find the line that says IP Address and write down the number.

Static and Dynamic IP Addresses

For a computer to participate in a network, it must follow a network protocol. The standard for most networks and the Internet is the Transmission Communication Protocol/Internet Protocol or TCP/IP. It must also have a node number or *network address*, so other devices on the network can communicate with it. For a computer to work on a TCP/IP network, the address is called an IP address. IP addresses are typically expressed in the *dotted-quad* format such as 192.168.0.152.

IP addresses can be *static* or *dynamic*. Static addresses never change, and are usually entered directly into your computer by hand. They are commonly issued by the system administrator so that no two computers have the same address and they follow a specific numbering scheme.

Dynamic addresses, on the other hand, can change. During boot-up, your computer looks for a *Dynamic Host Configuration Protocol* (DHCP) server. If it finds one, the DHCP server issues your computer an IP address, expiration date, and other information. The DHCP server can reside in your network connection equipment (a cable modem or router, for example), or reside at the Internet provider's location. Your Internet provider has the ability to change the IP address every time the DHCP server issues it, or allow you to reuse the same address.

Trying to use SimpleWeather's web server with a dynamic address that keeps changing isn't very practical. In fact, that's why some Internet service providers intentionally change your address; it keeps you from hosting a web server on your address. If this is the case, the next chapter shows you how to share your data using the Weather Underground web site.

Choosing a Port Number

Web pages are normally served from port 80. However, some Internet service providers block port 80 to discourage operating a web server at your location. To eliminate this potential problem, this project starts out using port 8080. Once everything seems to be working, you can try using port 80.

What's a Port Number?

A port number is a network connection to your computer. If an IP address is analogous to a street address, then the port number is like an apartment number. Just like many apartments can share a single address, there can be multiple connections sharing one IP address. Port numbers range from 1 to 65535. They are divided into three groups: the Well Known Ports, the Registered Ports, and the Dynamic and/or Private Ports.

The Well Known Ports range from 0 through 1023.

The Registered Ports range from 1024 through 49151.

The Dynamic and/or Private Ports range from 49152 through 65535, and are available for general use.

The Well Known Ports are pre-defined for certain tasks. For example, port 80 is defined as the default HTTP (or web-page) port. Other common ports are:

- 21 — File Transfer Protocol (FTP)

- 23 — Telnet

- 25 — Simple Mail Transfer Protocol (SMTP)

- 109 — Post Office Protocol 2 (POP2)

- 143 — Internet Message Access Protocol (IMAP)

If you are a Mac user, using port 80 also requires administrator privileges for security reasons. If you later switch to port 80, run SimpleWeather from the system admin account or make sure your account has admin rights.

Test it Locally

To initially test the web server, you are going to use the *local loopback* IP address. This is a specially defined IP address that is used for testing purposes only. Any network requests get directed right back to the computer originating the request without using the network. The local loopback address is always 127.0.0.1.

Go ahead and run SimpleWeather in NetBeans. If everything compiles and runs, then you won't notice too much difference. Look at NetBean's output window and shortly after the start-up messages, you should see a line stating the web server is starting:

```
Starting SimpleWeather 1.0
Found Adapter: DS9097U on Port /dev/tty.USA19QW4b44P1.1
Resetting 1-wire bus
Creating Log File WeatherLog.txt
Starting Web Server
Time = Fri Mar 10 14:56:01 MST 2006
...
```

Let SimpleWeather run until an even 10-minute time is reached and you should see the line stating the web page is being updated:

```
Time = Fri Mar 10 15:00:00 MST 2006
Temperature = 75.2 degs F
Wind Speed = 0.0 MPH from the NNE
Humidity = 24.107576 %
Dewpoint = 36.48154 degs F
Pressure = 29.715893 inHg
Rain = 0.0 in
Lightning = 0 SPM
Updating /Users/tbitson/Web/index.html
Updating Log WeatherLog.txt
```

If you see an error message instead, it means that most likely the Web folder path is incorrect. Check for typos and make sure SimpleWeather compiles before continuing.

Next, open your web browser and type in **http://127.0.0.1:8080**. This is the local loop-back address at port 8080. If all goes well, you should see the web page displayed, along with the ExtremeTech logo. If not, retrace your steps and double-check your work.

Test it on the Internet

The final test is to see if you can see your web page on the Internet. Start by checking on your computer first. Enter the computer's IP address in the web browser's address bar in the format `http://ip_address:8080` where the `ip_address` is the IP address discovered in this chapter. You should see the same weather page. Next, try using the same address on a remote computer, such as a friend's computer or from work. If all goes well, you should be able to see your weather data.

You can now switch to port 80 if you like, because this is the port defined for web pages. Mac users, remember you have to be an admin to use this port. Change the `PORT_NUMBER` constant in `WebWorker.java` to the new value and re-start NetBeans. When you type in the address, you don't need to include the 8080 anymore, because the web browser assumes port 80 by default. Remember, your ISP may be blocking port 80, so if port 8080 worked remotely and port 80 doesn't, this may be the problem.

Routers and Firewalls

If you can't see your weather data using your IP address on a remote computer, a firewall or a router may be blocking the network connection.

A firewall is designed to block network requests on all but a few selected ports. If the port you are using is on the list of blocked ports, then the web requests won't get through. Some operating systems like Mac OS X and Windows XP have a built-in firewall, while others can be added. Look through the user documentation for your firewall and enable or unblock the port you're using.

A router can perform several functions, but is commonly used to distribute IP addresses on a network. For example, your Internet provider may only supply a single network address. A router will allow you to connect many computers to that address. It does this by using Network Address Translation (NAT). NAT allows the computers it serves to masquerade as the only computer connected to the Internet. One way to tell if the network is using a router is by examining your IP address. If it is using a non-routable address like 192.168.x.x or 10.x.x.x then you are most likely behind a router. Look through your router's documentation to find out how to enable "port-forwarding" and set it to forward port 8080 (or whichever port you are using) to your computer's IP address. You now use the router's address to access the web page remotely. It, in turn, forwards the request to your computer.

Customize the Web Page

Now that you have the web page and server working, go back to the web page code and customize the user fields. The fields themselves can have embedded HTML so you can change the font sizes and colors within each line. Figure 12-2 shows an example of the completed web page.

Expanding Your Project

Before I wrap up this project, let me make a few suggestions to get you thinking about ways your can expand your project.

A Remote Weather Display Using an Old Laptop

Once the web server is up and running, there are several ways you can display your weather data. Do you have an older laptop sitting around? If so, you can use it as a remote display for displaying the weather web page. Just set it up, hook it to the network, launch the browser, and connect to your web page. Because the web page automatically refreshes every 10 minutes, you have up-to-date weather information at your fingertips. If you have a wireless network, even better! You can move the laptop around your house or business keeping tabs on the weather without wires.

Oro Valley, AZ
Weather Conditions
03/12/06 12:10 PM MST

Located in the Copper Creek neighborhood of Oro Valley, AZ
near the corner of La Canada & Naranja.
Weather data updated every 10 Minutes.
32.24.7 deg N 110.59.7 deg W

Temperature			
Current	Today's High	Today's Low	Trend
74.3 °F	74.3 °F	74.3 °F	0.00 °F/Hr
Wind			
Current	Today's High	Today's Peak	Trend
0.0 MPH NNE	0.0 MPH	0.0 MPH	0.00 MPH/Hr
Relative Humidity			
Current	Today's High	Today's Low	Trend
27.6 %	27.7 %	27.6 %	0.24 %/Hr
Dewpoint			
Current	Today's High	Today's Low	Trend
43.5 °F	43.5 °F	39.1 °F	0.22 °F/Hr
Barometric Pressure			
Current	Today's High	Today's Low	Trend
30.10 inHg	30.10 inHg	30.10 inHg	0.01 inHg/Hr
Rainfall			
Year to Date	Since Midnight	-	Rate
0.00 in	0.00 in	-	0.00 in/Hr
Lightning Activity			
Current	Today's High	-	Trend
0.0 SPM	0.0 SPM	-	0.00 SPM

Weather data collected using a 1-Wire Weather Station and SimpleWeather 1.0
JVM Free Mem = 1111 KB   Server Uptime: 0 Days, 0 Hours, 25 Mins   0 Hits since 06/01/06

This data is believed to be accurate at the station location.
Questions regarding this weather station may be sent to weathertoys@mac.com

FIGURE 12-2: Customized web page.

One 1-Wire weather hobbyist has re-packaged his laptop into a picture frame and mounted it on a wall. He now has an auto-updating weather electronic weather display right on his wall.

Switching over to the weather server side of your network for just a second, because the weather web server doesn't require a lot of processing power, that old 400 MHz P3 sitting in the back room might make a great weather server. Once it is configured, there is no need for the monitor to be on.

Get an Internet Domain Name

If you decided to share your weather data on the Internet and your ISP isn't blocking port 80, consider getting your own Internet domain name. It's kind of cool to type in something like www.timsweather.com on a friend's computer and see the weather in your own backyard.

To get a domain name, you need to contact a domain name provider, such as GoDaddy.com. Here you can check to see if the chosen domain name isn't already taken. If it is available, for just a few dollars a year, you can register the domain name. Once you have a domain name, then just follow the domain provider's instruction to set the Domain Name Service (DNS) to your IP address. Keep in mind, if your ISP blocks port 80 or you have a dynamic IP address and it keeps changing, this may not be a good option.

Expand Your Web Page

The web page presented in this project doesn't contain a lot of bells and whistles. Its primary goal is to show you how to do it. If you have never programmed in HTML before, there are some great books and free online articles showing how to create fancy web pages. Using some of the more advanced HTML techniques you can:

- Use graphical gauges showing weather data.

- Superimpose web data on top of a web cam photo showing up-to-the minute photos of the effects of the weather, like rain or snow on the mountains.

- Look into Cascading Style Sheets (CSS) to assemble your web pages in a free-form style.

- Change the look of the web page depending on weather conditions. For example, the temperature area could turn red when the temperature is over 100°, or display an icon showing a cloud with rain when you detect rain.

Changing the Update Frequency

If you want to update your weather page other than once every 10 minutes, you just need to change one line of code. In SimpleWeather.java near line 198 is the if statement that controls when the page is updated:

```
// and update the web page on the 10 minute mark
if (minute % 10 == 0)
{
```

The percent sign in Java is the Modulus function (or just "mod"). It returns the remainder of the first value (minute) divided by the second value (10). In this statement, if the remainder is 0, then do what's in the braces. To change the update time, simply change the 10 to the value you desire. Remember, this statement is only reached once a minute, so it can't update any faster than that.

If you are changing the value to a longer time, you may want to change the time of the trend calculation. The default value used here is 20 samples, with one sample a minute it equals 20 minutes. To change the trend length, modify the user constant TREND_SIZE in WeatherCruncher.java.

Add Error Checking

If an error occurs reading a sensor, the value returned is most likely incorrect. Some of the sensors return values like –999.9 if an error occurred. If this value is passed to the weather cruncher, it will throw off the average and your web page will display weird values.

One enhancement you can make is to check the values before adding it to the average, discarding bad values. But don't be deceived, this is harder than it looks.

Wrap Up

This was a pretty software-intensive chapter, and I hope you learned at lot. Getting the weather data "crunched" into usable data is an important step regardless of what you plan to do with the weather data. WeatherCruncher.java provides a good start and is fairly versatile, so you should be able to hack it to your needs.

The web server now built into SimpleWeather can serve more than just the weather data. Just build the HTML pages, and add them to your Web folder. You can even design your web pages in Microsoft Word and export the document as an HTML file. Remember, the web server used in this project only supports the web GET command, so users won't be able to post any data to your web site. By default, a web browser looks for the file index.html if an actual file isn't specified. If you're serving more than just the weather page, you may want to change the weather page's name to something other than index.html.

If you would prefer not to share the weather data from your computer, but still want to get it on the Net, in the next chapter you learn how to post it to the Weather Underground, where anyone on the Internet can view it.

Post Your Data to the Weather Underground

A nother great way to share your weather data is to post it to the Weather Underground. The Weather Underground, also known as Wunderground, is a comprehensive weather site that currently posts weather data for more than 60,000 cities internationally. It started as a project at the University of Michigan back in the 1990s and has evolved into one of the largest weather data clearinghouses in the world.

One of the cool features of the Weather Underground is the Personal Weather Station section. This is a section dedicated for hobbyists and weather enthusiasts to publish their weather data and share it with the weather community. It has a great user interface and a search feature that allows you to search personal weather stations worldwide.

To use SimpleWeather to post to the Weather Underground, you need an always-on Internet connection. Once you start posting your weather data, it is stored on the Weather Underground servers and can be viewed in several different formats. My favorite is the weather graphs. The Weather Underground displays plots of temperature, pressure, wind speed, wind direction, and rain rate. You can select the date and time span for the plotted data.

Another great feature is that weather statistics are created and displayed for your weather data. By default, daily, monthly, quarterly, and yearly stats are kept and available at the click of a button. The actual posted data is also displayed, allowing users to pull up historical data on your weather station. All this for free!

In this project you learn how to post your weather data to the Weather Underground using SimpleWeather. First, you'll create a Weather Underground account, getting a unique identifier for your weather station. Next you'll modify SimpleWeather to format your weather data into the Weather Underground format and send it via the Internet. Then you can point your web browser to your Weather Underground weather page and see the results.

How Do I Get a Weather Underground Account?

It's a snap to get a Weather Underground account, and it should only take about 10 minutes. First you'll log onto the Weather Underground and create a password so only you can post data to this account. Then you'll enter some data specific to your weather station. After that you'll receive an account identifier and a special link to a banner you can display on your web page.

Log on to the Weather Underground

The first step to setting up a Weather Underground account is to log on to the Weather Underground at www.wunderground.com. In the navigation menu, click the Personal Weather Stations link. There's some interesting information on various weather stations and weather station software on this page. Spend a few minutes looking around and examining the features. You may notice that some of these links are to 1-Wire weather station software. When you're done exploring, click the link "Signup or Modify Your Account."

Next a prompt will be presented to create a Weather Underground account. To complete the registration, you will need to supply a valid email address, a password, and a handle or username as shown in Figure 13-1.

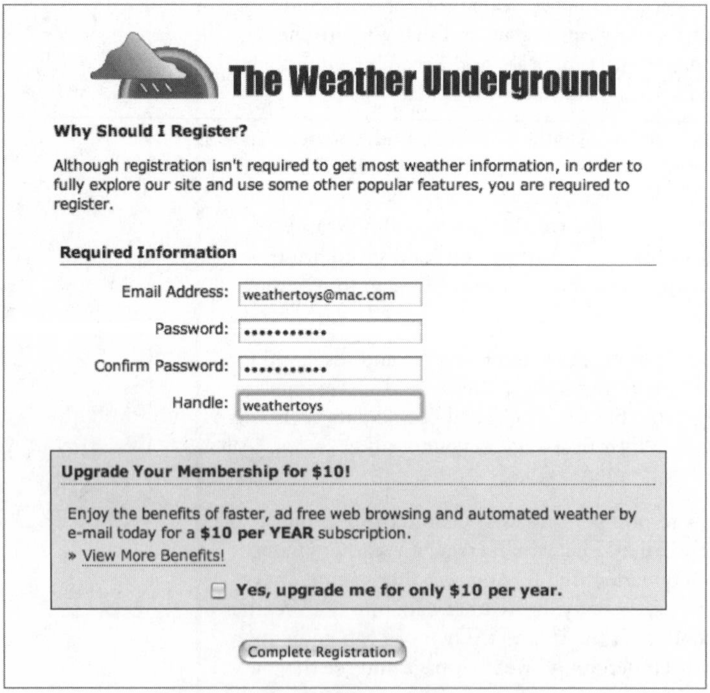

FIGURE **13-1:** The Weather Underground registration page.

Confirm Your Account

Once the data is entered, there will be a prompt to confirm the registration process and an agreement to the terms and conditions as shown in Figure 13-2. To complete this form, you will need to enter an authorization code that was just emailed to the account specified. If you want to come back later and complete this form, go ahead and close it. The email sent contains a link to enter the same data.

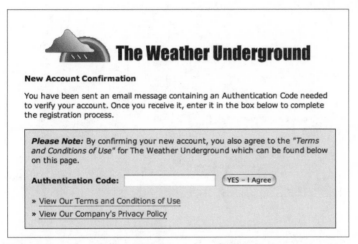

FIGURE 13-2: The Weather Underground confirmation page.

After you have confirmed your account, enter the station-specific data such as name, address, and time zone as shown in Figure 13-3. You can optionally enter a description of the weather station location and, if you are also using SimpleWeather's web server, a link to your weather web site. Click the Sign Up button to get to the weather station info page.

Enter Your Site's Information

After clicking the Sign Up button, the next page displayed shows the information for your current weather stations — yes, you can operate more than one. Enter the station-specific information as shown in Figure 13-4. Make sure to enter a longitude and latitude so users can find your station using the Maps feature. If you have your own weather web page, enter the URL in the designated field.

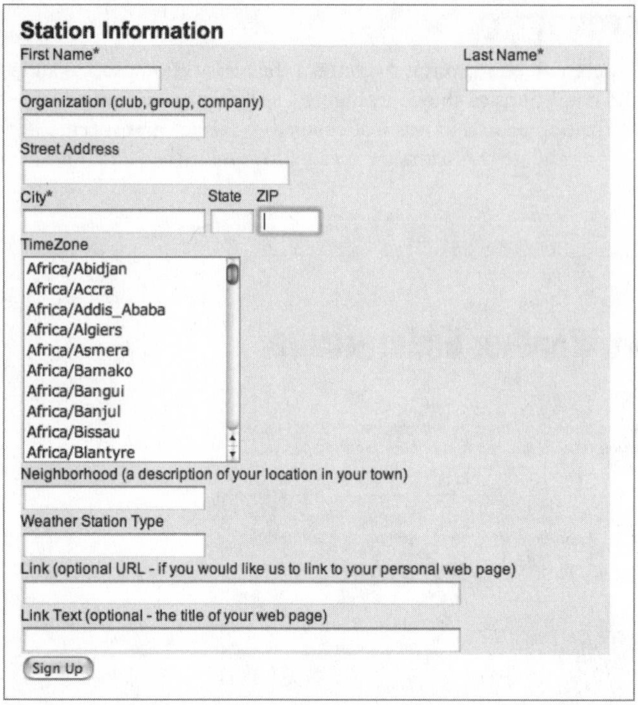

Station Information

First Name*   Last Name*

Organization (club, group, company)

Street Address

City*   State   ZIP

TimeZone

Africa/Abidjan
Africa/Accra
Africa/Addis_Ababa
Africa/Algiers
Africa/Asmera
Africa/Bamako
Africa/Bangui
Africa/Banjul
Africa/Bissau
Africa/Blantyre

Neighborhood (a description of your location in your town)

Weather Station Type

Link (optional URL - if you would like us to link to your personal web page)

Link Text (optional - the title of your web page)

(Sign Up)

FIGURE 13-3: General Station information page.

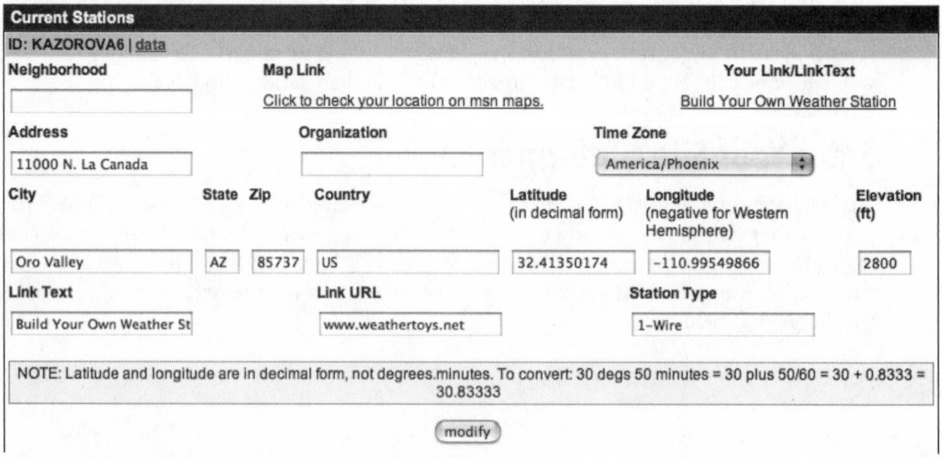

Current Stations

ID: KAZOROVA6 | data

Neighborhood				Map Link				Your Link/LinkText
				Click to check your location on msn maps.				Build Your Own Weather Station

Address			Organization			Time Zone		
11000 N. La Canada						America/Phoenix		

City	State	Zip	Country	Latitude (in decimal form)	Longitude (negative for Western Hemisphere)	Elevation (ft)
Oro Valley	AZ	85737	US	32.41350174	–110.99549866	2800

Link Text	Link URL	Station Type
Build Your Own Weather St	www.weathertoys.net	1-Wire

NOTE: Latitude and longitude are in decimal form, not degrees.minutes. To convert: 30 degs 50 minutes = 30 plus 50/60 = 30 + 0.8333 = 30.83333

(modify)

FIGURE 13-4: Station-specific data.

Copy Your Banner Information

Under the station data, there are two banner URLs that can be copied. As shown in Figure 13-5, the Membership HTML Code is used on the weather web page as a link to the Weather Underground page. The Weather Sticker HTML Code will display the current temperature and humidity values once you start posting data.

FIGURE 13-5: Banner URLs and examples.

For now, click the Modify button to save the data and leave this page open so you can copy the banner URL later.

Modify SimpleWeather to Post Your Data

Using the weather data cruncher used in the last project makes the modifications for this project easy. All you have to do is add the Wunderground class to your project, add a few lines to `SimpleWeather.java` to instantiate the class, and then call the `send()` method once every 10 minutes. You can also add the Weather Underground banner to your web page to provide a link to your weather station's data.

Add the Wunderground Class

Open the `WeatherToys` directory and find the project source code you downloaded from the companion web site. Next, locate the Chapter 13 directory. Inside, find the file `Wunderground.java` and copy it to your SimpleWeather ⇨ src ⇨ SimpleWeather folder.

Modify SimpleWeather.java

Launch NetBeans and open your SimpleWeather project. Check your list of source files to make sure `Wunderground.java` is listed. Open `SimpleWeather.java` in the NetBeans editor and scroll down to the `// class variables` comment near line 46. Add a class variable for the Wunderground class to the end of the list:

```
// class variables
public float temp;
...
private WebPage wp;
private Wunderground wu;
```

Scroll down to SimpleWeather's construction, and add a call to instantiate a new Wunderground class just after the call to initialize the web page class at line 110:

```
public SimpleWeather()
{
...
  // initialize the web page
  wp = new WebPage(wc);

  // initialize wunderground
  wu = new Wunderground();
}
```

Toward the bottom of the `mainLoop()` method near line 206, add the call to send your weather data to the Weather Underground just after the call to update the web page:

```
// and update the web page on the 10 minute mark
if (minute % 10 == 0)
{
  wp.updatePage(date);
  wu.send(date, wc);

  // reset the averages for the next set of data
  wc.resetAverages();
}
```

Save your file before continuing.

Add the Weather Underground Banner to Your Web Page

If you left the Weather Underground web page open to the Stations page, the banner information should still be displayed. If not, re-open the Weather Underground weather stations page and copy the text from the Membership banner.

Open `WebPage.java` in NetBeans, and paste the link text in the user constant `BANNER`:

```
// user constants
...
private static final String BANNER = "";
```

The banner link is a long URL. Be careful when pasting it into SimpleWeather that you don't inadvertently add a carriage return or line feed. It is acceptable to have very long lines in the

NetBeans editor, so you can leave it as one long String. Next, escape all quotes contained in the string (except for the first and last). Your results should look something like this, except your line won't be wrapped:

```
private static final String BANNER = "<a href=\"http://www.
wunderground.com/weatherstation/WXDailyHistory.asp?ID=
KAZOROVA5\"><img border=\"0\" src=\"http://banners.Wunderground.
com/cgi-bin/banner/ban/wxBanner?bannertype=WeatherStationCount
&weatherstationcount=KAZOROVA5\" width=\"160\"height=\"163\">";
```

In NetBeans, string text is colored red unless you have changed it. Use this as a guide to make sure the string is escaped correctly. Save the changes before continuing.

What's a URL?

URL stands for Uniform Resource Locater. You may think of it as simply a web location. It is a pointer to a resource on the Internet. It could be a web page, an image, or Internet mailbox. Officially, URLs have five parts:

```
protocol://address/path?parameter#fragment
```

- Protocol — This is the type of message. The most common are HTTP, FTP, and MAILTO. It defines the communications format of an Internet message.

- Address — The location or Internet address such as www.extremetech.com. This could also be an IP address, such as 63.87.252.160.

- Path — The path or file name to the desired web resource.

- Parameter — This follows the "?" symbol and is data sent to the host computer. A database or search engine can process this data. Multiple pieces of data are separated by "&".

- Fragment — The fragment comes after a "#" symbol and denotes a specific location within a web page, also known as an anchor.

Not all URLs use all five parts, but the first two are always required. Most browsers, including Internet Explorer and Safari, will assume the HTTP scheme if it is omitted and add it to the URL.

Here are a couple simple examples:

```
http://www.extremetech.com/update.php?item=5735
http://www.extremetech.com/index.html/#data
```

You can find a complete description of URL syntax and its big brother, the URI, at http://www.ietf.org/rfc/rfc3986.txt.

Walk Through the Wunderground Class

Take a quick look at the Wunderground class to see how your data gets posted before you compile and run the code. The weather data is posted by assembling a rather long URL that contains the Weather Underground address, the station ID, and parameters for each of the weather values.

Weather Underground Commands

These are the common commands that the Weather Underground accepts:

- `id` — ID as registered by wunderground.com.

- `password` — Password registered with this ID.

- `dateutc` — UTC date in the format YYYY-MM-DD HH:MM:SS (mysql format).

- `winddir` — The wind direction in degrees [0–360].

- `windspeedmph` — Wind speed in MPH.

- `windgustmph` — Wind gust in MPH.

- `humidity` — Relative humidity in %.

- `tempf` — Temperature in degrees F.

- `rainin` — Rain rate in inches/hour.

- `dailyrainin` — Daily rain amount in inches.

- `baromin` — Barometer pressure in inHg.

- `dewptf` — The dewpoint in degrees F.

- `weather` — The current weather in metar style text (+RA).

- `clouds` — Current cloud condition: SKC, FEW, SCT, BKN, OVC.

- `softwaretype` — The software used to collect data (SimpleWeather, ETWS, and so on).

- `action` — The type of action, such as update data (updateraw).

Each of the commands is appended to each other with an ampersand character. If Wunderground accepts your command, then a singe line reply of "success" is returned.

First are the two user constants for your account name and password. Edit these values for your station:

```
public class Wunderground
{
  // user constants
  public static final String USERNAME = "KAZOROVA5";
  public static final String PASSWORD = "password";
```

There are three class variables. The debugFlag you're already familiar with. The BufferedReader in and the PrintStream out are both streams, similar to the file stream used to write the log file data to disk. However, in this case these streams are connected to a Socket, which is Java's mechanism to connect to a network and the Internet:

```
  // user constants
  public static final String USERNAME = "KAZOROVA5";
  public static final String PASSWORD = "password";


  // class variables
  private BufferedReader in;
  private PrintStream out;
  private static boolean debugFlag = SimpleWeather.debugFlag;
```

Next is the send method, which posts your data. This may seem confusing at first, but although you are posting data to the Weather Underground, it is disguised as a GET command. The GET command actually has the weather data embedded in it and Weather Underground's computers know how to decipher it.

There are two key variables. url is the Weather Underground's address and is used to open the network connection. sendUrl is a StringBuffer and is used to assemble the URL. The sendUrl starts with the HTTP GET command. The username, password, and date parameters are added to the address. Next the individual weather parameters are appended. Notice that each parameter is preceded with an ampersand (&). If you don't have a particular sensor, then comment out the line for that sensor.

```
  public void send(Date d, WeatherCruncher wc)
  {
    String url = "weatherstation.wunderground.com";
    StringBuffer sendUrl = new StringBuffer();

    // build up wunderground message based on what sensors we have.
    // comment out the lines if you don't have that sensor

    sendUrl.append("GET /weatherstation/updateweatherstation.php?");
    sendUrl.append("ID=" + USERNAME);
    sendUrl.append("&PASSWORD=" + PASSWORD);
    sendUrl.append("&dateutc=" + utcDate(d));

    // Temperature
    sendUrl.append("&tempf=" + wc.getTemp());
```

```
// Humidity
sendUrl.append("&humidity=" + wc.getHum());

// Dewpoint
sendUrl.append("&dewptf=" + wc.getDP());

// Wind Speed and Direction
sendUrl.append("&windspeedmph=" + wc.getWind());
sendUrl.append("&windgustmph=" + wc.getWindPk());
sendUrl.append("&winddir=" + convertWindDir(wc.getWindDirAvg()));

// Baro Pressure
sendUrl.append("&baromin=" + wc.getBaro());

// Rain
sendUrl.append("&rainin=" + wc.getRainRate());
sendUrl.append("&dailyrainin=" + wc.getRain24());

// Software Type & action
sendUrl.append("&softwaretype=tws&action=updateraw
             HTTP//1.1\r\nConnection: keep-alive\r\n\r\n");
```

Note Make sure you comment out the sensors you don't use so that the Weather Underground plotting routines don't get confused. Simply put the comment symbol "//" in front of the line. For example, if you don't have a barometric pressure sensor, then the line would look like this:

```
// Baro Pressure
//sendUrl.append("&baromin=" + wc.getBaro());
```

Next, a socket (network connection) is opened to the Weather Underground on port 80. If the connection was successful, the PrintStream and BufferedReader are attached to the newly created socket stream. Next, the assembled URL is sent using the send() method:

```
try
    {
      if (debugFlag)
        System.out.println("Updating Wunderground...");

      // open a connection to wunderground
      Socket s = new Socket(url, 80);
      if (s != null)
      {
        // set up buffered readers & writers to the socket
        out = new PrintStream(s.getOutputStream());
        in = new BufferedReader(new InputStreamReader(s.getInputStream()));

        // send wunderground post string
        send(sendUrl.toString());
        readResponse();
        s.close();
```

```
      }
    }
    catch (IOException exception)
    {
      System.out.println("WU Error: " + exception);
    }
  }



  private void send(String s) throws IOException
  {
    if (s != null)
    {
      if (debugFlag)
        System.out.println("Sending " + s);

      out.print(s);
    }
  }
```

Then the software enters a loop, waiting up to 10 seconds for a response. If the URL was successfully contacted and the Weather Underground accepted the message string, then a single response of "success" is sent back. If any other response is received, the first line of the response is printed, the connection is closed, and the code exits the loop:

```
  private void readResponse() throws IOException
  {
    if (debugFlag)
      System.out.println("Getting Response...");

    // wait up to 10 seconds for reply
    int i = 0;
    while (i++ < 100)
    {
      if (in.ready())
      {
        String line = in.readLine();
        if (line.toLowerCase().indexOf("success", 0) > -1)
        {
          if (debugFlag)
          {
            System.out.println("Line = " + line);
            System.out.println("Weather Underground Updated");
          }
          return;
        }
        else
        {
          if (debugFlag)
            System.out.println("Line = " + line);
```

```
        }
      }
      else
      {
        try
        {
          Thread.sleep(100);
        }
        catch (InterruptedException e)
        {}
      }
    }
    System.out.println("WU No Response");
  }
```

If the Weather Underground could not be contacted, an Exception is thrown and the following error message will be printed. This could occur if your Internet connection is temporarily down or Weather Underground is busy:

```
  catch (UnknownHostException e)
  {
    System.out.println("Unable to Connect to Weather Underground");
  }
  catch (IOException e)
  {
    System.out.println("Wunderground: Error Connecting to Wunderground " + e);
  }

  return;
}
```

Weather Underground uses UTC (Universal Time Coordinated) dates and time, which used to be called Greenwich or Zulu Time. This routine converts the current time into UTC time and then formats the results into the date and time string specified by the Weather Underground:

```
public String utcDate(Date d)
{
  // convert to UTC in the format YYYY-MM-DD+HH:MM:SS
  long offset = Calendar.getInstance().get(Calendar.ZONE_OFFSET);

  if (offset != 0)
  {
    offset = offset / (3600*1000);
    d.setHours(d.getHours() - (int)offset);
  }

  int month = d.getMonth() + 1;
  int day = d.getDate();
  int hour = d.getHours();
```

```
    int minute = d.getMinutes();

    StringBuffer dateTime = new StringBuffer((d.getYear() + 1900) + "-");

    if (month < 10)
      dateTime.append("0");

    dateTime.append(Integer.toString(month) + "-");

    if (day < 10)
      dateTime.append("0");

    dateTime.append(Integer.toString(day) + "+");

    if (hour < 10)
      dateTime.append("0");

    dateTime.append(Integer.toString(hour) + "%3A");

    if (minute < 10)
      dateTime.append("0");

    dateTime.append(Integer.toString(minute) + "%3A00");

    // set date back to current timezone
    if (offset != 0)
      d.setHours(d.getHours() + (int)offset);

    return (dateTime.toString());
}
```

The last routine simply converts the wind direction value SimpleWeather user (0 to 15) into degrees:

```
private String convertWindDir(int windDir)
{
  return new Integer(windDir * 360/16).toString();
}
```

The results of the preceding routine convert your weather data into a long URL. If you have problems posting to the Weather Underground, try turning on debug mode and comparing the displayed URL string with this sample message:

```
Sending GET /weatherstation/updateweatherstation.php?
ID=KAZOROVA5&PASSWORD=wtoys&dateutc=2006-06-25+16%3A40%3A00
&tempf=79.1&humidity=30.3&dewptf=45.7&windspeedmph=0.0
&windgustmph=0.0&winddir=315&baromin=29.86&rainin=0.00
&dailyrainin=31931.82&softwaretype=tws&action=updateraw HTTP//1.1
Connection: keep-alive
```

With debug mode turned on, you'll also see the Weather Underground reply. If all goes well, you will see something like the following lines. The key is the word "success."

```
WU: Getting Response...
WU: HTTP/1.0 200 OK
WU: Content-type: text/html
WU: Content-Length: 8
WU: Connection: close
WU: success
```

If you are an experienced Java programmer, you may be asking why I didn't use Java's URL class to post the weather data, because it is much easier to use. I did originally, except I had a problem with the code occasionally hanging in the URL class. Maybe I was doing something wrong. If you would like to debug it for me, send me an email and I'd be happy to send you the code.

Run Your Project

Now that you understand how the code works, go ahead and run the project. Like the last project, you won't see any differences until the 10-minute mark. Then there should be a new line displayed:

```
Time = Mon Mar 20 08:40:00 MST 2006
Temperature = 72.5 degs F
Wind Speed = 0.0 MPH from the NNE
Humidity = 27.799574 %
Dewpoint = 37.747105 degs F
Pressure = 29.98122 inHg
Rain = 0.0 in
Lightning = 0 SPM
Updating /Users/tbitson/Web/index.html
Weather Underground Updated
Updating Log WeatherLog.txt
```

Once the weather data is posted to the Weather Underground, open your web browser and go to your Weather Underground weather station page. You may or may not be able to see any data on the plot yet, because a single data point doesn't show up well. Scroll down to the Tabular Data section and see if the posted data is displayed.

If your data didn't post, check to see if there was any error message displayed in SimpleWeather. Try turning on debug mode and check to see if the output looks like the preceding example. Also double-check your password and user ID.

Let SimpleWeather run for an hour or two and then check the data on the Weather Underground. Figure 13-6 shows an example of the weather station data page and Figure 13-7 shows the tabular data.

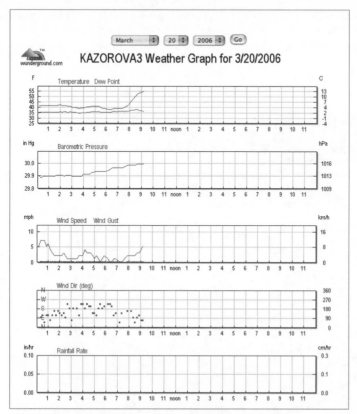

FIGURE 13-6: SimpleWeather data posted to the Weather Underground.

Tabular Data for March 20, 2006								
Time	Temperature	Dew Point	Pressure	Wind	Wind Speed	Wind Gust	Humidity	Rainfall Rate (Hourly)
00:00	41.0 °F / 5.0 °C	35.3 °F / 1.8 °C	29.88in / 1011.7hPa	East	5mph / 8.0km/h	-	80%	-
00:10	41.0 °F / 5.0 °C	34.9 °F / 1.6 °C	29.89in / 1012.1hPa	East	5mph / 8.0km/h	-	79%	-
00:20	41.0 °F / 5.0 °C	34.8 °F / 1.6 °C	29.88in / 1011.7hPa	East	7mph / 11.3km/h	-	78%	-
00:30	41.0 °F / 5.0 °C	34.7 °F / 1.5 °C	29.89in / 1012.1hPa	ESE	7mph / 11.3km/h	-	78%	-
00:40	41.0 °F / 5.0 °C	34.7 °F / 1.5 °C	29.89in / 1012.1hPa	NE	7mph / 11.3km/h	-	78%	-
00:50	41.0 °F / 5.0 °C	34.7 °F / 1.5 °C	29.89in / 1012.1hPa	ESE	5mph / 8.0km/h	-	78%	-
01:00	41.0 °F / 5.0 °C	34.7 °F / 1.5 °C	29.89in / 1012.1hPa	ESE	6mph / 9.7km/h	-	78%	-
01:10	41.0 °F / 5.0 °C	34.9 °F / 1.6 °C	29.89in / 1012.1hPa	ENE	4mph / 6.4km/h	-	79%	-
01:20	41.0 °F / 5.0 °C	35.0 °F / 1.7 °C	29.89in / 1012.1hPa	ESE	3mph / 4.8km/h	-	79%	-
01:30	41.0 °F / 5.0 °C	34.9 °F / 1.6 °C	29.89in / 1012.1hPa	SSE	2mph / 3.2km/h	-	79%	-

FIGURE 13-7: Weather Underground's tabular data.

If you are using the web server from the last chapter and entered the Weather Underground banner, open the weather web page in your browser. At the bottom of the page is the Weather Underground banner as shown in Figure 13-8. Click on it and it should take you to your station page.

Oro Valley, AZ
Weather Conditions
03/12/06 12:10 PM MST

Located in the Copper Creek neighborhood of Oro Valley, AZ
near the corner of La Canada & Naranja.
Weather data updated every 10 Minutes.
32.24.7 deg N 110.59.7 deg W

Temperature			
Current	Today's High	Today's Low	Trend
74.3 °F	74.3 °F	74.3 °F	0.00 °F/Hr
Wind			
Current	Today's High	Today's Peak	Trend
0.0 MPH NNE	0.0 MPH	0.0 MPH	0.00 MPH/Hr
Relative Humidity			
Current	Today's High	Today's Low	Trend
27.6 %	27.7 %	27.6 %	0.24 %/Hr
Dewpoint			
Current	Today's High	Today's Low	Trend
43.5 °F	43.5 °F	39.1 °F	0.22 °F/Hr
Barometric Pressure			
Current	Today's High	Today's Low	Trend
30.10 inHg	30.10 inHg	30.10 inHg	0.01 inHg/Hr
Rainfall			
Year to Date	Since Midnight	-	Rate
0.00 in	0.00 in	-	0.00 in/Hr
Lightning Activity			
Current	Today's High	-	Trend
0.0 SPM	0.0 SPM	-	0.00 SPM

Weather data collected using a 1-Wire Weather Station and SimpleWeather 1.0
JVM Free Mem = 1111 KB   Server Uptime: 0 Days, 0 Hours, 25 Mins   0 Hits since 06/01/06

Member of the

Weather Underground
Data Exchange
27282
members

This data is believed to be accurate at the station location.
Questions regarding this weather station may be sent to weathertoys@mac.com

FIGURE 13-8: SimpleWeather's web page with Weather Underground link.

Expanding Your Project

Once you get your project working reliably and you have had a chance to play, here are a few ideas to expand on this project.

Send a Second Time

The routine for sending data to the Weather Underground only tries once. If the Weather Underground is busy or the message doesn't get through, it could be lost. You can add some code around the sending routine so that if a failure occurs, it will try again. Make sure to put a limit on the number of times to retry, otherwise if your Internet connection goes down, the program would be stuck in a loop forever.

Write an Algorithm to Determine Cloud Cover

The Weather Underground accepts a post that describes cloud cover such as scattered ("SKC") or overcast ("OVC"). Can you think of a way to determine this using your weather sensors? With just a line or two of code, this could be added to the Weather Underground posting.

Adding a solar radiation sensor may help. If you've used the Hobby-Boards PC board for the humidity sensor, then by adding two more components and some software, solar intensity can be measured.

Save Your Data

The Weather Underground stores your weather data on its servers. I don't know how long it is stored, but you may want to keep a copy. The Weather Underground offers a Comma Separated Value (CSV) formatted file of the data posted for download. With a few variations of the Wunderground class, it is possible to request the data back from the Weather Underground once each day and save it to your hard drive.

The Weather Underground Quick Weather Link

In Figure 13-5 there were two banners. One of the links was copied to your weather web page. The other banner is also a link to the weather station page on the Weather Underground, but it displays the current (or last posted) temperature and humidity. If you or someone you know has a web site that would benefit from displaying the local weather conditions, just paste this link into the web page source code.

Post Your Data to the CWOP

Another great place to share your data is the Citizens Weather Observation Program, or CWOP. There are more than 2500 weather stations posting data in the US. Using the information presented in this chapter and just a few modifications, you can also post your data on CWOP. For more info, go to www.wxqa.com/.

Wrap Up

Posting weather data to the Weather Underground is a great way to get your data to the world. Using the Weather Cruncher from the last project made it easy to send it to the Weather Underground. Now you have current weather data plots and statistics to add to the list of cool things you can do with your weather station.

Now would be a great time to explore the many features of the Weather Underground. You'll find graphs of weather data and weather statistics for the week, month, and year. Check out some of the other contributing weather stations in your area and look at some of their historical data. Another fun item to play with is Weather Underground's Google Weather Station Map for viewing weather station locations.

This project, like the last chapter, was strictly a software project. In the next couple of chapters you'll get back to building some hardware. I don't want to give away too much, but in future chapters you will use the weather data to turn on or off lights, adjust the timing of an automated sprinkler timer, and more.

Add a Lightning Surge Suppressor

One drawback to a wired weather station is potential damage caused by the large amounts of Electro-Magnetic Energy (EME) radiated by lightning. The cables act like long antennas and pick up the radiated EME. If the lightning is sufficiently close and the cables long enough, damage to your weather station could occur.

This project presents a simple surge protector that can be installed across the 1-Wire bus to guide induced voltages to ground. In fact, you should install two suppressors: one on each end of the 1-Wire bus.

In this chapter, you also learn a little more about proper grounding. If you live in a lightning-prone area, this chapter is a must. If your location doesn't experience much lightning, then installing the surge suppressor will guard against Electro-Static Discharge (ESD) when working with your weather equipment.

Project Overview and Design

Because the 1-Wire bus is a simple twisted-pair, there are only two wires to protect: signal ground and data. Protecting the ground line is easy; just connect it to earth ground. Protecting the data line is a little harder. It needs to be clamped to a safe level so the 1-Wire devices don't get damaged from excessive voltage; however, adding any capacitance or inductance degrades 1-Wire bus performance. Typical surge suppressors will not work because they generally add too much capacitance to the line. Instead, a semiconductor Transient Voltage Suppressor (TVS) is employed.

Options

This project is not available as a kit. The individual parts must be purchased and some soldering may be required. Don't worry — there are only a handful of parts and they aren't easily damaged. You can build the surge suppressor on a small perf board or just solder the parts directly to each other. If you prefer to avoid soldering altogether, the parts can be connected in a modular telephone jack as well.

There are two versions of the surge suppressor: one is installed at the beginning of the 1-Wire bus, as close as possible to the DS9097U 1-Wire adapter. It also has a ground connection that gets connected to the single point ground. The other version is designed for installation at the end of the 1-Wire bus. Both use identical parts; the only difference is the connection method.

How It Works

The design featured in Figure 14-1 uses three diodes to suppress surge voltages. Diodes D1 and D2 are fast-response Schottky Barrier diodes that start conducting in the forward direction at about 0.25V. Because 1-Wire devices do not like to be reverse powered, diode D1 is installed across the 1-Wire such that it starts conducting if the data line becomes 0.25V lower than the ground line. This prevents the data line from going negative with respect to the ground.

Diode D3 is a 5V Transient Voltage Suppressor used to clamp the voltage on the data line to safe values. Because diode D3 has some capacitance, which may degrade bus performance, diode D2 is placed in series to decouple D3 from the bus during normal operation. The D2/D3 combination clamps the data line to under 7V, the maximum surge voltage for most of the 1-Wire devices.

If you are using the 100-ohm bus-matching resistor described in Chapter 11, the resistor should be installed in the same package.

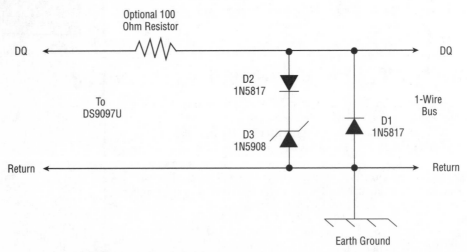

FIGURE 14-1: The in-line version schematic.

The bus-end version shown in Figure 14-2 operates identically but does not have the external ground connection. If you are using a Y-network and have multiple bus ends, you may want to install one on each leg. In general, don't install more than four total; the small amount of capacitance in this circuit will degrade 1-Wire bus performance. However, this is highly dependent on your configuration.

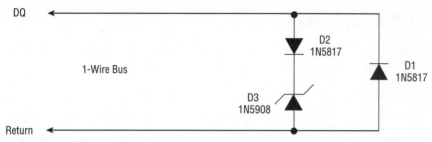

DQ

1-Wire Bus

D2
1N5817

D1
1N5817

D3
1N5908

Return

FIGURE 14-2: The bus-end version.

Figure 14-3 shows the semiconductor curve trace for the suppressor. In the reverse direction, the circuit clamps at about 0.25 volts with 100 milliamps (ma) of current. In the forward direction, clamping starts at 6.8V and reaches about 7V at 100ma.

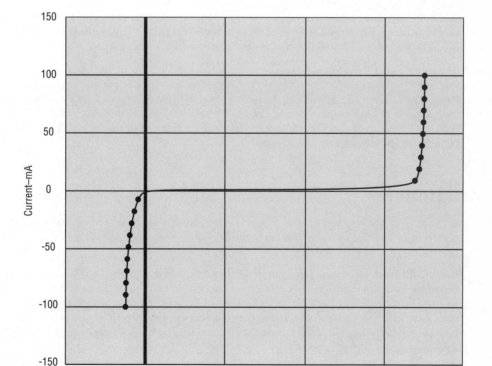

Surge Suppressor Clamping Voltage

FIGURE 14-3: Curve trace of surge suppressor performance.

Building the Hardware

Table 14-1 lists the parts needed for each surge suppressor, which can be purchased from Digikey or other online electrical parts suppliers.

Table 14-1   Parts List for 1 Surge Suppressor

Qty	Designator	Description	Vendor
2	D1, D2	1N5817 Schottky Diode	Digikey
1	D3	1N5908 Zener TVS Diode	Digikey
1	R1	100 ohm ¼ Watt Carbon Resistor (Optional)	Digikey, Radio Shack
1	—	Dual Modular Phone Jack	Radio Shack 279-448

Figure 14-4 shows the completed in-line version installed in a dual modular phone jack. The diodes are connected to the screw terminals, so no soldering is required. One of the RJ-14 jacks should be connected to the DS9097U 1-Wire adapter, and the 1-Wire bus to the other RJ-14 jack. The ground wire should be connected to a known good earth ground.

The end-of-line surge suppressor can be mounted on a small 0.1-inch hole pattern PC perf (perforated circuit) board as shown in Figure 14-5. Perf board can be purchased at Digikey or Radio Shack. The RJ-14 attaches to the circuit board with a single twisted-pair wire removed from a length of CAT5 cable. You will need to look at your last 1-Wire device on the bus to determine which type of connector to use. The surge suppressor can also be attached using the phone crimp connectors shown in Chapter 3.

Installation

If your 1-Wire bus is longer than 20 feet, two surge suppressors should be installed as shown in Figure 14-6. The first unit goes as close to the DS9097U as possible and the end unit should be installed close to the last device.

Because the surge suppressor described in this project effectively shorts the 1-Wire bus data line to the return/ground line whenever a transient occurs, the 1-Wire return or ground line needs to be connected to earth ground. This is typically accomplished through your computer's serial port, which should be connected to the computer's ground, which in turn is connected to your house's electrical ground. However, not all computers have a good connection between the serial port ground and the case.

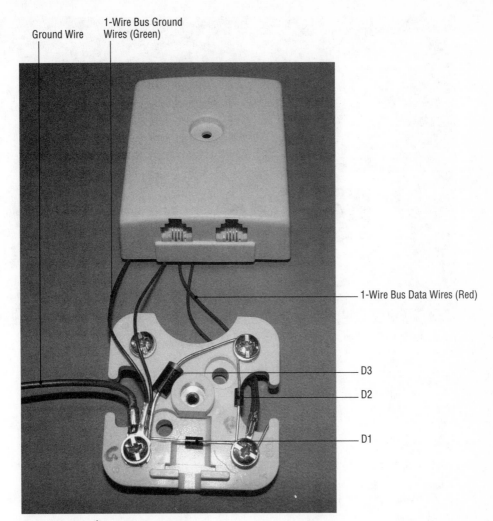

FIGURE 14-4: In-line surge suppressor.

Also, serial ports use pin 5 as the serial data return line, which, in this case, is the 1-Wire ground. Often, the serial return line is routed through the motherboard on a very small trace. It wouldn't take much of a static charge to do permanent damage.

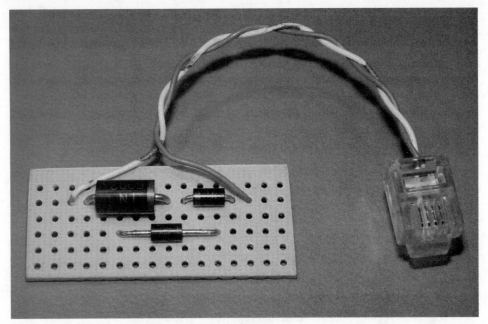

FIGURE 14-5: End-of-line suppressor with RJ-14 plug.

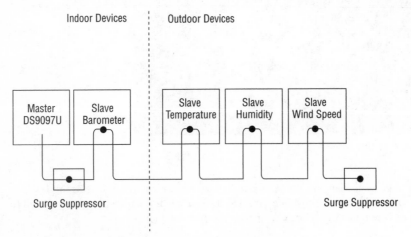

FIGURE 14-6: Suggested surge suppressor placement.

Ground Your Computer

To protect your computer, the 1-Wire bus ground line should be connected externally to a known good ground. Figure 14-7 shows the preferred grounding method. A separate earth ground, such as a water pipe or ground rod, is attached to the 1-Wire surge suppressor ground and the computer's case. Use 14-gauge or larger wire for this connection to minimize ground loop current and to couple the transient surges safely to ground.

If you live in an area that is susceptible to lightning, an AC power surge suppressor or Uninterruptible Power Supply (UPS) should also be used to prevent power line surges from reaching your computer.

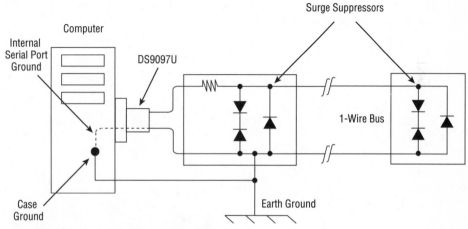

FIGURE **14-7: Recommended grounding scheme.**

Lightning Warning

No amount of surge suppression can completely guard against lightning damage. During lightning storms, always disconnect your weather equipment and stay away from any cabling that goes outdoors.

Wrap Up

Because every installation is a bit different, you should spend some time looking over your weather station to see if there are any other areas that need to be grounded. Any mast or pole that is above your roofline or higher than 10 feet in an open area should be grounded. If you haven't already done so, check to make sure that the AC outlet you are using is properly wired. You may want to have a qualified electrician check the main house ground in your power panel. You would be surprised at the number of houses that have a defective or damaged ground line.

Even if you don't live in an area that gets lightning storms, consider installing the in-line surge suppressor described in this chapter. It will help protect your 1-Wire weather equipment from damage caused by ESD. Believe me, it is *really* disappointing to have your entire weather station damaged by one little spark!

Add an LED Weather Display

All of the projects you've built so far require a computer to display the weather conditions. Wouldn't it be cool to have a display of some sort that allows you to get the time, temperature, and/or humidity without having to get up and go to your computer? Or maybe you would like to know the current outside temperature on your patio. Either way, having a weather display can be very useful.

Many LED and LCD displays are available on the market. Most are small and unreadable from just a few feet away. Some require electronics to drive the display and some way to interface it to your PC or 1-Wire bus. My personal favorite is the BetaBrite LED sign.

This chapter shows you how to connect the BetaBrite LED sign to your computer, and then modify SimpleWeather to display weather data. You'll learn the basics of how to communicate to the LED sign and how you can incorporate the many different features of the BetaBrite.

What's a BetaBrite?

A BetaBrite is an LED sign that displays text messages. Messages scroll across the screen, allowing lengthy messages to be displayed a dozen characters or so at a time. You may have seen one in a bank or a business displaying advertisements, promotions, and news.

The BetaBrite, manufactured by Adaptive Micro Systems, has some great features that make it a great hacker's toy:

- Large 8-color display using 2-color LEDs
- 2.1-inch high, 14-character display
- 32K memory for storing messages
- Multiple fonts
- Multiple scrolling options
- Serial port interface for remote programming
- Hand-held IR remote programmer

You can find out more by visiting the BetaBrite web site at www.betabrite.com. Figure 15-1 shows the BetaBrite sign displaying the current time and weather conditions. As you can see, the sign is easily readable from several feet away.

The biggest drawback to a BetaBrite is cost. Full retail price is more than $200, but by shopping around, you can find it cheaper. If you are looking for a new sign, one place to look is Sam's Club. As I write this, it is offering BetaBrite signs for less than $160 at www .samsclub.com. If you are looking for a less expensive way to get one, consider buying a used sign on one of the many online auctions. A quick look shows that they are currently going for around $80 used. There are several models of BetaBrites — make sure that one you are going to bid on is the newer one, the model 1036.

FIGURE 15-1: BetaBrite Model 1036 showing the current time, temperature, and humidity.

The BetaBrite sign is designed for indoor use only, so you can't mount it outdoors unless you protect it from the elements. When choosing a location, pick a spot that has AC power and is within 60 feet of the computer it will be connected to.

Tip   As this book was going to press, AMS announced the new BetaBrite Prism. Looking at the specifications, it appears that the Prism only supports USB and is not compatible with this project.

Setting Up the Sign

To display weather data, the BetaBrite is connected to a second serial port on your computer. The standard retail package does not include the serial cable needed to make the connection, but you can purchase one online through one of the many suppliers. Unfortunately, the cable only comes packaged with the BetaBrite Messaging Software, which retails for $99. Here again, several of the online auctions have new and used cables, currently going for between $10 and $25. Of course, there's always the option to build your own cable.

Building the Cable

Building your own serial cable isn't too hard and if you have been building your own 1-Wire cables, you have most of the tools and supplies already. Table 15-1 lists the parts necessary. You can use up to 20 feet of standard modular 6-conductor cable without having serial communication problems.

Table 15-1   BetaBrite Serial Cable Parts List

Qty	Description	Source
1	RJ-11 6-Position Modular Plug	Hardware Store
20 Feet	6-Conductor Modular Phone Cable (flat)	Radio Shack #278-864
1	9-Pin Female Solder Cup D Connector	Radio Shack #276-1538
1	9-Pin D Connector Backshell/Hood	Radio Shack #276-1508

Once you have collected the hardware, assemble the cable as shown in Figure 15-2. The RJ-11 gets crimped on one end, and the 9-pin D connector gets soldered on the other end. Don't rely on any specific color-code for the wires, because there is no standard. Instead, use the pin-out listed in Table 15-2. You may want to use a short length of 1/16-inch heat shrink sleeving on the 9-pin connector wires as a strain relief. Your completed cable should look similar to Figure 15-3.

Table 15-2   BetaBrite to 9-Pin Serial Port Connections

BetaBrite Pin	Function	DB-9	Comment
1	Serial Return	5	BetaBrite Signal Ground
2	N/C	N/C	No Connection
3	RXD	2	BetaBrite Serial Out
4	TXD	3	BetaBrite Serial In
5	N/C	N/C	No Connection
6	+5 Volts	N/C	BetaBrite 5V Output

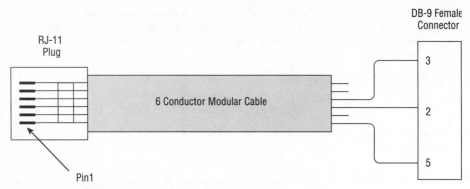

FIGURE 15-2: BetaBrite interface cable for short lengths.

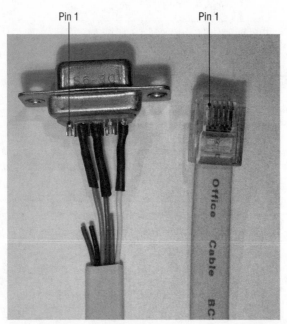

FIGURE 15-3: Completed cable without backshell installed.

Double-check your work with an ohmmeter to verify it's wired correctly. Also, make sure the ends of the unused wires don't have any exposed copper that could short to the other wires. If everything looks good, install the backshell.

If your sign is going to be located further than about 20 feet from your computer, you should consider using shielded cable, especially if you are mounting it outdoors. Just like the 1-Wire bus, long cable lengths are subject to interference and damage from nearby lightning strikes. Using good quality shielded cable minimizes this risk and you can place your sign up to 60 feet away. You can find 3- or 4-conductor 22-gauge shielded cable at many hardware stores. It is usually sold by the foot as security or alarm cable.

To use the shielded cable, you will need to build a 1-foot version of the RJ-11 to 9-pin D connector cable described earlier. Then attach a 9-pin D connector to each end of the shielded cable, one end male and one end female making an extension cable. Connect the shield on the indoor end to your computer's case (preferably the single point ground described in the previous chapter) and connect the other end to an electrical ground, such as the ground in an electrical outlet as shown in the schematic in Figure 15-4.

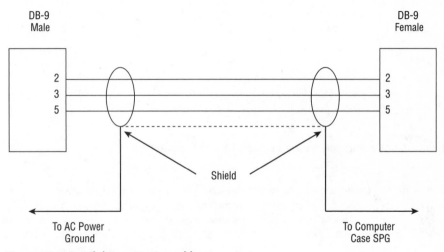

FIGURE 15-4: Serial data extension cable.

Installing the Sign and Running the Cable

The BetaBrite sign comes with two types of mounting hardware: L-brackets for wall mounting and two small loops for hanging it from a chain. You will also need an AC outlet to power your sign. The power supply is a brick with an AC cord on one end and a coaxial power connector on the other.

One solution for mounting the power supply is to epoxy it to the back of the BetaBrite sign. This makes the sign easier to install. If you do, you'll have to purchase longer L-brackets for mounting. Standard 4-inch brackets will work if you grind the ends round as shown in Figure 15-5.

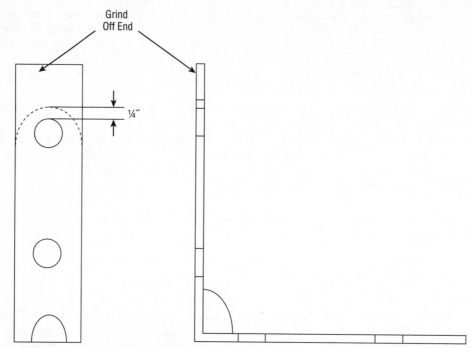

FIGURE 15-5: Modifications to use a standard 4-inch L-bracket.

Follow the same guidelines for installing the serial cable as you did for running the 1-Wire cable. Stay away from noise sources such as fluorescent lights and motors. Unlike the 1-Wire bus, the serial cable is not as susceptible to capacitance so you can run the wire through metal conduit or along a metal pipe.

Warning

During the development of this chapter, I managed to kill two BetaBrites. I believe that static electricity was the culprit; however, I don't have any hard evidence. The BetaBrite is isolated from the AC power using a "brick" transformer. This means the BetaBrite is "floating" and not connected to ground. At the time, the serial cable was connected to the BetaBrite but not the computer, so I think it developed a charge, even though I didn't touch it. I suggest that you treat the BetaBrite as static sensitive. Once you have it installed, connect the case to an earth ground.

Connect to Your Computer

Connect the BetaBrite to the second serial port on your computer. This may be a built-in port or a USB-to-Serial adapter. Just like the 1-Wire port, you will need to know the serial port identifier or name. You can use the OneWireLister program to list the names of all the ports available.

Modify SimpleWeather

After you have your sign installed, the next step is to modify SimpleWeather. Just as you have done in the past, you start by adding the class file that provides support for the BetaBrite sign. Then modify SimpleWeather to instantiate the class, initialize the sign, and send weather data to it.

Add the BetaBrite Class

In your downloaded source code, locate the Chapter 15 folder. It should contain a single Java source file, BetaBrite.java, and a completed project folder. Copy BetaBrite.java to your SimpleWeather project source code folder. Fire up NetBeans and verify that the new file shows up in the list of source files.

Modify SimpleWeather

The first step is to add a new user constant defining the serial port used to communicate with the sign. Add a new user constant BETABRITE_SERIAL_PORT after the existing ONE_WIRE_SERIAL_PORT near line 29:

```
// user constants
  public static final String VERSION = "SimpleWeather 1.0";
  public static final String ONE_WIRE_SERIAL_PORT = "COM1";
  public static final String BETABRITE_SERIAL_PORT = "COM2";
  public static final String LOG_PATHNAME = "WeatherLog.txt";
```

Add a new class constant near line 46. SPACE is simply a space character used in formatting the display.

```
// class constants
public static final String ADAPTER_TYPE = "DS9097U";
public static final char SPACE = ' ';
```

In the list of class variables, declare a new BetaBrite class variable:

```
// class variables
  public float temp;
  ...
  private Wunderground wu;
  private BetaBrite sign;
```

When SimpleWeather starts up, it would be nice to have the BetaBrite display a "splash screen" indicating that the program is indeed running. Near line 123, after the call to instantiate the Wunderground class, add the following code to display "SimpleWeather" on the sign. This section of code is an example of the fundamental operation of the BetaBrite class. The following six steps are used to create and display a message:

1. Create new instance of the BetaBrite class, passing it the maximum size of the message; in this case it is 1,024 characters (far more than needed).

2. Call openSerialPort(), passing the designated serial port for the BetaBrite. This will attempt to open the serial port, and if successful, returns TRUE. If this call fails, the remaining steps are skipped.

3. Build the message header with buildHeader() using the message file name (A), the scroll type (HOLD_MODE, meaning don't scroll), and the desired color (YELLOW_COLOR).

4. Call the addText() method passing it the scroll speed (NO_HOLD_SPEED), the desired font (FIVE_HIGH_STD), and the actual text to display (SimpleWeather).

5. Call sendMsg() to assemble the message and send it via the serial port to the BetaBrite sign.

6. If there is no more text to send, call the closeSerialPort() method.

You'll follow these same six steps later when the actual weather data is sent to the sign. You will learn more about the various parameters when the BetaBrite class is discussed.

```
public SimpleWeather()
  {
  ...
    // initialize wunderground
    wu = new Wunderground();

    // initialize the BetaBrite sign & display program's name
    sign = new BetaBrite(1024);

    if (sign.openSerialPort(BETABRITE_SERIAL_PORT))
    {
      sign.buildHeader('A', sign.HOLD_MODE, sign.YELLOW_COLOR);
      sign.addText(sign.NO_HOLD_SPEED, sign.FIVE_HIGH_STD, "SimpleWeather");
      sign.sendMsg();
      sign.closeSerialPort();
    }
  }
```

What's Garbage Collection?

When coding software, many programming languages require the developer to explicitly dispose of old variables and objects no longer needed. This prevents "memory leaks" where memory slowly gets eaten up by leftover objects called garbage. Java is smart enough to know when objects and variables are no longer needed, and will automatically dispose of them for you. This process is called Garbage Collection, and runs automatically depending on complex rules. If the running program has a main loop that executes periodically, I prefer to manually invoke the garbage collection routine to keep the program from becoming bloated.

In the `mainLoop()` near line 231, add the call to update the BetaBrite prior to the existing call to log the data. This call will be executed once every minute. Just after the call to the data logger, add the call to invoke garbage collection:

```
// and update the web page on the 10 minute mark
if (minute % 10 == 0)
{
  wp.updatePage(date);
  wu.send(date, wc);

  // reset the averages for the next set of data
  wc.resetAverages();
}

// update BetaBrite sign
updateBetaBrite(sign, date, temp, humidity);

// log the data
logger.logData(date, this);

// force garbage collection
System.gc();

System.out.println("\n");
```

Lastly, add the following method for sending the weather data to the sign at the end of the program near line 326, just before the ending bracket. This is the routine you can modify to display the weather parameters of your choice:

```
private void updateBetaBrite(BetaBrite sign, Date d, float temp, float hum)
  {
    String message;

    // convert date to a String
    SimpleDateFormat formatter = new SimpleDateFormat("h:mm");
    message = formatter.format(d) + SPACE + SPACE;

    // display the current temperature & humidity
    message += (int)temp + sign.DEG_SYMBOL + SPACE + (int)hum + "%";

    // select color based on temperature
    char color = sign.YELLOW_COLOR;
    if (temp < 70)
      color = sign.GREEN_COLOR;
    else if (temp > 90)
      color = sign.RED_COLOR;

    // put it all together and send to sign
    if (sign.openSerialPort(BETABRITE_SERIAL_PORT))
    {
```

```
      sign.buildHeader('A', sign.HOLD_MODE, color);
      sign.addText(sign.NO_HOLD_SPEED, sign.SEVEN_HIGH_STD, message);
      sign.sendMsg();
      sign.closeSerialPort();
   }
}
```

In this code, the first few lines of code build a `String` message containing the time in a simple "hh:mm" format. After the time, temperature is added by taking the integer value of the temperature value passed in, and converting it to a String. Humidity is added to the String in the same manner.

To show one of the many ways you can hack the code to create a customized display, the color of the display is changed depending on the temperature. By default, color is set to equal the constant value YELLOW_COLOR. If the temperature value is less than 70 degrees, the variable `color` is set to GREEN_COLOR, and if the value is greater than 90 degrees, the variable is set to RED_COLOR. Once the message is assembled, it is sent to the sign using the same steps used to display the splash screen.

Try It

Make sure your BetaBrite sign is connected and enter the serial port name in `SimpleWeather.java`'s user constant BETABRITE_SERIAL_PORT. Run the program in NetBeans. You will see the splash screen message displayed briefly followed by the current time and temperature. The display should update once a minute with current values. Depending on the temperature, the displayed message should be red, yellow, or green. If the temperature sensor is accessible, gently blow on it with a heat gun or a hair dryer to just over 100 degrees. The displayed message should turn red on the next update. Be careful to not overheat the sensor.

If you get an error in NetBeans, check the message and fix the problem. If you get stuck, the completed code for this chapter is included in the source code folder. You can also open the `BetaBrite.java` class and set the variable `DebugFlag` to `true` to enable the debug messages.

If the code runs, but the display doesn't show the splash screen, check your cables and check that you specified the correct serial port. Check that the BetaBrite is turned on using the remote control. Also, double-check the code to make sure you specified the correct serial port parameters listed in the code.

Walk Through the Code

The BetaBrite class contains many user constants and a lot of code. To save space, I will just cover the important ones. Look through the BetaBrite class in NetBeans to see all of the constants and methods.

First you will find almost 100 user constants for building messages. These include fonts, colors, transition modes, transition speeds, and more. The following listing just lists a portion:

```
public class BetaBrite
{
```

```
...
// Character Sets
public static final char SELECT_CHAR_SET = 26;      // character selection code
public static final char FIVE_HIGH_STD = '1';       // 5 dots high standard
public static final char FIVE_HIGH_FANCY = '2';     // 5 dots high Fancy
public static final char SEVEN_HIGH_STD = '3';      // 7 dots high standard
public static final char SEVEN_HIGH_FANCY = '5';    // 7 dots high fancy

// Transition speed
public static final char NO_HOLD_SPEED = 9;         // no hold speed
public static final char SPEED_1 = 21;              // slowest speed
public static final char SPEED_2 = 22;              // medium slow
public static final char SPEED_3 = 23;              // medium
public static final char SPEED_4 = 24;              // medium fast
public static final char SPEED_5 = 25;              // fast

// Colors
public static final char SELECT_COLOR = 28;         // color selection char
public static final char RED_COLOR = '1';           // red display
public static final char GREEN_COLOR = '2';         // green display
public static final char AMBER_COLOR = '3';         // amber display
public static final char YELLOW_COLOR = '8';        // yellow display
public static final char RAINBOW1_COLOR = '9';      // rainbow color display
public static final char RAINBOW2_COLOR = 'A';      // rainbow color display
public static final char MIXED_COLOR = 'B';         // mixed color display
public static final char AUTO_COLOR = 'C';          // auto color display

// modes
public static final char ROTATE_MODE = 'a';     // message travels right to left
public static final char HOLD_MODE = 'b';       // message stays stationary
public static final char FLASH_MODE = 'c';      // message flashes
...
```

There is one special character defined for this application: the degree symbol used when displaying temperature or dewpoint:

```
// special characters
public static final char BACKSPACE = 0x08;
public static final String DEG_SYMBOL = BACKSPACE + "I";
```

The class construction is fairly straightforward. It allocates a new char buffer `sendBuffer` based on the size passed in. The variable `num` is used to keep track of the total number of bytes to send. In this routine, it is initialized to zero:

```
public BetaBrite(int bufferSize)
{
   this.debugFlag = debugFlag;

   if (debugFlag)
```

```
      System.out.println("Constructing new BetaBrite");

   // create message buffer
   sendBuffer = new char[bufferSize];

   // clear message buffer
   for(int i = 0; i < bufferSize; i++)
     sendBuffer[i] = '0';

   num = 0;
 }
```

The `openSerialPort()` method attempts to open the designated serial port for output. If the port exists and is not in use by another program, it is set for 9600 baud, 7 data bits, 1 stop bit, and even parity. A serial port output stream is created to allow data "writes" to the port:

```
public boolean openSerialPort(String portName)
{
  CommPortIdentifier commPort;

  if (debugFlag)
    System.out.println("Attempting to Open Serial Port " + portName);

  // try to get the port with a 5 second timeout
  try
  {
    commPort = CommPortIdentifier.getPortIdentifier(portName);

    // is port busy?
    if (commPort.isCurrentlyOwned())
      return false;

    // now open it
    port = (SerialPort)commPort.open("BetaBrite", 5000);

    // set to 9600 7E1, no flow control
    port.setSerialPortParams(9600, SerialPort.DATABITS_7,
                    SerialPort.STOPBITS_1, SerialPort.PARITY_EVEN);
    port.setFlowControlMode(0);

  }
  catch (NoSuchPortException e)
  {
    System.out.println("Specified Serial Port Does Not Exist: " + e);
    return false;
  }
  catch (PortInUseException e)
  {
    System.out.println("Serial Port In Use: " + e);
    return false;
  }
```

```
catch (UnsupportedCommOperationException e)
{
  System.out.println("Error Setting Serial Port Parameters: " + e);
  return false;
}

// get the output stream
try
{
  outputStream = port.getOutputStream();
  return true;
}
catch (IOException e)
{
  System.out.println("Unable to get serial port for Output: " + e);
  return false;
}
}
```

If the serial port was opened successfully, the next step is to assemble the message. The BetaBrite uses Adaptive Micro's Alpha Sign Protocol. Referring to Figure 15-6, there are eight parts to a standard transmission packet. The first six synchronize the serial transmissions and address the sign. Because there is only one sign in this configuration, the "all signs" address is used. Part 7 contains the desired message and message formatting, and part 8 closes the transmission.

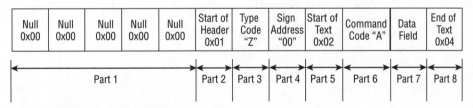

FIGURE 15-6: Standard Alpha Protocol format.

The buildHeader() method builds a character array containing the first six parts of the protocol and part of the seventh, based on the values passed. The BetaBrite can hold several message "files." A single letter designates each file. In this application, all messages are stored in file 'A.' The mode parameter indicates the desired scrolling mode. It should be one of the values listed in the user constants. The third parameter, color, designates the desired color and should also be one of the values in the user constants list:

```
public void buildHeader(char file, char mode, char color)
{
  if (debugFlag)
  {
    System.out.println("Adding Header Info:");
```

```
      System.out.println("File : " + file);
      System.out.println("Mode : " + mode);
      System.out.println("Color: " + color);
    }

    // send 5 nulls to sync serial port
    for (num = 0; num < 5; num++)
      sendBuffer[num] = 0;

    // send betabrite header info
    sendBuffer[num++] = SOH;
    sendBuffer[num++] = ALL_SIGNS_TYPE;
    sendBuffer[num++] = '0';                   // all addresses
    sendBuffer[num++] = '0';

    // start of text char
    sendBuffer[num++] = STX;

    // file name
    sendBuffer[num++] = WRITE_TXT_FILE;
    sendBuffer[num++] = file;

    // send line code
    sendBuffer[num++] = ESC;
    sendBuffer[num++] = TOP_LINE;

    // send mode code
    sendBuffer[num++] = mode;

    // select color
    sendBuffer[num++] = SELECT_COLOR;
    sendBuffer[num++] = color;

    return;
  }
```

After the header is assembled, the desired text to display on the sign is appended to the data buffer using the addText() method. Three parameters are passed: the desired scroll speed, the desired font, and the actual text to display. The message is followed with an End-of-Text (EOT) character as specified in the protocol for part 8:

```
  public void addText(char speed, char font, String text)
  {
    // select character set
    sendBuffer[num++] = SELECT_CHAR_SET;
    sendBuffer[num++] = font;

    // set update speed
    sendBuffer[num++] = speed;

    // insert text
```

```
    int len = text.length();

    text.getChars(0, len, sendBuffer, num);
    num += len;

    sendBuffer[num++] = EOT;
}
```

Once the message has been assembled in the preceding two methods, the call to sendMsg() attempts to send the message buffer to the sign via the serial port:

```
public void sendMsg()
{
  if (debugFlag)
    System.out.println("Attempting to Send " + num + " Bytes");

  try
  {
    for (int i = 0; i < num; i++)
    {
      outputStream.write((byte)(sendBuffer[i] & 0x00ff));

      if (debugFlag)
        System.out.println("Byte " + i + " = " + (byte)(sendBuffer[i] &
                      0x00ff) + " (" + sendBuffer[i] + ")");
    }

    // make sure all data is sent
    outputStream.flush();

    if (debugFlag)
      System.out.println("Sent " + num + " bytes");

  }
  catch (IOException e)
  {
    System.out.println("Error Getting Output Stream: " + e);
  }
}
```

After the message has been sent, the serial port can be closed using the closeSerialPort() method. Because SimpleWeather may share this serial port with another device, it is closed after each message:

```
public void closeSerialPort()
{
  if (debugFlag)
    System.out.println("Closing Port");

  port.close();
}
```

Expanding Your Project

There is a lot to learn about the BetaBrite and its programming. The master decoder ring for BetaBrite serial communications is Adaptive Micro's *Alpha Sign Communications Protocol* document, available on Adaptive Micro's web site at www.ams-i.com/Pages/97088061.htm. It offers several programming examples and a discussion forum for programming support.

Modify the Message Mode

There are dozens of different ways to display your weather. You can try different fonts, colors, and scrolling modes. To help get you started, try this: Edit SimpleWeather.java's updateBetaBrite() method. Change sign.HOLD_MODE to sign.ROTATE_MODE and sign.NO_HOLD_SPEED to sign.SPEED_1 like this:

```
if (sign.openSerialPort(BETABRITE_SERIAL_PORT))
    {
        sign.buildHeader('A', sign.ROTATE_MODE, color);
        sign.addText(sign.SPEED_1, sign.SEVEN_HIGH_STD, message);
        sign.sendMsg();
        sign.closeSerialPort();
    }
```

Re-run SimpleWeather and the weather data is now displayed in a scrolling fashion.

Stop That Flashing

When a serial message is detected on the BetaBrite, it stops displaying and reads the message. This causes a noticeable flash in the display once a minute when SimpleWeather updates the display. You might find this somewhat annoying.

The Alpha Protocol supports what are called String Files. A single message is sent to the BetaBrite that contains special characters, or variables, for the text or numeric values in the message that change. Then, on each update cycle, just the value for that variable is sent to the BetaBrite. Because the entire message does not have to be rebuilt in the BetaBrite, the value on the display is updated without a noticeable flash.

If you're interested, look through the Alpha Protocol documents under String Files and modify SimpleWeather to use this method. It's not as easy as it sounds. If you need a little help, look at the ExtremeTech Weather Server source code on the companion web site.

Change the Display Based on Time

In the code for this chapter, the sign changed color based on temperature. There are many other possibilities. If the sign is indoors, maybe you want it to turn off at night. Based on the time value passed to the updateBetaBrite() method, you could send a blank (empty) message, effectively turning it off. Or maybe you want the sign to switch to display the rain count only when it is raining. Another option is to display the time, temperature, and humidity on the even minute, and time, temperature, and dewpoint on the odd minute. There are lots of options and choices.

Warning Messages

Another option for your sign is to display warning messages when your weather data indicates bad weather approaching. For example, if you built the lightning detector, a warning message could be displayed when the lightning count exceeds some threshold. Or, based on the barometric pressure trend, a "storm is approaching" message could be displayed.

Pool Temperature

My favorite use for the LED sign is to display swimming pool temperature. To do this, you'll have to add another temperature sensor that goes in the pool water. Use an outdoor UV-resistant length of CAT3 or CAT5 cable and run it from your 1-Wire bus to your pool. The tricky part is completely waterproofing the temperature sensor. You can do this by soldering the 1-Wire temperature sensor directly to the end of the cable, referring to Chapter 5 for guidelines. Next, take about a 6-inch length of poly tubing and fill it with RTV, avoiding any bubbles. Then slide the temperature sensor into the tube so that it's completely embedded in the RTV and let it cure overnight. Once the RTV is cured, simply drop the sensor in your pool.

In SimpleWeather, add another instance of the temperature sensor class and modify the code to use the new sensor to display on the BetaBrite. Now you'll know your swimming pool temperature year 'round without having to go outside.

Wrap Up

As you wrap up this project, there are a couple of things to think about. Just like the 1-Wire bus, long cable lengths can pick up radiated energy from nearby lightning. You may want to consider using a serial port surge suppresser on the PC side or both ends of the serial cables. There are several suppliers of 9-pin serial port surge protectors, including APC at www.apcc.com.

As I mentioned earlier, the case on the BetaBrite is not connected to ground. I suggest attaching a wire to one of the case screws and connecting it to an AC power ground, such as the screw on an outlet plate.

Adding a BetaBrite sign adds a certain "wow" factor to your weather station, and is really quite useful. The large display is readable many feet away, so there are no more trips to the computer to check the weather. A quick glance, and you know the outdoor temperature when you're getting dressed in the morning or thinking about a dip in the pool.

Turn Appliances On or Off Based on Weather

There are times when you probably wish you could turn something on or off based on the current weather conditions. Maybe you have a whole-house fan and want it to run when it is hotter inside the house than outside. Or maybe your citrus trees have a warming device and it needs to be turned on when the temperature drops below freezing. Once you start thinking about it, you may come up with all kinds of custom applications.

In this project, you learn how to turn on or off just about any electrical appliance you have. If it plugs into a 120V AC outlet, then it's a snap to control. You can even dim lights to a certain setting based on time and weather conditions.

The software application in this chapter is a bit more advanced in terms of coding. To turn an appliance on or off, you will have to develop your own code, writing the conditions in Java. It isn't that hard to learn, and in fact, you may find it is pretty easy. Options like these are why you own your weather station.

There is a safety concern when working directly with 120V to control AC-powered lights and appliances, exposing you to shock hazards. Plus, working with AC power near other low-voltage electronics like 1-Wire and your computer also poses a risk. One wrong connection and your computer is history. Furthermore, running wires all over your house is not only a safety hazard itself, but is unsightly. What you need is some sort of remote control system. That's where X10 comes in.

What Is X10?

X10 is a remote-control system that allows you to control AC-powered devices remotely without additional wiring. It uses your existing house wiring to transmit signals from the X10 transmitter to one of the X10 receivers or modules. You simply plug the transmitter-controller into an AC outlet, plug in a lamp or appliance module into another outlet, and then a lamp or appliance into the module. By sending commands from the controller, the module will turn on or off. X10 can also dim lights on command.

X10 was first introduced back in 1978 for the Sear's Home Control System and Radio Shack's Plug n' Play Power System. Since then, many other companies jumped on the X10 bandwagon, offering many different types of modules and accessories.

How Does X10 Work?

X10 sends its commands over your house's power lines using a high-frequency signal right on the "hot" line of the AC outlet. As shown in Figure 16-1, the signal is a 120-kHz "burst" that occurs as the AC voltage crosses zero. By sending out a series of bursts and no-bursts, the X10 transmitter essentially is sending out a sequence of 1s and 0s. The bursts shown in Figure 16-1 have been exaggerated; the real levels are much lower.

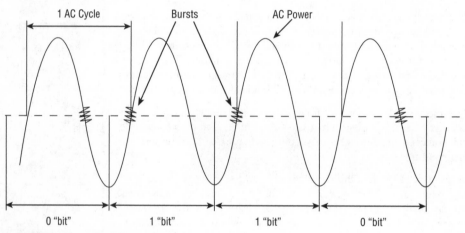

FIGURE 16-1: X10 signal basics.

Because the signal is transmitted over the power line, it could go through your main power panel and right into your neighbor's house. To prevent you from controlling your neighbor's X10 modules, each X10 transmitter and receiver has a switch to set the *house code*. The house code is a letter from 'A' to 'P' that you choose for your house. The receiver only process commands for the house code it is set for.

Each receiving device or module also has a switch to set the *unit number* as shown in Figure 16-2. The unit number is a value between 1 and 16 (inclusive), and allows you to have up to 16 different devices on a single house code. As long as your neighbors aren't using X10, you could theoretically have 16 house codes x 16 unit codes, or 256 individually controlled devices.

FIGURE **16-2: X10 modules showing the house and unit code switches.**

X10 isn't without its flaws. Because it shares the house wiring with many appliances, noise on the AC power line can have a serious effect on signal reliability, essentially "drowning out" the signal. Here are a few examples of noisy devices:

- **Motors** — Many appliances, such as hand-held drills, vacuum cleaners, and blenders have electric motors that use brushes to supply power to the rotating armature. As the armature rotates, the brushes make and break contact with different coils. This causes large noise spikes on the power line.

- **Fluorescent lights** — Fluorescent lights use an arc to generate light. This arc switches on and off at 120 Hz causing considerable noise on the power line. The newer compact fluorescent light uses a high-frequency switching power supply that generates even more noise.

- **Computers** — Computers and many computer accessories use switching power supplies that generate high-frequency noise as the power is switched on and off. Additionally, the inputs to these power supplies have high capacitance that tends to filter or reduce the X10 signal.

Another issue with X10 is two-phase power. As shown in Figure 16-3, your house is powered by two different sets of coils in the power transformer that supplies power to your house. Because the power transformer is designed for 60 Hz, 120-kHz X10 signals don't travel too well through the coils. In most houses, half of your house is connected to Phase X, and the other half to Phase Y. This means that X10 modules connected to a different phase than the transmitter is connected to may not get sufficient signal to operate. There are several ways to mitigate this; the most common is a 2-phase coupler. This device connects between the two phases, coupling X10 signals (but not power). Check the resources list in Appendix A for some suppliers of X10 power phase couplers.

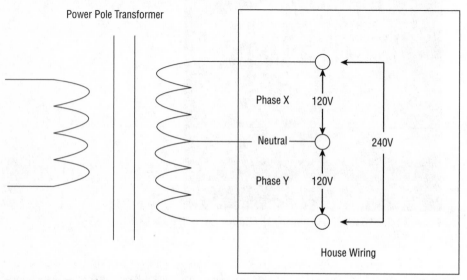

FIGURE **16-3: Typical two-phase home power distribution.**

Even though it has its drawbacks, X10 is still quite useful. For a more detailed discussion on X10, specific house code and unit code bit patterns, see ExtremeTech's *Geek House* by Barry and Marcia Press. There's a whole chapter dedicated to X10, modules, and theory. There's also more details on the X10 support web site at www.x10.com/support/technology1.htm.

What Can I Do with It?

You have many different X10 module controllers and modules to choose from. Some are designed to control lights, and also have the ability to dim incandescent lights. Others are designed to control appliances, such as TVs and motorized devices.

Some of the X10 modules are designed for installation in a wall switch or outlet. As shown in Figure 16-4, these modules resemble a wall outlet or light switch, and contain the X10 electronics to allow remote control. They are very handy if the light you want to control is installed as part of your existing house wiring.

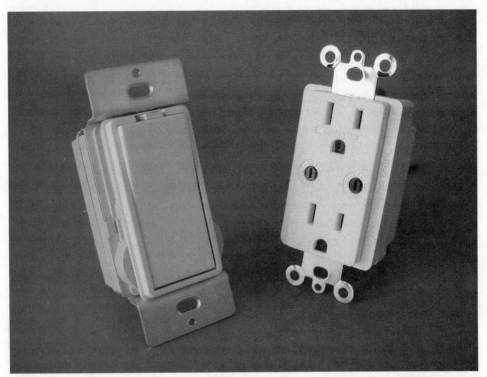

FIGURE 16-4: X10 wall switch and outlet modules.

The X10 transmitter or controller must be connected to AC power to work; otherwise it couldn't get its signal onto the power lines. The folks at X10 realized this was a limitation, and developed a "remote control" for the remote control. The remote unit transmits its signals to the X10 receiver/controller using Radio Frequency (RF) signals. The X10 receiver/controller, in turn, converts the commands it receives from RF into X10 signals that are transmitted over your house wiring to your modules. You can now sit in your easy chair and turn on the reading light or control your ceiling fan.

What Basic Hardware Do I Need?

To get started with X10, you need at least one controller and one module. However, this project isn't about using X10 manually, it's about using your computer to control X10. What you really need is a device to connect your computer to X10.

Several X10 computer interfaces are available on the market. Some connect an X10 controller directly to your computer using an opto-isolator to isolate your computer from the AC power line. Another option, the one chosen for this project, uses a small module that sends out RF signals to the X10 RF receiver/controller module. The module, called a FireCracker, is about half the size of a pack of chewing gum. It is designed to plug right into a serial port on your computer. It has a mating serial port on the other end, allowing it to share a serial port with another device.

The X10 FireCracker is available at www.x10.com/automation/firecracker.htm as part of the FireCracker 4-Piece Home Automation Starter System. Shown in Figure 16-5, the kit contains the FireCracker module, the receiver/controller, a lamp module, and a hand-held remote controller. The receiver/controller module also has an X10 controlled output permanently set to Unit Code 1.

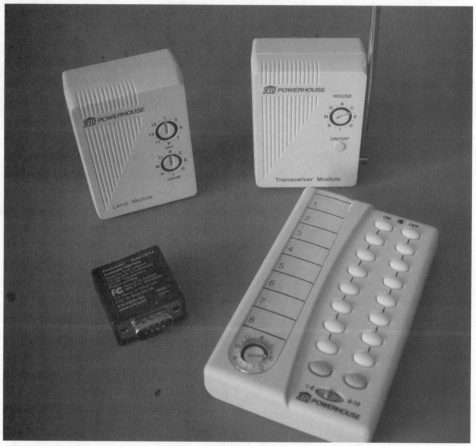

FIGURE 16-5: X10 FireCracker 4-Piece Home Automation Starter System.

Testing It Out Using the Remote

Before you try to control an appliance with SimpleWeather, set up the X10 system and test it out using the hand-held remote. First, pick a house code. It doesn't matter which letter you pick, just make sure that both the remote and the receiver module are set to the same code. Install the receiver/controller module within 10 or 20 feet from your computer so the FireCracker's signal will reach it, and install fresh batteries into the remote. Next, plug a lamp into the receiver's outlet and make sure the lamp is turned on. Because the receiver module is set for Unit Code 1, press the unit 1 "ON" button on the remote and the lamp should turn on. If it doesn't, re-check the house code settings and try again.

If you can't seem to get things working, check the following:

- Don't use a surge suppressor outlet strip. Some of these have an internal noise filter that blocks X10 signals.

- Make sure that the lamp you are controlling is not a fluorescent lamp. These energy-saving bulbs use an electronic power supply to generate the high voltages necessary for the lamp to light. If you want to control fluorescent lamps, use appliance modules only.

- Try a different outlet. There may be a noise-producing appliance on the same line corrupting the X10 signals.

If you experience the problem where a light or appliance turns on unexpectedly, maybe your neighbor is using X10. Try switching house codes to see if the problem goes away. The X10 web site has several technical support pages, online chat, and more. Find it at www.x10.com/support/support.htm.

Connecting the FireCracker

Once you have X10 working with the hand-held remote, you are ready to install the FireCracker. If you are using a BetaBrite sign as described in the previous chapter, disconnect it from your computer's serial port and connect the FireCracker. Re-connect the BetaBrite to the other end of the FireCracker as shown in Figure 16-6.

X10.com offers free automation software as part of the FireCracker starter package. If you are interested in experimenting with the software, now is a good time to try it out. Make sure SimpleWeather is not running, because it may be using the serial port that FireCracker is connected to. This gives you the opportunity to experiment with X10's capabilities.

Add the X10 Software Module

In your downloaded source code, locate the Chapter 16 folder. It should contain a single Java source file, X10FireCracker.java, and a completed project folder. Copy X10FireCracker.java to your SimpleWeather project source code folder. Launch NetBeans, open your SimpleWeather project, and verify that the new file shows up in the list of source files.

FIGURE 16-6: BetaBrite cable plugged into the FireCracker using the second serial port.

Modify SimpleWeather

Start the modification by adding a user constant for the X10 serial port. In this project, the second serial port will be used. If you are using a BetaBrite, the port will be shared between the two, or if your computer has a third serial port, you can use it. In the `user constants` near line 27, add the following line, entering the name of your serial port:

```
public class SimpleWeather
{
    // user constants
    public static final String VERSION = "SimpleWeather 1.0";
    public static final String ONE_WIRE_SERIAL_PORT  = "COM1";
    public static final String BETABRITE_SERIAL_PORT = "COM2";
    public static final String X10_SERIAL_PORT       = "COM2";
    public static final String LOG_PATHNAME = "WeatherLog.txt";
```

In the list of class variables near line 52, add a new class variable for the X10 class:

```
// class variables
  ...
  private DSPortAdapter adapter;
  private DataLogger logger;
  private WeatherCruncher wc;
  private WebPage wp;
  private Wunderground wu;
  private BetaBrite sign;
  private X10FireCracker x10;
```

Near line 136, after the BetaBrite Splash Screen code, instantiate a new X10 class:

```
public SimpleWeather()
  {
   ...
    // initialize the BetaBrite sign & display program's version
    sign = new BetaBrite(1024);

    if (sign.openSerialPort(BETABRITE_SERIAL_PORT))
    {
      sign.buildHeader('A', sign.HOLD_MODE, sign.YELLOW_COLOR);
      sign.addText(sign.NO_HOLD_SPEED, sign.FIVE_HIGH_STD,
                  "SimpleWeather");
      sign.sendMsg();
      sign.closeSerialPort();
    }

    // initialize the X10 interface
    x10 = new X10FireCracker();
```

In the `mainLoop()` near line 142, add a new method String variable used to hold the X10 command:

```
public void mainLoop()
{
  Date date = new Date();
  int minute, lastMinute = -99;
  int hour;
  boolean quit = false;
  InputStreamReader in = new InputStreamReader(System.in);
  String command;
```

At the end of the `SimpleWeather.java` code, just before the closing brace (near line 370), add a new method called `x10Command()`. This is the method that accesses the X10 class. To use this new method, you need to pass it through three parameters: the house code, the unit code, and the command. The house code is a `char` variable 'A' (use a single quote or specify a `char`) through 'P'. The unit code is an `int` value 1 through 16 representing the unit to access. The command is a string and can be OFF, ON, BRIGHT, or DIM.

In this method, the X10 serial port is opened, the X10 command is sent, and then the port is closed. This leaves the serial port available for the BetaBrite sign if you are using one:

```
private void x10Command(char houseCode, int unitCode, String command)
{
    try
    {
      x10.open(X10_SERIAL_PORT);
      x10.sendCommand(houseCode, unitCode, command);
      x10.close();
    }
    catch(Exception e)
    {
      System.out.println("X10 FireCracker: " + e);
    }
  }
}
```

Here's where things get a bit tricky. In the `mainLoop()`, just after the call to log the data, add a few extra lines to mark the spot. This is where you will add your code to define the rules that turn on or off the X10 modules. This is going to be unique to your application, so you have to do the actual coding. Start off with something simple like the following code example. In this example, the module at house code 'A' turns on if the temperature is greater than 73 degrees, and turns off if the temperature is less than 73 degrees:

```
// log the data
logger.logData(date, this);

// Send X10 Command
if (temp > 73)
{
  command = "ON";
}
else
{
  command = "OFF";
}

x10Command('A', 1, command);

// force garbage collection
System.gc();
```

Before compiling the code, make sure the house code and unit code in the preceding example match the house code and unit you are attempting to activate. Also, make sure the X10 appliance you are trying to control works with the X10 remote first. It can be quite frustrating trying to turn on a lamp in your code if the lamp itself is turned off (don't ask how I know this).

Select Run ⇨ Run Main Project on the NetBeans menu. If you get a compile error, check the error message. If the error is in the X10 command code, look at the preceding example to make sure you have semicolons in the right spot; they don't go on the `if` statement lines.

When you get your project running, check the temperature. If the temperature is greater than 73 degrees (or whatever temperature you chose), the light or appliance should turn on. Cool the temperature sensor below 73, and the light should turn off. Look at the NetBeans output window to see the X10 commands being sent to the X10 class. There should be one every minute when `mainLoop()` executes:

```
Time = Fri Apr 07 07:05:13 MST 2006
Temperature = 69.9125 degs F
Wind Speed = 0.0 MPH from the  E
Humidity = 22.485159 %
Dewpoint = 30.207876 degs F
Pressure = 30.009905 inHg
Rain = 0.0 in
Lightning = 0 SPM
Updating Log WeatherLog.txt
X10 Housecode = A  Unitcode = 1  Cmd = OFF
X10 Command Sent
```

Once you have successfully switched on or off an appliance, you are ready to begin experimenting and hacking your own commands. For a few tips, read the sidebar on Java's if/else statement.

A Quick Look at the X10 Class

The X10 class is quite long and a little complex. Rather than walk you through the code, I'm going to explain how the X10 FireCracker operates. If you want to get into the details of the class, read this to understand how the FireCracker works, and then open the source file in the NetBeans editor and look through the code. By now, you should be familiar enough with Java code that you can read it and get a good understanding of what the code is doing.

The FireCracker sends X10 commands to the receiver by a sequence of bits. The exact bit sequence in defined in the FireCracker's protocol manual available from X10.com at `ftp://ftp.x10.com/pub/manuals/cm17a_protocol.txt`.

The FireCracker is not really a serial device, even though it plugs into your serial port. Instead, it uses two serial port signal output lines to control its operation: Data Terminal Ready (DTR) and Request To Send (RTS). The FireCracker also uses these two lines to steal its power, similar to the 1-Wire serial port adapter. At least one of these two lines must be high at all times to keep the FireCracker powered.

Java's if/else Statement

Although you have been using it throughout most of the SimpleWeather project, here's a little more about how Java's if/else statement works. If you're an experienced programmer, you can skip this section.

The if/else statement controls program flow. It enables your program to selectively execute certain program statements based on a condition.

The standard syntax for the if/else statement is as follows:

```
if (logical condition)
{
    do this block if the condition is true
}
else
{
    do this block if the condition is false
}
```

In the `if` statement, a logical expression is contained in parentheses. If that expression evaluates to `true`, then the first block is executed. If it is `false`, then the code in the block following the `else` is executed. If there is only one statement line in the blocks, the brackets are not required and if there is no `else` block, then it should be omitted.

The logical expression can use any of the Java logical operators, such as `<, >, ==, !=`.

a < b: a is less than b

a > b: a is greater than b

a == b: a equals b

a != b: a does not equal b

For example:

```
If (temperature > 70)
```

If the variable `temperature` is greater than 70, then this expression evaluates to `true` and the first block would be executed. If it is less than (or equal to) 70, then the block following the `else` will be executed.

You can combine several logical expressions together using the logical and symbol `&&` to develop complex rules. For example:

```
If (temperature > 70 && humidity < 50)
```

In this case temperature has to be greater than 70 *and* humidity has to be less than 50 for this expression to be true.

Java's if/else Statement *(continued)*

From SimpleWeather, this is the list of the weather variables that you can use in your expression to generate the rules for controller X10 modules:

- temp
- Speed
- windDir
- humidity
- dewpoint
- pressure
- rain
- lightning
- hour
- minute

The FireCracker uses these two lines to generate one of four possible conditions as listed in Table 16-1.

Table 16-1   X10 FireCracker Mode Selection

Mode	RTS	DTR
Reset	Off	Off
Send Logical '1'	On	Off
Send Logical '0'	Off	On
Standby	On	On

To keep the device from resetting, it is important to enable the one line prior to disabling the other. For example, if the FireCracker sent a logical '1' and the next bit is a logical '0', enable the DTR line before disabling the RTS line. The minimum wait time at each state is 0.5 milliseconds.

Each X10 command sequence is made up of 40 bits or 5 bytes. The first 16 bits are the header, followed by 16 command bits that correspond to the desired house code, unit code, and command word. This is followed by an 8-bit footer. The header and footer never change; they are always the same sequence of bits shown in Table 16-2.

Table 16-2   Header and Footer Bit Sequence

Part	Bit Sequence
Header	11010101 10101010
Footer	10101101

The most complicated part of the X10 class is developing the 16-bit command. If you look through the X10 protocol document, you'll see that the house code, unit code, and command are combined in kind of an oddball fashion. To convert the codes into the necessary bit pattern, the X10 class uses a lookup table similar to that shown in following tables. The house code, shown in Table 16-3, and the unit code, shown in Table 16-4, are combined using a bitwise OR to form the device code. If the command, as shown in Table 16-5, is OFF, BRIGHT, or DIM, the command is combined with the device code using a bitwise OR. If the command is ON, the command is combined to the device code using a bitwise AND, forming the 16-bit pattern that is sent to the FireCracker.

Table 16-3   X10 House Code Bit Patterns

House Code	Bit Sequence	Hex Value
A	01100000 00000000	0x6000
B	01110000 00000000	0x7000
C	01000000 00000000	0x4000
D	01010000 00000000	0x5000
E	10000000 00000000	0x8000
F	10010000 00000000	0x9000
G	10100000 00000000	0xA000
H	10110000 00000000	0xB000
I	11100000 00000000	0xE000
J	11110000 00000000	0xF000
K	11000000 00000000	0xC000
L	11010000 00000000	0xD000
M	00000000 00000000	0x0000
N	00010000 00000000	0x1000
O	00100000 00000000	0x2000
P	00110000 00000000	0x3000

Table 16-4   X10 Unit Code Bit Patterns

Unit Code	Bit Sequence	Hex Value
1	00000000 00000000	0x0000
2	00000000 00010000	0x0010
3	00000000 00001000	0x0008
4	00000000 00011000	0x0018
5	00000000 01000000	0x0040
6	00000000 01010000	0x0050
7	00000000 01001000	0x0048
8	00000000 01010001	0x0051
9	00000100 00000000	0x0400
10	00000100 00010000	0x0410
11	00000100 00001000	0x0408
12	00000100 00011000	0x0418
13	00000100 01010000	0x0450
14	00000100 01110000	0x0470
15	00000100 01001000	0x0448
16	00000100 01011000	0x0458

Table 16-5   X10 Command Code Bit Patterns

Command	Bit Sequence	Hex Value
Off	00000000 00100000	0x0020
On	11111111 11011111	0xFFDF
Bright	00000000 10001000	0x0088
Dim	00000000 10011000	0x0098

For example, to turn house code C unit 5 ON:

```
House Code C: 01000000 00000000
- bitwise OR with-
Unit Code 5:  00000000 01000000
- bitwise AND with -
ON Command:   11111111 11011111
Result is:    01000000 01000000
```

Once the 40-bit sequence is assembled, the FireCracker is reset, and then set to standby mode using the RTS and DTR lines as shown in Table 16-1. Then the 40-bit sequence is sent 1 bit at a time. At the end of the sequence, the FireCracker is set back to standby mode for a few seconds while it processes the command and sends it to the X10 receiver.

If this doesn't make sense to you, don't worry; you don't need to know the inner workings of the X10 class to use it. All that is necessary to use it is in the x10Command() method you added to SimpleWeather.

Expanding Your Project

Take a look at some of the ways you can use X10 to control appliances and lights based on weather.

Freeze Alarm

Living in the Southwest, it occasionally drops below freezing at night. I have a citrus tree that has warming lights (actually, Christmas lights strung in the trees) that I plug in when I think the temperature is going to drop below freezing. Using the X10 weather controller, the lights now turn on automatically when the temperature the drops below freezing. This keeps the fruit from freezing if I forget and saves electricity.

You can use the same idea to keep your pipes from freezing in the winter. Use the X10 weather controller to activate the pipe heaters or circulation system when the temperature drops below freezing.

Close Your Blinds When the Sun Is Shining

To conserver energy, it might be advantageous to open or close your blinds when the sun is shining on the windows. Smarthome offers several drapery controllers that are X10 adaptable. Check out the drapery controllers at www.smarthome.com/3142.html.

There are several ways to detect if the sun is shining. One method is to place a temperature sensor where the sun would strike it. When the sun hits it, there is a significant increase in temperature. Using the WeatherCrucher's trend calculation, you can detect when the sun hits the sensor (positive trend) and when the sun stops shining through the window (negative trend). Painting the sensor flat black or mounting it to a small flat black plate would increase the sensitivity.

Although it is not covered in this book, Hobby-Boards offers a solar sensor that detects light levels. You can use this as the sun detector. Example code that can be integrated into SimpleWeather is included in the Bonus Code section on the companion web site.

Turn Off Your A/C

There's a lot you can do here. Many household central air conditioning (A/C) indoor evaporator units (the part that's indoors) plug into a standard 120V AC outlet. Using one of the X10's heavy-duty appliance modules, you can use the weather as the master A/C controller, enabling or disabling your A/C based on outdoor conditions. For example, if it's cooler outside than in, you can disable the A/C and turn on a fan. You could even go a step further, and display a message on your BetaBrite sign suggesting you open your windows.

If your A/C unit is hard-wired or runs on 220 VAC, then the smart thermostat in Chapter 18 may be a better solution.

Do not try to switch an appliance or other equipment with a lamp module. The circuitry for dimming the lamp in the module may permanently damage both the equipment and the module.

Build a Thermostat Set-Back Device

Rather than turn off your A/C, another option is to build and install a device in your home thermostat that biases the temperature up a few degrees. This provides a "poor-man's" remote programming capability.

It works like this: A small heater (very small) is mounted next to the temperature sensor or coil in your thermostat. When it is turned on, the heater makes the thermostat think the temperature is actually a few degrees warmer than it really is. Therefore, the house heater won't turn on until it gets colder, or the A/C will turn on sooner.

Building the "mini-heater" for your thermostat is actually easier than it sounds. Simply install a small resistor across the output of an AC "wall-wart" transformer. In Figure 16-7, a 12V AC adapter (such as the Radio Shack 273-1773) is plugged in to an X10 appliance module. A small ½-watt resistor is connected across the wires and acts as the heater. You can experiment with different resistor values and placement in the thermostat until you get the results you want. I suggest starting out with a 560- to 1000-ohm ½ watt resistor. Increase the resistance for less heating, or lower the resistance for more. Don't go below 270 ohms or you risk burning up the resistor.

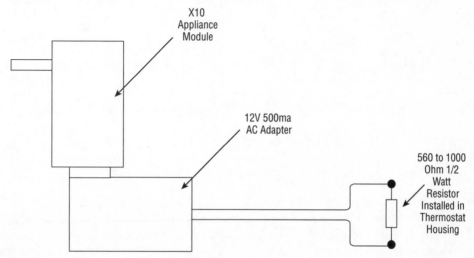

FIGURE 16-7: Thermostat set-back heater design.

Wrap Up

Adding X10 to your weather station opens the door to many possibilities. Not only can you control devices based on weather, but you can use time as well. With a little programming, you can create sophisticated commands that control lights and appliances based on many factors.

As you dream up new ways to hack this project, keep in mind there are many ways to use non-weather-related data to control equipment. For example, a temperature sensor mounted on the top of your TV could detect the heat when it's turned on, closing the blinds behind you and dimming the lights. Or a sensor installed in your cigar humidor could turn on a light warning you that the humidity is out of range.

Up until now, only the ON and OFF commands have been discussed. Using an X10 lamp module, you also have a DIM and BRIGHT command. Each command brightens or dims the lamp by approximately 5%.

As I wrap up this chapter, I want to remind you that X10 is not an infallible system. Several things can cause improper action, such as noise on the power line or a neighbor experimenting with house codes. Don't use X10 to control anything that is life-critical or could cause a loss of property. X10 is a cool weather toy, but be careful how you use it.

Build a Moisture Sensor

There may be times that you need to know the moisture content of a substance. Maybe you want to check the moisture level in soil for your award-winning violets, or want to detect water leaks in the basement. Although it's not really a weather sensor in the strictest sense, this project shows you how to build a simple 1-Wire moisture detector.

One way moisture can be detected is by measuring the resistance of a medium. This could be accomplished as easily as sticking two nails in the soil, attaching a couple of wires, and measuring the resistance. The challenge is that the resistance measured varies greatly depending on the mineral and salt content of the water. For example, distilled water has a very high resistance (almost undetectable), whereas salt water is fairly low. Fertilizers and other soil conditioners can have a major impact on the soil moisture readings.

The solution chosen for this project is to use a "gypsum block" moisture sensor. A basic gypsum block contains two electrodes embedded in a block or cylinder of gypsum. For most soil and water types, the gypsum block provides a fairly repeatable resistance value dependent on the moisture it absorbs, yet independent of the salt content. By measuring the electrical resistance between the electrodes in the block, you can get a good idea of the moisture content.

Project Overview

This project has three parts: building and testing the moisture sensor circuit, building or buying a gypsum block, and adding moisture measurement capability to SimpleWeather.

Options

Building the moister sensor circuit has several options:

- **Option 1: Buy a Complete Assembly.** Hobby-Boards offers the moisture sensor as a completed circuit. You can also buy a gypsum soil moisture sensor.

- **Option 2: Buy a Kit.** Hobby-Boards offers a complete kit for the moisture sensor circuit. If you choose this step, you'll have to solder several surface-mount components, so you need the necessary equipment.

- **Option 3: Build it Yourself.** A schematic, list of parts, and sources are shown later in this chapter.

How It Works

Measuring the resistance of the gypsum block is not as simple as it sounds. If you use a standard DC resistance check, in a short time the gypsum becomes polarized (also known as "gassing"), leading to faulty measurements. Instead, an AC measurement method must be employed. The schematic shown in Figure 17-1 does just that. It is a frequency-to-voltage converter where the frequency is dependent on the moisture level in the block.

U1 is a standard 555 timer IC. In this circuit it is configured as a "monostable multivibrator," or in other words, a simple oscillator. The resistance of the gypsum block connected to the sensor terminals determines the frequency of the oscillation. The more moisture the block holds, the lower the resistance. Lower resistance allows C4 to charge faster, causing the frequency to go up. Conversely, as the gypsum block dries up, its resistance increases, causing the oscillator to run slower. The gypsum block is protected from any DC voltages by capacitors C5 and C6.

To convert the frequency to a voltage that can be measured with a 1-Wire device, the oscillator current is measured. As with most CMOS parts, the current is proportional to circuit "activity," which in this case is frequency. The higher the frequency, the more current it draws. The current is converted to a voltage via resistor R2.

U2 is a DS2760 1-Wire battery monitor device. In this application, the battery functions are not employed, just the precision voltage analog-to-digital converter. The voltage drop across R2 is measured and converted to a 1-Wire voltage reading.

U3 is a simple regulator IC that provides 5V to run the circuit. Note that this circuit is not parasitically powered from the 1-Wire bus. It must be supplied with a separate +7V to +18V source.

How Does the Gypsum Block Work?

Normally, the gypsum block is buried several inches below the soil surface. When the soil is wet, the pores in the gypsum absorb the surrounding water, which dissolves some of the gypsum, and makes a saturated solution of calcium sulphate (a salt). The saturated solution is very conductive and conducts electrical current mostly independent of soil water salinity. In wet conditions, the resistance measures around 100 to 200 ohms.

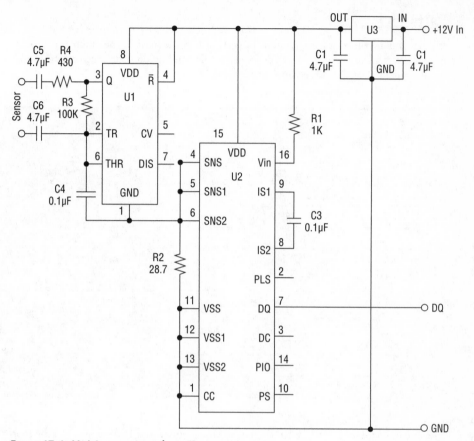

Figure 17-1: Moisture sensor schematic.

As the soil around the block dries, it absorbs water from the block, causing the block to dry out. As the block dries out, the moisture in the larger pores dries out first, causing the AC current flowing between the electrodes in the block to travel a longer path through the smaller pores, causing the resistance to increase. In the 1-Wire circuit described, this causes the frequency of the oscillator to drop, resulting in a lower measured voltage.

Building the Hardware

If you have chosen to build your own circuit, I suspect you have built surface-mount circuits before and have the required tools (and patience!).

Warning | The moisture module is ESD sensitive. Always use an anti-static wrist strap when handling your module or parts.

Parts List

If you're building the kit from scratch, Table 17-1 lists the parts in the schematic. Hobby-Boards offers a PC board to make the job easier. Depending on your construction technique, surface mount or axial lead resistors and capacitors may be used.

Table 17-1   Basic Parts List

Qty	ID	Description	Vendor
1	U1	DS2760 1-Wire Battery Monitor IC	Digikey, Maxim, Hobby-Boards
1	U2	uA555D Timer IC	Digikey
1	U3	LM78L05 5-Volt Regulator	Digikey
1	R1	1K Ohm Resistor	Digikey
1	R2	28.7 Ohm Resistor	Digikey
1	R3	100K Ohm Resistor	Digikey
1	R4	430 Ohm Resistor	Digikey
4	C1, C2, C5, C6	4.7 μF 20V Capacitor	Digikey
2	C3, C4	0.1μF Capacitor	Digikey
1	—	Terminal Strip, Circuit Board	Various
1	—	Gypsum Block	See Text

The circuit board uses five screw terminals to attach power, the 1-Wire bus, and the sensor block. Figure 17-2 shows the completed circuit installed on the Hobby-Boards PC board.

FIGURE 17-2: Completed moisture sensor circuit.

Building the Gypsum Block

You can build your own gypsum block out of nothing more than plaster of paris and two wires. Pour the plaster of paris in a small mold approximately 1- to 2-inches in diameter and insert the stripped wires into the mixture as shown in Figure 17-3. You can use an old paper towel roll as the mold. To get consistent readings from sensor to sensor, the wires must be evenly spaced and the stripped portion must be the same. The leads must be held steady during the drying process to avoid fractures in the gypsum.

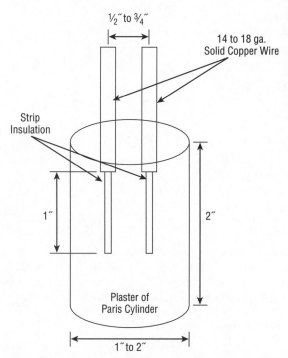

FIGURE 17-3: Gypsum block construction.

Figure 17-4 shows the Hobby-Boards gypsum block. Notice that to keep the wire spacing consistent, Hobby-Boards uses 300-ohm TV "flat" antenna cable. This can be purchased at Radio Shack and most hardware stores. Because the antenna cable only comes with stranded wires, make sure to twist the exposed copper tightly.

FIGURE 17-4: Hobby-Boards gypsum block.

Completing the Hardware

Once you have assembled or purchased the moisture sensor circuit, you will need a housing to protect the circuit if you're going to mount it outdoors. Figure 17-5 shows one method; the circuit is housed in PVC pipe using two 1-inch end caps and a 1¾-inch length of 1-inch pipe. Drill a hole in one end cap for the 1-Wire, power, and sensor cables, similar to the rain gauge circuit board enclosure in Chapter 9. When you're ready to install it outdoors, seal the end with RTV and wrap electrical tape around the joint to keep water out.

FIGURE 17-5: Weatherproof housing for the circuit board.

You can attach the gypsum block to the moisture sensor circuit using up to five feet of standard "zip" cord (also know as lamp cord). If you go any longer than this, you may get strange results due to the oscillation frequency "wandering."

You'll also need to supply +7 to +18 volts DC to the circuit. This can be a simple 12V DC wall-wart transformer. If the cable that connects the circuit to your 1-Wire bus is less than about five feet long, then you can use this cable for both 1-Wire and power. Figure 17-6 shows a block diagram of the overall setup.

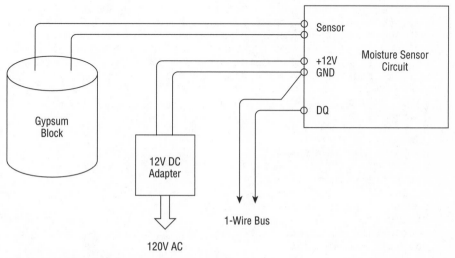

FIGURE **17-6: Moisture detector interconnection.**

Add Moisture Detection to Your Software

Before you place your gypsum block in soil, add the MoistureSensor class to the SimpleWeather project so you can test your circuit.

Add the MoistureSensor Class

Open your `WeatherToys` directory and find the project source code you downloaded from the companion web site. Locate the Chapter 17 directory and the file `MoistureSensor.java`. Copy it to your SimpleWeather ➪ src ➪ SimpleWeather folder.

Modify SimpleWeather

Launch NetBeans and open your SimpleWeather project. Check the list of source files to make sure `MoistureSensor.java` is listed. Next, open `SimpleWeather.java` in the NetBeans editor and scroll down to the `// 1-Wire Devices` comment near line 35. At the end of the list, add a new string to hold the serial number of the moisture sensor device:

```
// 1-Wire Devices
private final String TEMP_SENSOR_ID      = "78000800E97BA210";
private final String WIND_SPD_ID         = "C20000000143E01D";
private final String WIND_DIR_ID         = "3A00000000CE6920";
private final String HUMIDITY_SENSOR_ID  = "3C00000073B6F426";
private final String BARO_SENSOR_ID      = "6D00000073A34F26";
private final String RAIN_SENSOR_ID      = "C20000000143E01D";
private final String LIGHTNING_SENSOR_ID = "C20000000143E01D";
private final String MOISTURE_SENSOR_ID  = "";
```

In the list of class variables, add a new integer variable for the moisture measurement:

```
// class variables
public float temp;
public float windSpeed;
public int windDir;
public float humidity;
public float dewpoint;
public float pressure;
public float rain;
public int lightning;
public int moisture;
```

Add a new MoistureSensor class variable to the list of sensors:

```
// sensors
private TempSensor ts1;
private WindSensor ws1;
private HumiditySensor hs1;
private BaroSensor bs1;
private RainSensor rs1;
private LightningSensor ls1;
private MoistureSensor  ms1;
```

In SimpleWeather's constructor, add the following line to instantiate the MoistureSensor class:

```
// initialize sensors
ts1 = new TempSensor(adapter, TEMP_SENSOR_ID);
ws1 = new WindSensor(adapter, WIND_SPD_ID, WIND_DIR_ID);
hs1 = new HumiditySensor(adapter, HUMIDITY_SENSOR_ID);
bs1 = new BaroSensor(adapter, BARO_SENSOR_ID);
rs1 = new RainSensor(adapter, RAIN_COUNTER_ID);
ls1 = new LightningSensor(adapter, LIGHTNING_SENSOR_ID);
ms1 = new MoistureSensor(adapter, MOISTURE_SENSOR_ID);
```

Finally, near line 223, add the code that requests the moisture value from the MoistureSensor class:

```
// get lightning count
lightning = ls1.getLightning();
System.out.println("Lightning = " + lightning + " SPM");

// get moisture percent
moisture = ms1.getMoisture();
System.out.println("Moisture = " + moisture + "%");

// if it's midnight, reset the highs & lows
if (hour == 0 && minute == 0)
wc.resetHighsAndLows();
```

Find Your Device Serial Number

To complete the SimpleWeather code additions, connect the moisture sensor circuit, gypsum block, and power supply as described earlier, and connect the circuit to your 1-Wire bus. Launch OneWireLister and determine the 1-Wire serial number of the DS2760. Look for the family code '30'. Add the serial number to 1-Wire device list you updated earlier:

```
private final String MOISTURE_SENSOR_ID    = "52000011201CFB30";
```

Testing the Project

With the moisture sensor circuit connected and power applied, run SimpleWeather. There should be an additional output printed on the screen:

```
Time = Sat Apr 22 08:48:07 MST 2006
Temperature = 82.0625 degs F
Wind Speed = 0.0 MPH from the  N
Humidity = 15.80268 %
Dewpoint = 31.695833 degs F
Pressure = 29.887997 inHg
Rain = 0.0 in
Lightning = 0 SPM
Moisture = 6%
```

Don't worry if the value you get isn't the same as shown here; you'll calibrate it later. If your code doesn't compile and run, check your typing and the serial number for the 1-Wire device. If the code runs, but you get strange results, try turning on debug mode to see what's happening in the code.

Walk Through the Code

The code for the moisture sensor is similar to the humidity sensor in that they both use a 1-Wire battery monitor device. The code starts off with two user constants: MOISTURE_OFFSET and MOISTURE_GAIN, which are used to perform a simple calibration (which is described in the next section). Next is the definition of several class variables and constants:

```
public class MoistureSensor
{
  // calibration constants
  private final float MOISTURE_OFFSET = 0.0f;
  private final float MOISTURE_GAIN   = 1.0f;

  // class variables
  private DSPortAdapter adapter;
  private OneWireContainer30 moistureDevice = null;
  private static boolean debugFlag = SimpleWeather.debugFlag;

  // class constants
  private final int VDD_SENSE_AD = 0;
  private final int CURRENT_SENSE_AD = 1;
```

The class constructor makes a new 1-Wire container for the DS2760 using the adapter object and serial number value:

```
  public MoistureSensor(DSPortAdapter adapter, String deviceID)
  {
    // get an instance of the 1-wire device
    moistureDevice = new OneWireContainer30(adapter, deviceID);
  }
```

The getMoisture() method starts by reading the current state of the DS2760 and storing it in the byte array state. Next, the DS2760 is commanded to do a temperature conversion. This isn't required to measure the sensor voltage, but I've included it so that you can also use the DS2760 as a temperature sensor.

Next, an A-to-D conversion is initiated on the current sense channel (the one you are using to measure the current drawn by the oscillator). The result is stored in the class variable Vad. The supply voltage is also measured, and is only used when debug mode is enabled so that you can check the supply voltage remotely:

```
  public int getMoisture()
  {
    double moisture = -999.9;

    if (moistureDevice != null)
    {
      if (debugFlag)
      {
        System.out.print("Moisture: Device = " + moistureDevice.getName());
        System.out.print("  ID = " + moistureDevice.getAddressAsString() +
                                  "\n");
```

```
  }
  try
  {
    // read 1-wire device's interal temperature sensor
    byte[] state = moistureDevice.readDevice();
    moistureDevice.doTemperatureConvert(state);
    double temp = moistureDevice.getTemperature(state);

    // Read moisture sensor's output voltage
    moistureDevice.doADConvert(CURRENT_SENSE_AD, state);
    double Vad = moistureDevice.getADVoltage(CURRENT_SENSE_AD, state);

    // Read the moisture sensor's power supply voltage
    moistureDevice.doADConvert(VDD_SENSE_AD,state);
    double Vdd = moistureDevice.getADVoltage(VDD_SENSE_AD, state);
```

The DS2706 is designed to monitor battery charge and discharge current. In your application, the current required by the oscillator is considered to be a current "draw," so the DS2760's measured value is negative, and the full-scale output is 2.56 volts. Dividing the measured output by –2.56 and then multiplying by 100, the A-to-D output is converted to a 0 to 100% reading:

```
    // Convert to percentage of full scale (2.56v)
    moisture = Vad / -2.56  * 100.0;
```

To compensate for circuit variations, an offset and gain factor is applied to the result that provides a simple calibration:

```
    // apply the calibration scale factor and offset
    moisture = (moisture + MOISTURE_OFFSET) * MOISTURE_GAIN;
```

As with most of the sensor classes, all intermediate measurement data can be viewed by enabling debug mode. If you forgot how to turn on debug mode, check out Chapter 7.

```
    if (debugFlag)
    {
      System.out.println("Supply Voltage = " + Vdd + " Volts");
      System.out.println("Sensor Output  = " + Vad + " Volts");
      System.out.println("Temperature    = " + temp + " C / " +
                         ((temp * 9/5) + 32) + " F");
      System.out.println("Gain           = " + MOISTURE_GAIN);
      System.out.println("Offset         = " + MOISTURE_OFFSET);
      System.out.println("Comp Moisture  = " + moisture + " %\n");
    }
  }
  catch (OneWireException e)
  {
    System.out.println("Error Reading Moisture Sensor: " + e);
  }
}
```

After the conversion to a percent, the integer value of the percent is returned to the calling method:

```
    // return integer value
    return (int)moisture;
  }
}
```

Calibration and Installation

Before burying your gypsum block, you can do a simple calibration and test. This will verify that the soil moisture circuit is working properly and provide a rough calibration so you can compare readings from multiple sensors.

Calibration

A simple calibration can be performed to eliminate variations caused by the moisture sensor circuitry. First, make sure SimpleWeather runs and displays a value of 0 +/− 20%. If your starting value is higher that this, check the circuitry and wiring connections. If you built your own board, try cleaning it with alcohol and drying it well.

Disconnect your gypsum block from the sensor terminals. With open sensor terminals, the oscillator will run at its slowest speed, simulating a fully dry gypsum block.

Tip

If you need debug mode turned on for a single file only, you can modify the code to set the `debugFlag` to `true` in the file itself. For example, to enable debug mode in `MoistureSensor.java` only, scroll to the top of the file and set `debugFlag` to `true` like this: `private static boolean debugFlag = true`. Don't forget to change it back when you're done.

Enable debug mode for your project and run SimpleWeather. Note the compensated moisture value. In the following example, it measures about 6.0%:

```
Moisture: Device = DS2760   ID = D5000011201CEF30
Supply Voltage = 4.82144 Volts
Sensor Output  = -0.15375 Volts
Temperature    = 28.375 C / 83.075 F
Gain           = 1.0
Offset         = -0.0
Comp Moisture  = 6.005859375%
```

Stop SimpleWeather and open `MoistureSensor.java` in the NetBeans editor. Enter the *negative* value of the measured moisture in the user constant `MOISTURE_OFFSET`. In this example, −6.0 is entered:

```
// calibration constants
private final float MOISTURE_OFFSET = -6.0f;
private final float MOISTURE_GAIN   = 1.0f;
```

Next, short the sensor terminals on the circuit board with a small length of wire. This simulates full scale, resulting in the oscillator running at max speed. Re-run SimpleWeather and note the new compensated moisture value:

```
Moisture: Device = DS2760  ID = D5000011201CEF30
Supply Voltage = 4.82144 Volts
Sensor Output  = -2.56 Volts
Temperature    = 30.875 C / 87.575 F
Gain           = 1.0
Offset         = -6.0
Comp Moisture  = 94.0%
```

Divide 100% by the new moisture value to get the gain calibration factor. In this example, 100% / 94.0% = 1.064. Enter the calculated number in the user constant MOISTURE_GAIN:

```
// calibration constants
private final float MOISTURE_OFFSET = -6.0f;
private final float MOISTURE_GAIN   = 1.064f;
```

Stop and restart SimpleWeather. The moisture value should now read close to 100%:

```
Moisture: Device = DS2760  ID = D5000011201CEF30
Supply Voltage = 4.82144 Volts
Sensor Output  = -2.56 Volts
Temperature    = 29.875 C / 85.775 F
Gain           = 1.064
Offset         = -6.0
Comp Moisture  = 100.01600098609924%
```

You can now remove the jumper and re-attach the gypsum block.

Installation

The gypsum block is typically installed several inches to 1 foot below the surface, at the root level of the plants for which you want to monitor the moisture level. Follow these steps to get good readings:

1. Soak the gypsum block overnight in clean tap water to fully saturate the block.

2. Dig a hole to the desired depth, saving the removed soil in a bucket.

3. Place the gypsum block at the bottom of the hole.

4. Add water to the removed soil until it is wet enough to pour.

5. Pour a little of the mud into the hole around the block and lift the block up a little bit to allow the mud to get underneath.

6. Pour the remaining mud around the block, using a small stick or screwdriver to stir the mud to make sure there are no bubbles. The goal is to have intimate contact between the gypsum and the surrounding soil.

7. Allow the installation to sit for a few hours, and then add some additional soil to the top to make sure there is not a depression where the block is installed. Otherwise, water may pool and give false readings.

8. Connect the block to the moisture sensor electronics.

It may take several days for the gypsum block to settle in and give stable readings. Experimenting with my sensor, I get readings in the 20% to 40% range for "moist" soil. My plants begin to wilt when the moisture level drops below about 10%. Your results may vary.

Expanding Your Project

Although the gypsum block is used primarily for measuring ground moisture, there are other applications for which it will work equally well. This section presents a few ideas to get you thinking.

Soil Moisture Detection

To expand on soil moisture, you could install multiple blocks in your potted plants, both indoor or outdoors. Set up SimpleWeather to monitor each sensor and display a warning when the plants need watering. You could use the X10 system described in the previous chapter to turn on a watering system.

Rainfall Detector

Another application is to install the block above ground where it is subject to rain. Because it takes a while for the sensor to dry out (hours to days, depending on the humidity), you can use this as a rainfall detector "with memory." Based on the moisture level, you can decide if you should turn on your sprinklers.

Basement Water Detector

A common problem in some areas is basement flooding. By installing the gypsum block on top of the basement floor, or slightly below, you have a moisture detector. Once water hits the block, it quickly wicks up, causing the moisture sensor to detect the presence of water. It could sound an alarm or using X10 from Chapter 16, turn on a pump.

Wrap Up

Here are a couple of last minute notes on using the gypsum block. When exposed to moisture, over time the gypsum will slowly dissolve. In a typical soil installation, it will last from one to five years. With this in mind, you should periodically check the readings to make sure the block hasn't failed.

Another potential issue is mounting the gypsum block too close to a ground rod. Because the current flows between the electrodes in the block, if you have high conductivity soil, some of the current could bypass the block and flow back to your 1-Wire ground through the ground rod. The solution is to install the gypsum block as far away from the ground rod as possible.

You now have a full complement of weather sensors to measure many aspects of the weather. In the next chapter, you read about another option to control devices based on your measurements.

Build a Smart Sprinkler Timer or Home Thermostat

I'm sure you've seen it more than once: it's pouring rain and you see someone's sprinklers are running full blast, with water running down the street. In the drier parts of the country, water is starting to become a precious resource. Minimizing your water usage not only makes good sense, but your lawn and garden will appreciate it too.

I have to adjust the timer for my irrigation system several times a year. In the winter, I shorten the duration the sprinklers run. If I don't remember to set it back in the summer months, my grass reminds me by turning brown. It sure would be nice have this done automatically, based on the time of year and current weather conditions. Enter 1-Wire.

In this project you learn how to control low-voltage equipment like your sprinkler timer and home Heating, Ventilation, and Air Conditioning system (HVAC). You will use 1-Wire controlled relays capable of replacing your sprinkler timer or thermostat. And because it is controlled by 1-Wire, it is easy to connect it to your existing 1-Wire weather station.

Once you have the relay board interfaced to your system, you then use SimpleWeather to develop the "rules" that turn on or off your heater or set how long your sprinklers will run.

Project Overview

To interface your 1-Wire bus to your home sprinklers or HVAC, you will need to build a 1-Wire relay control module that will replace your thermostat or sprinkler timer. Then using SimpleWeather and a new relay controller class, you can activate or deactivate the relays programmatically. To set the conditions to turn on or off the relays, you will develop rules for activating the device similar to the rules you developed for the X10 system in Chapter 16.

For each sprinkler station, you will need a relay channel. If your current sprinkler system has eight stations, then you will need eight channels. For HVAC applications, most home and small office thermostats require three channels, one for heat, one for cooling, and one to control the fan.

How It Works

If you're not familiar with relays, you can look at them as simply an electrically activated switch: apply power to the "coil" terminals, and the switch terminals will be connected together. Remove the power, and the switch opens. What's actually happening is pretty simple. The power to the coil terminals is actually powering a small coil that acts as an electro-magnet. The magnetism pulls on a metal arm that moves and connects a set of contacts. When power is removed, the magnet is de-energized and a small spring pulls the relay arm away from the contacts, opening the circuit.

In the circuit shown in Figure 18-1, one channel of the relay module is shown plus the common power supply. If you're building your own, you can have almost as many channels as you want. The only limitation is the number of relays that U1 can power.

U1 is a 5-volt regulator that takes unregulated DC power and regulates it to a constant 5 volts at 1 amp to power the rest of the circuitry. You can use a power supply that supplies +7 to 18 volts.

U2 is a DS2406 1-Wire Addressable Switch. The 1-Wire connection to pin 2 provides power and bi-directional communication. By sending the correct command sequence, the output pin 3 can be enabled or disabled. This is an active low device. When commanded on, it pulls the base of Q1 to ground, activating the transistor and turning on the relay. When commanded off, the open collector output is pulled high by resistor R1, which deactivates Q1.

Relay RLY1 is a single-pole single-throw (SPST) relay with a 5V coil. It can switch up to 3 amps at 250V AC or 3 amps at 30V DC. Using a relay, the output is completely isolated from the 1-Wire bus. LED1 is used as a visual indication that the relay is activated.

Building the Hardware

Before you start to build the relay board, look at your intended application and decide how many relays you need. Will you need more relays later? What other devices could you control with relays? Once you decide on the number of channels, you can look at the options you have.

Options

You have three options with the relay board:

- **Option 1: Buy a Complete Assembly.** Hobby-Boards offer a 4-channel relay board as a completed module. If you need more than four channels, you can buy additional 4-channel modules and connect them together.

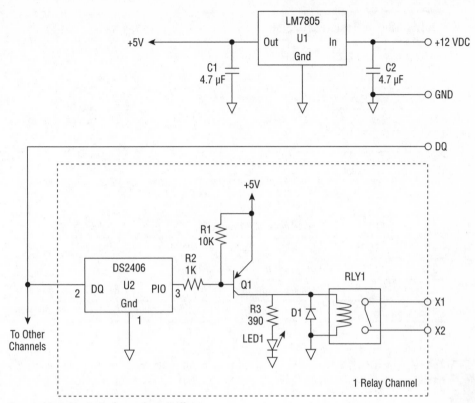

FIGURE 18-1: Basic 1-channel controller.

- **Option 2: Build the Kit.** Hobby-Boards offers the 4-channel relay board as a kit if you'd like to build it yourself.

- **Option 3: Build it from Scratch.** Using the schematic in Figure 18-1 and the parts list in Table 18-1, you can build your own relay card from scratch. The advantage is that you can add as many channels as you'd like, up to 10, all on one circuit card. This project can be built without surface-mount components, so it is possible to use perf board and point-to-point wiring.

Parts List

If you're building the kit from scratch, Table 18-1 lists the parts in the schematic for one channel. The items marked with an asterisk (*) need to be duplicated for each additional channel. Using screw terminals to connect the external wiring makes connecting it much easier. You'll also need a data cable to connect your relay board to your 1-Wire bus.

Table 18-1 Basic Parts List

Qty	ID	Description	Vendor
1	U1	LM7805 5Volt Regulator	Digikey
1*	C1, C2	4.7 μF 20V Electrolytic Capacitor	Digikey
1*	U2	DS2406 1-Wire Addressable Switch	Digikey, Maxim
1*	R1	10K Ohm Resistor 1/10w	Digikey
1*	R2	1K Ohm Resistor 1/10w	Digikey
1*	R3	390 Ohm Resistor 1/10w	Digikey
1*	Q1	2N4403 Transistor	Digikey
1*	D1	1N914 or BAS16 Diode	Digikey
1*	LED1	Red LED, Axial Lead or Surface Mount	Digikey
1*	RLY1	Potter and Brumfield T77S1D3-05	Digikey
3*	—	Terminal Screw Connectors	Radio Shack 276-1388
1	—	Heat Sink for U1	Digikey, Radio Shack

Warning The relay module is ESD sensitive. Always use an anti-static wrist strap when handling your module.

Building the Board

Because this design can be built without surface-mount components, you can build it on experimenter's perf board. If you're experienced at soldering surface-mount components, you can use the Hobby-Boards PC board.

Each relay pulls about 50 milliamps when activated, so if you add some margin, the regulator should be able to power at least 10 relays simultaneously. Make sure you install a heat sink on the regulator to prevent overheating.

The relays specified in the parts list are capable of switching up to 3 amps at 250V AC, so you could actually switch AC line loads. I strongly suggest you use the X10 system described in Chapter 16 if you need to control AC-powered devices. Because the relays are in close proximity to the DS2406, the switching noise generated by the relay opening and closing could be coupled into the 1-Wire bus.

Warning

Because the wiring on the board is exposed, it is not recommend that you use the relay board to switch 120V AC equipment.

Testing the Board

Once you have completed building your relay board, you can test each channel by connecting +12 volts DC to the power inputs. At this point, no relays should activate and all LEDs should remain off. Carefully connect pin 3 of the DS2406 for each channel to ground. The corresponding relay and LED should activate. If not, check your work. If none of the relays will activate, check the power supply and regulator with a voltmeter.

Figure 18-2 shows the Hobby-Boards 4-channel relay module. Notice the use of heavy-duty screw terminals for attaching to the relay outputs and +12V power supply.

FIGURE 18-2: Completed Hobby-Boards 4-channel relay board.

Add Relay Control to Your Software

Once you have your board complete and tested, you're ready to add the code to SimpleWeather to control the relays. In this section, I'm going to assume that you are using a 4-channel relay board. If you've built your own board with more channels, or are using more than one of the 4-channel boards, you will have to adjust the code for the total number of channels you are using.

Note This project assumes that you are either building the sprinkler timer *or* the HAVC controller. The code provided will do both, but you'll have to have enough relay channels to control both and combining the code is left up to you.

Add the OWSwitch Class

Open your `WeatherToys` directory and find the project source code you downloaded from the companion web site. Locate the Chapter 18 directory and the file `OWSwitch.java`. Copy it to your SimpleWeather ⇨ src ⇨ SimpleWeather folder.

Modify SimpleWeather

Launch NetBeans and open your SimpleWeather project. Check the list of source files to make sure `OWSwitch.java` is listed. Next, open `SimpleWeather.java` in the NetBeans editor and scroll down to the `// 1-Wire Devices` comment near line 35. At the end of the list, add four new String variables to hold the serial numbers of the four 1-Wire switches:

```
// 1-Wire Devices
private final String TEMP_SENSOR_ID       = "3F00080008F42810";
private final String WIND_SPD_ID          = "87000000013E301D";
private final String WIND_DIR_ID          = "B200000000F3B620";
private final String HUMIDITY_SENSOR_ID   = "8A00000072E88D26";
private final String BARO_SENSOR_ID       = "8A00000072E88D26";
private final String RAIN_COUNTER_ID      = "87000000013E301D";
private final String LIGHTNING_COUNTER_ID = "87000000013E301D";
private final String MOISTURE_SENSOR_ID   = "D5000011201CEF30";
private final String RELAY_1_ID           = "";
private final String RELAY_2_ID           = "";
private final String RELAY_3_ID           = "";
private final String RELAY_4_ID           = "";
```

Although it isn't really a sensor, in the list of sensors near line 84, add one `OWSwitch` variable that you will use to access the relays:

```
// sensors
private TempSensor ts1;
private WindSensor ws1;
private HumiditySensor hs1;
private BaroSensor bs1;
private RainSensor rs1;
```

```
private LightningSensor ls1;
private MoistureSensor ms1;
private OWSwitch relay;
```

In the `// initialize sensors` block near line 163, add the code to instantiate the class. Notice that the 1-Wire device serial number is not passed like the rest of the sensors. Instead, the 1-Wire serial number for the relay you want to activate will be passed in the call that activates or deactivates the relay:

```
// initialize sensors
ts1 = new TempSensor(adapter, TEMP_SENSOR_ID);
ws1 = new WindSensor(adapter, WIND_SPD_ID, WIND_DIR_ID);
hs1 = new HumiditySensor(adapter, HUMIDITY_SENSOR_ID);
bs1 = new BaroSensor(adapter, BARO_SENSOR_ID);
rs1 = new RainSensor(adapter, RAIN_SENSOR_ID);
ls1 = new LightningSensor(adapter, LIGHTNING_SENSOR_ID);
ms1 = new MoistureSensor(adapter, MOISTURE_SENSOR_ID);
relay = new OWSwitch(adapter);
```

In the `mainLoop()`, after the call to the last weather sensor and just before the code to reset the highs and lows, add the code that you will use to test your relays:

```
// get moisture percent
moisture = ms1.getMoisture();
System.out.println("Moisture = " + moisture + "%");

// turn on a relay
relay.setSwitchState(RELAY_1_ID, false);
relay.setSwitchState(RELAY_2_ID, false);
relay.setSwitchState(RELAY_3_ID, false);
relay.setSwitchState(RELAY_4_ID, false);

// if it's midnight, reset the highs & lows
if (hour == 0 && minute == 0)
  wc.resetHighsAndLows();
```

In the preceding code, `setSwitchState()` is called passing the device ID of the desired 1-Wire switch and a Boolean that sets the state: `false` for off and `true` for on. By doing it this way, you only need one instance of the relay class.

Find Your Device Serial Numbers

Launch OneWireLister and look for devices with the family code of '05'. There should be one for each of the relay channels. Copy each serial number to your list of 1-Wire devices. At this point, you won't know which device controls which relay; you'll find that in the next section:

```
private final String RELAY_1_ID = "190000002A2A3E05";
private final String RELAY_2_ID = "180000002A30C805";
private final String RELAY_3_ID = "BC0000002A2A1005";
private final String RELAY_4_ID = "550000002A324705";
```

Testing the Project

Connect power to the relay board and connect it to your 1-Wire bus. In your code, set the first relay to turn on by setting the "on/off" variable to `true`:

```
// turn on a relay
relay.setSwitchState(RELAY_1_ID, true);
relay.setSwitchState(RELAY_2_ID, false);
relay.setSwitchState(RELAY_3_ID, false);
relay.setSwitchState(RELAY_4_ID, false);
```

Run SimpleWeather. One of the LEDs should turn on and you should be able to hear the corresponding relay click as it activates. Note which channel LED comes on. For example, if LED 3 turns on, you know that the ID for relay 1 is really relay 3. Stop SimpleWeather and modify the code to turn each relay on one-by-one, noting which relay actually turns on. After you have done all the relay channels, re-arrange the device ID numbers so that `RELAY_1_ID` is actually relay 1. If one or more of your relay channels doesn't activate, check the device ID.

Controlling Your HVAC

With your relay module working and checked out, you are ready to connect it to your HVAC system. If you aren't familiar with thermostat wiring, go slow and double-check your work. Also, many home HVAC systems have a wiring diagram on the indoor unit or just inside the front cover.

Connecting Your HVAC System

Most home and small office HVAC systems use a 4-wire low-voltage thermostat. One wire is for power, one wire is for heat, one for cooling, and one is to activate the fan. If your system does not have cooling capability, you may only have two or three wires. As shown in Table 18-2, there is a standard color scheme for wiring thermostats. However, before you start disconnecting any wires, make sure that you write down which color goes where. Many thermostats also have labels, so compare the wire color and label to the table to make sure your HAVC system is wired to the standard.

Table 18-2   Common Thermostat Wiring and Color Code

Function	Wire Color	Thermostat Marking
Control Power (Transformer Hot)	Red	R, RH, RC
Heat	White	W
Cooling	Yellow	Y
Fan	Green	G
Transformer Common	Blue, Brown, or Black	C

Don't assume the wire colors are correct. If you have a heat pump system or an uncommon heating/cooling system, make sure you refer to the system's manual before disconnecting the wiring or making any changes. Always disconnect power to your HVAC system before disconnecting wires. Even though 24 volts won't kill you, you could get a surprising shock.

Referring to Figure 18-3, to activate the heater, the red wire (control power) is connected to the white wire (heater). To activate cooling, the red wire (control power) is connected to the yellow wire (cooling). The fan is activated by connecting the red wire to the green wire. Note that on some systems, the cooling (yellow) wire only turns on the A/C compressor and the fan must be turned on separately. Many installations do not include the transformer common wire.

 Running your A/C compressor without the fan could damage the A/C compressor. When activating the cooling system and writing your rules, make sure the fan turns on with the A/C.

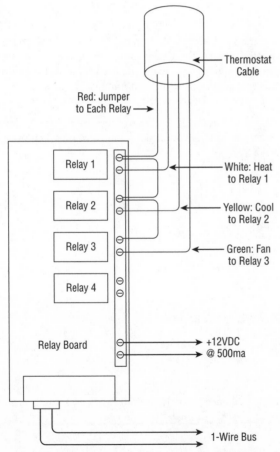

FIGURE **18-3: Relay board wiring for a typical thermostat.**

Testing Your Wiring

After connecting the thermostat wire to your relay board, it's time to test your wiring. Activate one relay at a time like you did when you tested the relay card:

```
// turn on a relay
relay.setSwitchState(RELAY_1_ID, true);  // Heating
relay.setSwitchState(RELAY_2_ID, false); // Cooling
relay.setSwitchState(RELAY_3_ID, false); // Fan
relay.setSwitchState(RELAY_4_ID, false); // Spare
```

Run SimpleWeather and the relay set to true should turn on, activating the desired function. To help remember what relay does what, add a comment to each relay line as you go. When testing the cooling relay, if only the A/C compressor turns on, you will have to activate both the cooling and fan relay simultaneously when turning on the A/C. Some of the newer HVAC systems are wired to do this for you.

Add the Code

Once the relay board and proper operation of your HVAC system have been tested, you're ready to start writing the rules. But before you do, there's one more issue to cover. Because most HVAC systems don't like to be toggled on and off repeatedly, you need some sort of protection to keep SimpleWeather from switching faster than once every 5 minutes. Otherwise, normal variations in the measurements could cause the A/C to toggle on and off once per minute.

To fix this problem, you simply "wrap" your rules and relay control statements inside a Java mod statement that is set for 5-minute intervals. Modify your code as shown, also adding three new variables you will use for the relay rules:

```
// thermostat control
if (minute % 5 == 0)
{
  boolean heaterOn = false;
  boolean coolingOn = false;
  boolean fanOn = false;

  // rules go here


  // turn on a relay
  relay.setSwitchState(RELAY_1_ID, heaterOn);  // Heater
  relay.setSwitchState(RELAY_2_ID, coolingOn); // Cooling
  relay.setSwitchState(RELAY_3_ID, fanOn);     // Fan
  relay.setSwitchState(RELAY_4_ID, false);     // Spare
}
```

With this code, the rules are only evaluated once every five minutes. You may want to consult your HVAC system instruction manual to see if this delay is sufficient.

Writing the Thermostat Rules

The "rules" are formed similar to the rules used in the X10 project in Chapter 16. If you need a refresher or skipped that project, go back and read the section now. Pay particular attention to the sidebar about Java's if/else statement.

When you develop your rules, be careful of conflicting rules! You don't want to command both your heater and your cooling to turn on at the same time. Also, it is best to have a "dead" zone; a temperature range between heat and cool where neither are turned on. Otherwise, you will switch from heat to cool and then back to heat over and over again.

Here's an example of a simple rule that turns on the heat if the temperature is less than 68 degrees, turns on just the fan if the temperature is greater that 75 degrees but less than 78, or turns on the cooling if the temperature is greater than 78:

```
// thermostat control
if (minute % 5 == 0)
{
  boolean heaterOn = false;
  boolean coolingOn = false;
  boolean fanOn = false;

  // rules go here
  if (temp < 68)
    heaterOn = true;

  if (temp > 75 && temp < 78)
    fanOn = true;

  if (temp > 78)
    coolingOn = true;

  // turn on a relay
  relay.setSwitchState(RELAY_1_ID, heaterOn);   // Heater
  relay.setSwitchState(RELAY_2_ID, coolingOn);  // Cooling
  relay.setSwitchState(RELAY_3_ID, fanOn);      // Fan
  relay.setSwitchState(RELAY_4_ID, false);      // Spare
}
```

If you have a separate indoor temp sensor, you can include that in the rules too. For example, the following rules turn on the cooling at a different inside temperature based on the outside temperature:

```
if (in_temp > 82)
  coolingOn = true;

if (in_temp > 78 && out-temp > 85)
  coolingOn = true;
```

Like the X10 project, you can use any of the weather data variables in your rules. For example, some houses in drier climates have both a "swamp" cooler and refrigerated air. Based on the current dewpoint, your smart thermostat could pick the proper cooling mode for the day. Because everything is controlled by your code, there are virtually no limits to what you can do.

Caution If your computer should happen to freeze with the heating or cooling on, it could stay on. Until you develop full confidence in your system, don't run it when you're not home. During system development (and maybe permanently), consider wiring your old thermostat in series with the relay board as a "backup" to turn it off if your house exceeds some threshold. If you live where it could freeze, you may want to wire the backup thermostat to activate your heater to keep your pipes from freezing.

Controlling Your Sprinklers

Low-voltage controlled sprinklers operate on 24 volts AC. By applying power to the valve, the sprinklers turn on. Depending on the kind of irrigation system and the soil type, most sprinklers run for 5 to 60 minutes, and then the next set of sprinklers or *station* is turned on. This sequence continues until all the stations have run for the programmed time.

The method that was used in the HVAC application to control the relays would require keeping track of which sprinkler is on, how long it has been on, and the next station that needs to run. This can make the code complicated and messy. So for this application, you will use a different approach. Once the designated start time is reached, the time for each sprinkler will be determined and stored in an array. Once all the values are stored, a separate timer "program" will be launched in its own thread, freeing up SimpleWeather to do its weather collection. The timer thread will sequence through each station for the desired on time, and then automatically step to the next one.

Connecting the Sprinklers

Figure 18-4 shows the wiring for a typical 4-station sprinkler timer. To power the valves, you will need a source of 24 volts AC at a minimum of 500 milliamps. If you're adapting an existing system, you may be able to use the old transformer. If not, many hardware stores and irrigation supply stores offer 24V wall-wart style power supplies specifically for this application.

Add the Code

First, open the `OWSwitch.java` file in the NetBeans editor and set the user constant `NUM_SWITCHES` to the number of stations you have (obviously, you will have that many relay channels):

```
public static final int NUM_SWITCHES = 4;
```

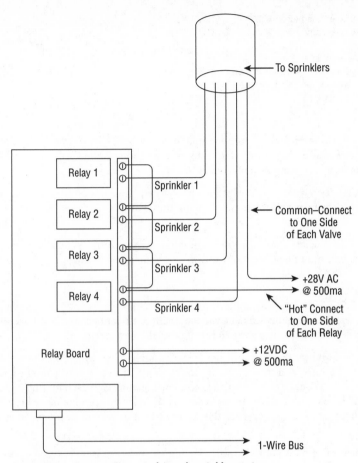

FIGURE 18-4: Connecting a 4-channel sprinkler system.

Next, add the following basic sprinkler timer code after the last weather sensor (in this example, moisture):

```
// get moisture percent
moisture = ms1.getMoisture();
System.out.println("Moisture = " + moisture + "%");


// sprinkler timer
if (hour == 8 && minute == 45) // start time
{
  // setup the switch program
  relay.setDeviceTime(0, RELAY_1_ID, 1);
  relay.setDeviceTime(1, RELAY_2_ID, 1);
```

```
relay.setDeviceTime(2, RELAY_3_ID, 1);
relay.setDeviceTime(3, RELAY_4_ID, 1);

// start the sprinklers
Thread relayRunner = new Thread(relay);
relayRunner.setName("timer");
relayRunner.start();
}
```

In the preceding code, the start time is set in the statement `if(hours == 8 && minutes == 45)`. This is where you set the desired start time in hours and minutes. Because there is no AM or PM, the hours are set using 24-hour values.

Next you'll see four `setDeviceTime()` calls. This is what sets the duration for each station. Pass the device ID for the relay and the duration in minutes. For the initial test, the values are set for 1 minute.

The next block of code is what creates a new thread, starts it, and returns. When started, the sprinkler timer code will run independent of the SimpleWeather `mainLoop()`.

Testing the Code

If you have already connected your sprinklers, you may want to disconnect the 24V transformer from the wall for this test to keep your sprinklers from actually turning on.

Check the current time and set the start time hours and minutes values for a time that will occur in the next few minutes, and then run your project. When the start time is reached, the LEDs on the relay board should light up in succession starting at relay 1, each for one minute. If nothing turns on, double-check the start time and that the relay board is connected to +12V. If the LEDs don't light up in sequence, check the device serial number as presented earlier in this chapter.

You now have a functioning basic sprinkler timer. By adjusting the duration for each station ("hard-coding the value"), you can start using your sprinkler timer. You may want to try it out for a few days with hard-coded values to make sure it works as expected before going to the next step.

Writing the Sprinkler Rules

Once you have the relays working correctly, you are ready to start developing the rules. If you skipped the rules discussion for the HVAC controller, go back and read them now. There are so many options for your rules, I can't begin to list them all here. So, instead I'll show you an example and you can experiment with others.

In the following code example, each station shows a different option for setting the duration:

```
// sprinkler timer
      if (hour == 10 && minute == 30)
      {
```

```
System.out.println("Sprinkler Start Time Reached");
// station 1
int station1 = 10;      // minimum of 10 minutes
if (temp > 70)
  station1 += 5;        // if temp is > 70, add 5 minutes
else if (temp > 85)
  station1 += 10;       // if temp is > 85, add 10 minutes

// station 2
int station2 = 30;      // minimum is 30 minutes
if (date.getMonth() > 5 && date.getMonth() < 10)   //summer months
  station2 += 30;        // during summer, add 30 minutes

// station 3
int station3 = (int)(20.0f + (temp - 50.0f)/2.0f);

// station 4
int station4 = 5;       // normal is 5 minutes
if (moisture < 20)
  station4 = 15;        // if soil is dry, run 15 minutes


// setup the switch program
relay.setDeviceTime(0, RELAY_1_ID, station1);
relay.setDeviceTime(1, RELAY_2_ID, station2);
relay.setDeviceTime(2, RELAY_3_ID, station3);
relay.setDeviceTime(3, RELAY_4_ID, station4);

if (wc.getRain24Float() < 0.1)
{
  System.out.println("Starting Sprinklers");

  // start the web server
  Thread relayRunner = new Thread(relay);
  relayRunner.setName("relayRunner");
  relayRunner.start();
}
}
```

The method for station 1 uses a simple if statement. First, the default duration of 10 minutes is assigned to an int variable station1. If the current temperature is greater than 70 degrees, then 5 minutes is added to station1, or if the temperature is greater than 85, then 10 minutes is added.

Station 2 is adjusted by the month. The variable station2 is set for default value of 30 minutes. If the month is greater than 5 and less than 10 (June through September), it is assumed to be a summer month and station2 is set for 60 minutes.

Station 3 calculates the duration by taking the current temperature minus 50 degrees, adding 20, and dividing it by 2. This causes the duration to increase proportionally as the temperature rises. This calculates to 0 minutes at 30 degrees and 35 minutes at 100 degrees. The result of the calculation must be cast as an integer before assigning its value to station 3.

Station 4 uses the measured value from the soil moisture sensor. A default value of 5 minutes is assigned to station 4. If the soil is very dry, the station duration is set for 15 minutes.

In this example, the rain is checked before starting the timer thread. If the rain in the last 24 hours is greater than 0.1 inches, the call to start the timer thread is skipped, preventing the sprinklers from running for the day.

Because a method doesn't currently exist in the WeatherCruncher class that returns a float value for the 24-hour rain amount, a new method has to be added. Open `WeatherCruncher.java` in the NetBeans editor, and in the list of rain "getters" near line 494, add the following:

```
// Rain -------------------------------------
public String getRain()
{
  return formatValue(rain, 2);
}

public String getRain24()
{
  return formatValue(rain24, 2);
}

// added for sprinkler timer
public float getRain24Float()
{
  return rain24;
}
```

This is just one example of the many ways you can develop the rules. They can be as long and complex as you want, and each station can have a completely separate rule. This example only shows one start time; with a little coding, you can have multiple start times, or even different start times for different days or stations.

Walk Through the Code

Before I wrap up this chapter, take a quick look at the OWSwitch class. The first thing you may notice is that this class implements the Runnable class. This is the mechanism that allows you to launch a separate thread to control the relays in the sprinkler timer.

There is one USER_CONSTANT to set the number of relays your system uses. This is used later in the setDeviceTime() and run() method:

```
public class OWSwitch implements Runnable
{
  // user constants
  public static final int NUM_SWITCHES = 4;
```

```
// class variables
private DSPortAdapter adapter;
private OneWireContainer05 switchDevice = null;
private String[] deviceID;
private int[] runTime;
private static boolean debugFlag = SimpleWeather.debugFlag;
```

The class constructor receives an instance of the 1-Wire adapter when it is constructed so it will know how to communicate with the 1-Wire bus. It also declares two arrays, one for the device IDs and one for the desired run time for each of the relays. Both of these arrays are used in the `run()` method:

```
public OWSwitch(DSPortAdapter adapter)
{
  this.adapter = adapter;
  deviceID = new String[NUM_SWITCHES];
  runTime = new int[NUM_SWITCHES];
}
```

The `setSwitchState()` method is how you turn on or off a particular relay. This is what you used to control the relays in the thermostat code. To switch a relay, the 1-Wire device is passed in along with a Boolean that determines whether you want the switch to turn on or off the relay. The 1-Wire Java API method `readDevice()` gets the current device state, `setLatchState()` sets the state to on or off, and `writeDevice()` sends the command to the desired 1-Wire device. For more information on these methods, look at `OneWireContainer05` in the 1-Wire API.

```
public void setSwitchState(String deviceID, boolean latchState)
{
  // get instances of the 1-wire devices
  switchDevice = new OneWireContainer05(adapter, deviceID);

  try
  {
    ...

    byte[] state = switchDevice.readDevice();
    switchDevice.setLatchState(0, latchState, false, state);
    switchDevice.writeDevice(state);
  }
  catch (OneWireException e)
  {
    System.out.println("Error Setting Switch: " + e);
  }
}
```

The `getSwitchState()` method wasn't used in this chapter, but is provided for applications where you might need to read the current switch state. Simply pass in the 1-Wire switch's device ID and this method will return a Boolean indicating the switch state, `true` for on or `false` for off:

```
public boolean getSwitchState(String deviceID)
{
  // get instances of the 1-wire devices
  switchDevice = new OneWireContainer05(adapter, deviceID);

  try
  {
    ...
    byte[] state = switchDevice.readDevice();

    boolean latchState = switchDevice.getLatchState(0, state);

     if (debugFlag)
       System.out.println("The Latch Reads: " + latchState);

    return latchState;
  }
  catch (OneWireException e)
  ...
}
```

In the sprinkler timer code, you set the desired duration for each sprinkler by calling the `setDeviceTime()` method. It simply stores the device ID and corresponding time in an array for the `Run()` method. There is also a check to make sure that the duration is not less than zero. If it is, the value is set to zero:

```
public void setDeviceTime(int deviceNum, String serialNum, int minutes)
{
  if (minutes < 0)
    minutes = 0;

  deviceID[deviceNum] = serialNum;
  runTime[deviceNum] = minutes;
}
```

The `run()` method in this class can be launched in its own separate thread. This is how SimpleWeather kept running while the sprinklers continued with their program. Once the durations are loaded using the `setDeviceTime()` method, the `run()` method is called. This starts a separate thread that loops through the array, turning on each relay one at a time. In the loop, after activating the relay, the method sleeps for the duration value stored in the array converted to milliseconds:

```
public void run()
{
  int i;

  for (i=0; i< NUM_SWITCHES; i++)
```

```
    {
      setSwitchState(deviceID[i], true);

      try
      {
        //convert time to minutes and sleep
        Thread.sleep(runTime[i] * 1000 * 60);
      }
      catch (InterruptedException e)
      {}

      setSwitchState(deviceID[i], false);
    }
  }
```

Wrap Up

This chapter focused on building either the HVAC system controller or the sprinkler timer. It isn't too hard to integrate the two projects and control both sprinklers and your HVAC system.

With the ease of 1-Wire, you can easily adapt this project to multiple systems. For example, some of the larger two-story houses have separate HVAC systems for upstairs and downstairs. You could have a separate relay board replace each thermostat and have one computer act as the master controller. It could activate one HVAC system based on data from both sets of sensors, taking into account the time of day and which day it was, making a very intelligent system. If you have a whole-house fan, you can use the relay controller to control another high-power relay that turns the fan on or off, adding another option to your intelligent controller. The possibilities are endless!

There is a lot to digest in this chapter and you really covered a lot of topics. Your code is starting to become more complex, and I bet you're starting to see the possibilities. Even though this chapter focused on HVAC and sprinklers, you can adapt this project to about anything that can be controlled with a relay. Let your imagination run wild and see what else you can control.

Add a User Interface to SimpleWeather

in this chapter

- ☑ Add a GUI to Your Code
- ☑ Customize It
- ☑ Build a Stand-Alone Version of SimpleWeather
- ☑ Double-Click Your Application
- ☑ Wrap Up SimpleWeather

In all of the projects so far, you've been running SimpleWeather inside the NetBeans IDE. Though this is great during development work, now that you're close to being done, having a stand-alone, double-clickable application would be very useful.

In Chapter 5, you learned how to build the OneWireLister program into a stand-alone "jar" program. If you tried that same process with SimpleWeather, you may have discovered that it doesn't work the same. Double-clicking the executable jar file seems to launch the program, but then nothing happens. In actuality, the program *is* running; you just can't see it because it doesn't have a user interface.

In this chapter you add a simple user Graphical User Interface (GUI) to SimpleWeather. It won't have a lot of bells and whistles, but it will provide a way to run SimpleWeather as a stand-alone application. By keeping it simple, even if you haven't done any Java GUI programming you should be able to hack it to customize it to your needs.

Project Overview and Design

This is strictly a software project. You don't need to buy any hardware to complete it. You will add the GUI class, modify the SimpleWeather code to launch the GUI on startup, and then tailor the GUI to what sensors you have.

To keep the design simple, the GUI will have just a single window. It just needs to contain the items that give program status and current data, similar to what you see in text form in the NetBeans IDE. With this in mind, here's a list of the specific items:

- Show the program status so you will know if the program is running and what it's doing.

- Display the current time so that you'll know if a weather collection is about to start or if a web page update is about to occur.

- Display the current data for each of the active sensors, updating it once a minute. This will let you know the sensors are working and get current weather data.

- Provide a quit button to let you stop the program.

- Display a logo or banner image that can be customized to add a personal touch.

Add the GUI to Your Software

The example GUI code is contained in a file in your downloaded source code. All you have to do is add the class to your project and then modify the `SimpleWeather.java` file to call the GUI class.

Add the UserWindow Class

Copy the `UserWindow.java` file to your SimpleWeather ➪ src ➪ SimpleWeather folder. Open your SimpleWeather project in NetBeans and check the list of project files to make sure it is included.

Also copy the `logo.png` file to your SimpleWeather project folder. This is an example logo that you can replace with your own logo or photo to customize your GUI.

Modify SimpleWeather

Open `SimpleWeather.java` in the NetBeans editor. Scroll down to the list of class variables near line 60. At the end of the list, add a new `UserWindow` class variable `myWindow`:

```
// class variables
public float temp;
...
private BetaBrite sign;
private X10FireCracker x10;
private UserWindow myWindow;
```

Scroll down to the SimpleWeather constructor and add the code to instantiate the new `UserWindow` class before the code that initializes the DataLogger. This will open the GUI and display the window:

```
catch (OneWireException e)
{
  System.out.println("Error Finding Adapter: "+ e);
  System.exit(1);
}
// init the user interface window
myWindow = new UserWindow();

// initialize the data logger
logger = new DataLogger(LOG_PATHNAME);
```

To update the window, you need to call UserWindow's updateGUI() method, passing three parameters:

```
public void updateGUI(String status, Date date, SimpleWeather sw)
```

Here's what the three parameters do:

- status is a short text message that gets copied to the Status field in the GUI. This is how you display program status and messages.

- The date updates the current time field in the GUI.

- The reference to the SimpleWeather class sw is how the method gets the weather data. If the reference to SimpleWeather is null, the weather data update is skipped. This provides the ability to update the status or the time without changing the displayed weather data.

In the mainLoop(), just after the code to check the current time near line 193, add a GUI update call that displays "Idle..." in the status display and updates the current time in the GUI. Because this call is before the code that checks if one minute has elapsed, it gets called once a second:

```
// check current time
date.setTime(System.currentTimeMillis());
minute = date.getMinutes();
hour = date.getHours();

// update user interface
myWindow.updateGUI("Idle...", date, null);
```

Just prior to the code that collects the weather sensor data, add another call to updateGUI() method to change the status message to display "Collecting..." to let you know that SimpleWeather is reading the various sensors:

```
// only loop once a minute
if (minute != lastMinute)
{
    System.out.println("Time = " + date);

    // get the weather
    // update user interface
    myWindow.updateGUI("Collecting...", date, null);
```

Next, before the code that resets the highs and lows near line 274, add a call to updateGUI() that changes the status to "Updating Display" and updates the weather sensor data in the GUI. Notice this is the only call that passes the updateGUI() method a reference to the SimpleWeather class (the "this" keyword refers to the current object, which in this case is the SimpleWeather object itself):

```
// update user interface
myWindow.updateGUI("Updating Display", date, this);
```

```
// if it's midnight, reset the highs & lows
if (hour == 0 && minute == 0)
  wc.resetHighsAndLows();
```

Finally, add one more call to `updateGUI()` near line 288 that lets you know when the web page is being updated:

```
// and update the web page on the 10 minute mark
if (minute % 10 == 0)
{
  // update the web page
  myWindow.updateGUI("Updating Web Page", date, null);
  wp.updatePage(date);
  wu.send(date, wc);

  // reset the averages for the next set of data
  wc.resetAverages();
}
```

Customize It for Your Sensors

If you haven't built all the sensors, the GUI is easily modified to remove the fields for a specific sensor. In the following code snippet, you can see how two fields are added to the window:

```
// temperature
y += height;
lable[4] = addText(myPanel, "Temperature:", x1, y, width, height, null);
lable[5] = addText(myPanel, " ", x2, y, width, height, myBorder);

// humidity
y += height;
lable[6] = addText(myPanel, "Humidity:", x1, y, width, height, null);
lable[7] = addText(myPanel, " ", x2, y, width, height, myBorder);
```

Looking at the code for humidity in the preceding code, there are three lines that do the work. The first line adds the height offset to the height position variable "y" to move the text fields down to the next position. The next line calls a method in the UserWindow class that adds a Java text label to the window, displaying the String passed in quotes. The remaining values set the starting horizontal position, starting vertical position, width, and height, respectively. The last parameter is used to add a border around the text label if desired.

If you are missing one of the sensors, simply "comment out" the three lines for that sensor like this:

```
// humidity
// y += height;
// status[6] = addText(myPanel, status[6], …
// status[7] = addText(myPanel, status[7], …
```

If you later add that sensor, simply uncomment the code.

Comment Out All Print Statements

Now that the values are displayed in the GUI, you no longer need to print them to the screen in NetBeans. In fact, depending on your computer's operating system, there are reasons why you shouldn't print the data. Simply comment out any code that prints the sensor data or other non-error data to the screen like this:

```
// update user interface
myWindow.updateGUI("Collecting...", date, null);

// get temperature
temp = ts1.getTemperature();
// System.out.println("Temperature = " + temp + " degs F");

// get wind speed & direction
windSpeed = ws1.getWindSpeed();
windDir = ws1.getWindDirection();
// System.out.println("Wind Speed = " + windSpeed + " MPH " +
//                    "from the " + ws1.getWindDirStr(windDir));

// get humidity
humidity = hs1.getHumidity();
// System.out.println("Humidity = " + humidity + " %");
...

//      System.out.println("\n");
```

Testing Your Project

Are you ready to see if it works? Run your project like you have done in the previous chapters. This time, you should see a new window pop up in front of NetBeans. The status box shows "Starting…" and a few seconds later switches to "Collecting…" indicating that SimpleWeather is reading the sensors. After a few more seconds, the status box should switch to "Updating Display…". The GUI will then update the sensor fields with the current weather data as shown in Figure 19-1.

If all goes well, you are ready to build the stand-alone version. If not, recheck your code for typos. As usual, there is a completed project in the source code folder for this chapter in case you need it for reference.

Here's how to build the stand-alone Jar file:

1. Click the Quit button in SimpleWeather's new GUI window. You will see a confirmation dialog making sure you really want to quit. Click Yes.

2. From the NetBeans menu, select Build ➪ Clean and Build Main Project. This will build the Jar file.

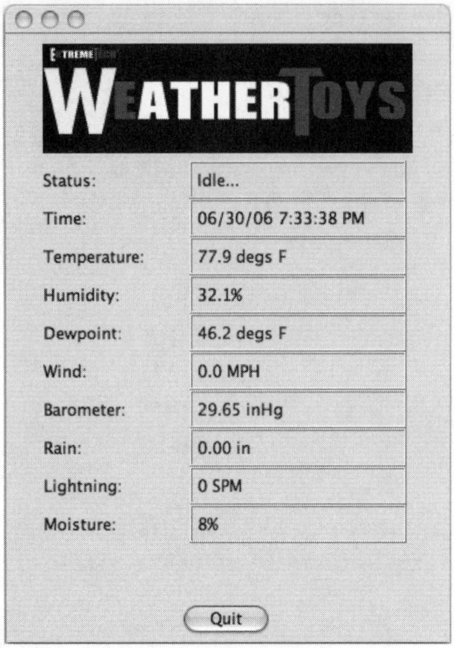

FIGURE 19-1: The new SimpleWeather GUI.

3. Navigate to your SimpleWeather project folder and locate the folder name "dist."

4. Copy the `logo.png` file to the dist folder.

5. Open the dist folder. You should see several files. Make sure you have `SimpleWeather.jar` and `logo.png`.

6. Launch the `SimpleWeather.jar` file like you would any other application. It should open and display the SimpleWeather window just like it did when you ran it inside the NetBeans IDE.

Congratulations! You've just finished a complete, stand-alone application.

Walk Through the Code

Take a quick look through the UserWindow class code. First, you might notice the long list of additional packages that have been imported. These provide the libraries for the various Java GUI supports. Java's built-in GUI package is called Swing. It provides numerous buttons, text boxes, and just about any kind of GUI object you've ever seen:

```
import java.util.*;
import java.util.Date;
```

```
import java.awt.*;
import java.awt.event.*;
import java.util.Locale;
import java.util.ResourceBundle;
import javax.swing.*;
import javax.swing.BorderFactory;
import javax.swing.border.Border;
import javax.swing.border.EtchedBorder;
import java.text.SimpleDateFormat;
```

What Do You Do if Your Jar File Won't Run?

You double-click the Jar file; it seems to run for a while, then nothing. What do you do? One method for troubleshooting a Jar file issue is to run it from the command line. Mac users need to launch the Terminal application, and Windows users open a command window by clicking Start ⇨ Run and typing `cmd`. Once you have a command window open, use the `cd` command to switch to the directory where the jar file resides the `dist` folder). Next, in the command window, type

```
Java -jar SimpleWeather.jar
```
This will run your program from the command line, and you will see your program output just like it did inside the NetBeans IDE. Look at the output messages to see if you can determine the problem.

```
Starting SimpleWeather 1.0
Stable Library
=========================================
Native lib Version = RXTX-2.1-7
Java lib Version   = RXTX-2.1-7
Found Adapter: DS9097U on Port /dev/tty.USA19QW4b44P1.1
Resetting 1-wire bus
```

If your code runs fine inside the NetBeans IDE, but you get an error that mentions the manifest file when you try to run the jar file, such as

```
java -jar SimpleWeather.jar
Failed to load Main-Class manifest attribute from
SimpleWeather.jar
```

This indicates that the manifest file is not being built properly or has been corrupted. Copy the `manifest.mf` file from the Chapter 19 project source code folder to your SimpleWeather project folder, and select Clean and Build from the menu again to rebuild your project with the new manifest.

Next is the list of class variables. Notice the `label` array. This is the array of 20 text "labels" that are displayed in the window. If you want to add additional sensors or labels, you will have to increase the size of this array to accommodate your new items:

```
public class UserWindow extends JFrame
{
  private JFrame myFrame;
  private static JLabel[] label = new JLabel[20];
  private static Border myBorder;
```

The next section of code is the class constructor. Most of the work of creating the window and showing it is done here. First, a frame is created. A Java Frame is the fundamental "empty window" where the various GUI objects are drawn. The frame is set so that it is not resizable. Inside the frame, a Java Panel is added that will contain the 20 text labels that make up the majority of the window. The logo image is loaded and added to the panel:

```
// set up the window
   setResizable(false);  // user can't resize it
   setSize(360, 500);    // width, height
   setLocation(60, 100); // x, y

   JPanel myPanel = new JPanel();
   myPanel.setLayout(null);
   this.getContentPane().add(myPanel, BorderLayout.CENTER);

   myBorder = BorderFactory.createEtchedBorder(EtchedBorder.RAISED);

   // get logo image and add it to panel
   ImageIcon icon = new ImageIcon("./logo.png");
   if (icon != null)
   {
     JLabel logoLabel = new JLabel(icon);
     logoLabel.setBounds(20, 5, 320, 92);
     myPanel.add(logoLabel);
   }
   else
     System.out.println("Couldn't Load Icon");
```

The next section of code is where the text labels are added to the panel. There are several variables that define the size, starting position, and spacing. After the variables, you will see a long list of calls to the UserWindow's `addText()` method. Each of these calls adds the text to the panel in the locations specified.

Notice that for the sensor names, the last parameter is null, and the text labels for the sensor values have the variable `myBorder`. Earlier in the code, a simple Java "Etched Border" was created and stored in the `myBorder` object. If this is passed to the `AddText()` method, the border is added. If null is passed, a border isn't added:

```
// text label positions
int y = 100;       // y is vertical (y) position
int width = 175;   // text box width
```

```
int height = 30;     // text box height
int x1 = 30;         // lable text box horiz start
int x2 = 150;        // data text box horiz start

// status
label[0] = addText(myPanel, "Status:",      x1, y, width, height, null);
label[1] = addText(myPanel, " Starting...", x2, y, width, height, myBorder);

// time
y += height;
label[2] = addText(myPanel, "Time: ",       x1, y, width, height, null);
label[3] = addText(myPanel, " ",            x2, y, width, height, myBorder);

// temperature
y += height;
label[4] = addText(myPanel, "Temperature:", x1, y, width, height, null);
label[5] = addText(myPanel, " ",            x2, y, width, height, myBorder);
...
```

A second panel is created and added to the frame in the following code. Inside this new panel, the "Quit" button is added, with its code that detects the button click, and displays the "Are You Sure?" dialog. If the "Yes" button is clicked, System.exit(0) is called, quitting the program:

```
// set up the main panel
JPanel buttonPanel = new JPanel();
this.getContentPane().add(buttonPanel, BorderLayout.SOUTH);

JButton quitButton = new JButton("Quit");
buttonPanel.add(quitButton);

quitButton.addActionListener(new ActionListener()
{
  public void actionPerformed(ActionEvent newEvent)
  {
    int n = JOptionPane.showConfirmDialog(myFrame,
            "Are You Sure You Want to Quit?",
            "Confirm Quit", JOptionPane.YES_NO_OPTION);

    if (n != JOptionPane.YES_OPTION)
      return;

    System.out.println("User Selected Quit");
    System.exit(0);
    }
  });

  setVisible(true);
}
```

The addText() method simplifies adding the 20 text labels. Each time the method is called, it creates a new Java Label, adds it to the panel, and then sets the location in the panel. If the border variable is not null, the border is added to the text label:

```
private JLabel addText(JPanel panel, String text, int x1, int y1,
                       int x2, int y2, Border border)
{
  JLabel label = new JLabel(text);
  panel.add(label);
  label.setBounds(x1, y1, x2, y2);

  if (border != null)
    label.setBorder(border);

  return label;
}
```

The final method in this class is the mechanism that updates the window. The String weatherStatus is copied to the status label. The date d is converted into a date/time format specified in the SimpleDateFormat() code, and then copied to the time label.

If the call to updateGUI() does not have a null value for the SimpleWeather parameter, each of the current weather measurements are copied to the corresponding text label using the FormatValue() method to round the data and convert it to a String. The Java SetText() method is used to set the label text to the String value:

```
public void updateGUI(String weatherStatus, Date d, SimpleWeather sw)
{
  SimpleDateFormat formatter = new SimpleDateFormat("MM/dd/yy h:mm:ss a");


  label[1].setText(" " + weatherStatus);
  label[3].setText(" " + formatter.format(d));

  if (sw != null)
  {
    label[ 5].setText(formatValue(sw.temp, 1) + " degs F");
    label[ 7].setText(formatValue(sw.humidity, 1) + "%");
    label[ 9].setText(formatValue(sw.dewpoint, 1) + " degs F");
    label[11].setText(formatValue(sw.windSpeed, 1) + " MPH");
    label[13].setText(formatValue(sw.pressure, 2) + " inHg");
    label[15].setText(formatValue(sw.rain,2) + " in");
    label[17].setText(" " + sw.lightning + " SPM");
    label[19].setText(" " + sw.moisture + "%");
  }
}
```

Expanding Your Project

You can make many additions to your SimpleWeather GUI window. Here's just a sampling of what you can do:

- If you're using SimpleWeather as a sprinkler timer, you can create little sprinkler icon-sized images. Make one for each station, putting the station number in the corner. Using the method used to display the logo, add an image icon to show when each sprinkler station is on.

- If you're using the HVAC controller from the previous chapter, add little text labels on the bottom of the GUI window showing "Heat," "Cool," or "Fan" when the corresponding relay is activated. Here's a hint: look at the `getSwitchState()` method in `OWSwitch.java` to check if a relay is on.

- If you have been using debug mode to help debug your projects, you can't turn on debug mode when running from the Jar file. Consider adding another button that turns on or off debug mode.

- Add a new text box, possibly in a second window, and print all error messages (like the ones in the `catch()` statements) to the error text window.

- If you're really adventurous, add a button that opens a new window. Inside the new window, draw plots of the weather data for the last 24 hours. This is not a trivial task. You'll have to find a way to save the data for the last 24 hours and then develop a plotting routine. For a hint, look at the code for the ExtremeTech Weather Server in Appendix B.

Wrap Up

In this chapter, you added the necessary code to allow you to run SimpleWeather as a stand-alone application. You can now copy the Jar file to your desktop and launch the program without the NetBeans environment. You can also move the program to a different computer and it will run, as long as you install Java and RxTx.

Even though you have built a robust weather program, SimpleWeather's biggest shortcoming is the lack of error checking. If you have a sensor error, which you will get occasionally, the incorrect value gets averaged and posted on the web page or sent to the Weather Underground. To round out your software, look at ways that you might be able to detect the errors and omit them from further processing.

This is the last project in this book for SimpleWeather, but the fun shouldn't stop here. You have developed a highly capable weather station program and can continue to expand upon it. Look at some of the other weather station software packages listed in the "Additional Resources" in Appendix A. See if there are any features that interest you and see if you can add that feature to SimpleWeather.

Eliminate Your
PC with TINI

You've probably spent several weeks, if not months, building your weather station and developing the code. Now that the code is working, projects like the weather web server require that your computer be on 24 hours a day, 7 days a week. Having SimpleWeather tie up your computer all the time can be a pain, and I bet there are times when you wish you had your computer back.

In this chapter you learn how to build a separate computer to run your weather station. It won't have a monitor or a keyboard, and a simple wall-wart AC adapter powers it. But it will free up your computer and at the same time, save you money on electricity.

The "computer" you can switch to is a small single-board device called TINI (pronounced "teeny"). In the basic weather station application, you will have essentially three connections to TINI: a 1-Wire port to connect to your weather station, an Ethernet port to connect to your network, and power. It's really that simple.

Project Overview

This is the first project that doesn't use your PC or SimpleWeather. Instead, you will be using a TINI and a new software program called TiniWeatherServer. You will look at the software a little later in this chapter, but first, focus on the hardware.

This chapter is considerably more complex than previous chapters. I could have filled up an entire book on TINI and interfacing it to your weather station, but I condensed it down to a single chapter. You will be using new applications like FTP and Telnet to copy files and communicate with TINI. Because many of the parts in this project are intertwined, I suggest you read the entire chapter first before you start construction. I also suggest you download or purchase a hardcopy of Don Loomis's book *The TINI Specification and Developer's Guide* published by Addison Wesley. You can download the online version at www.maxim-ic.com/products/tini/devguide.cfm. This is a comprehensive book on TINI and what it can and can't do. This is an invaluable guide for getting started with TINI.

What Is a TINI?

TINI is an acronym for Tiny Inter-Net Interface. It is a single-board computer (technically, it's a single board *microcontroller*), about the size of a computer RAM module. Dallas/Maxim manufactures TINI, the same folks who build the 1-Wire devices you have used in your weather sensor projects. As shown in Figure 20-1, you can think of TINI as a "black-box" that connects devices on one side to an Ethernet network on the other.

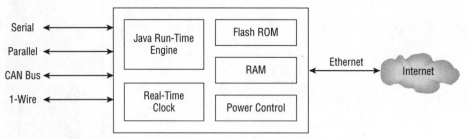

FIGURE 20-1: TINI interfaces your devices to the Internet.

Inside the "black-box" is a powerful Java engine that allows you to write sophisticated code to interface the device to the network. Looking at your weather application, the full Java 1-Wire API is implemented allowing full use of all the 1-Wire devices.

The TINI platform is a chipset manufactured by Dallas/Maxim designed for engineers to embed in their application to "Internet-enable" everything from coffee pots to refrigerators. Just like your desktop PC, different versions of the processor are available: the older DS80C390 and the newer DS80C400. Both will work fine for the weather application described in this chapter.

TINI has many features that make it ideal for running your weather station. Here's a quick summary:

- DS80C400 or DS80C390 processor
- 15-MHz processor
- Built-in Ethernet controller
- Flash ROM (512K on the 390 or 1 MB on the 400)
- Battery backed RAM (512K or 1 MB on the 390, 1 MB on the 400)
- Two integrated serial ports and support for two external serial ports
- Internal and external 1-Wire buses
- Real-time clock
- JDK 1.1 API and firmware providing a multi-tasking, multi-threaded run-time engine

- TINI OS with a full IP4/IP6 TCP/IP stack, garbage collection, serial port and 1-Wire drivers, and PPP support

- Native interface that offers developers access to hardware using assembly language

- Slush, a system shell giving a UNIX-like interface with TTY, Telnet, and FTP servers

To facilitate system development, a complete TINI development module using the DS80C400 processor is offered by Dallas/Maxim as an evaluation board that has everything you need to use the TINI chipset, including ROM, RAM, Ethernet interface, serial ports, a 1-Wire port, and more. The Dallas/Maxim TINI board is shown in Figure 20-2. Several other manufacturers, such as Systronix (www.systronix.com) offer their own versions of the TINI board. The TINI board requires a separate socket, or "motherboard," to ease connections to the various ports and power. As with the TINI module, there are several suppliers to choose from. The Dallas/Maxim version is shown in Figure 20-3.

FIGURE 20-2: The Dallas/Maxim DSTINIM400 TINI evaluation board.

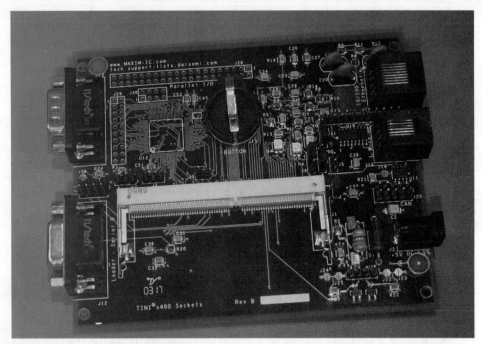

FIGURE 20-3: The DSTINIS400 TINI socket.

What TINI Isn't

Looking through some of the features, you may be wondering how you can do anything without a hard drive and only 1 MB of RAM. Let me assure you, you can do a lot! Of course, with only 1 MB of RAM you won't be able to store more than a few days' worth of weather data and you won't be able to store any graphic images, except for very small ones. This means you can't host an entire web site (but you can integrate TINI's web pages with your hosted web site).

TINI's processor isn't very fast, which means it may not be the right choice for you if you have a high-traffic site. Simple weather pages load fairly quickly one at a time. But TINI can get bogged down between serving web pages and collecting weather data if you start getting many simultaneous requests. TINI's 15-megahertz microcontroller is no match for your 3-*giga*hertz dual-processor computer!

How Do I Connect It?

The TINI board and socket simply require power, a network connection, and the 1-Wire cable to your weather station. If you're using one of the options that require a serial port, then you'll have a serial connection too. Figure 20-4 shows the basic connections to TINI. Because you can't do much with just the TINI module, from now on, whenever I refer to TINI, I'm referring to the TINI module connected to a TINI socket.

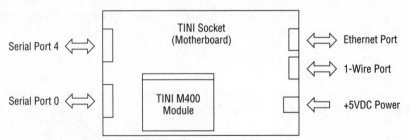

FIGURE 20-4: Basic TINI connections.

If There's No Monitor, How Do I See the Weather Data?

TINI doesn't have a way to connect to a computer monitor. To view the weather data it collects, you have to convert it into a form that can be accessed by a remote computer. This chapter provides several choices, and as you explore TINI, you may come up with some additional options.

In Chapter 12 you added a web server to share your data. The same web server is built into the software project for TINI. Just point your web browser to TINI's IP address and view the weather web page. Of course, you'll have to have your computer networked to TINI using Ethernet, but you most likely have a network in place already. Using the web server is the preferred method because you can see all your weather data, plus if you allow HTTP access through your firewall, anyone else on the Internet can access your weather data. To see an example of TINI in action, visit my TINI weather page at www.timbitson.com/weather/weather.html.

Another option is to post your weather data to the Weather Underground and view the weather data there. You can also use the BetaBrite LED sign from Chapter 15 and connect it to TINI's serial port.

Building the Project

You need several items to build your TINI-powered weather station. You'll need a TINI, a TINI socket, and a power supply. You will also need some sort of enclosure to package your TINI to keep it safe.

Buying a TINI

There are several TINI suppliers, the primary being Dallas/Maxim. Search the Internet to see who is offering TINI boards and sockets, and check the list of resources in Appendix A. This project focuses on the Dallas/Maxim DSTINIM400 version, but it should be pretty easy to translate the project to the flavor of TINI you have.

Table 20-1 lists the key components required to build your own TINI weather server. There are several options for the power supply, and you'll read more about it later. The TINI evaluation board and socket can be ordered directly from the Dallas/Maxim web site at www.maxim-ic .com/TINIplatform.cfm.

Table 20-1   TINI Web Server Parts List

Qty	Part Number	Description	Source
1	DSTINIM400	TINI Evaluation Board	Dallas/Maxim
1	DSTINIS400	TINI Socket Board	Dallas/Maxim
1	See Text	5VDC Regulated Power Supply	See Text
1	See Text	Enclosure	Various

Setting Up Your TINI

The TINI comes without the TINI Operating System loaded, so you'll have to download the OS and install it yourself. As I'm writing this, the current version of the TINI OS is 1.17. You'll also need Java installed on the computer that you will be loading the OS from. If you are going to use a different computer than the one you have been using for SimpleWeather development, go back to Chapter 4 and follow the instructions to load the Java JDK and install the RxTx package.

Caution TINI is ESD sensitive. Always use an anti-static wrist strap when handling your TINI.

Download the Necessary Files

If you have not already done so, download the TINI.zip package from the companion web site and expand it on your hard drive. This will be your TINI development folder.

Next, navigate your web browser to www.maxim-ic.com/TINIplatform.cfm and click the link to the TINI Software. Follow the on-screen prompts to download the latest TINI 1.1 SDK. Expand the SDK in your TINI development folder.

There should be four folders in your TINI development folder:

- **tini1.17:** The TINI OS downloaded from Maxim's web site
- **TWS Install:** The TiniWeatherServer application program
- **TWS400 Source:** The TiniWeatherServer source code
- **JavaKitMac:** The JavaKit program for Mac Users

Load the TINIOS

To install the OS, you'll need to connect TINI's serial port 0 to your computer's serial port. Then using a special serial communications package called JavaKit, you will download the OS to TINI and configure TINI for network communication. After that, all communication and installation will be performed using the network connection.

Detailed steps for loading the OS are provided in the README.txt file located in the top folder of the TINIOS package. Here's a quick look at the steps for getting the OS loaded along with a few notes to help make the process easier:

1. Connect TINI to power. The DSTINIM400 requires a source of regulated 5V power with a minimum of 250 milliamps. Use a good quality 5% regulated wall-wart style power supply, such as the Digikey T312-P13P-ND. (If you have an old external "ZIP" drive, its power supply is a regulated 5V). You can also use the 5V regulator described in the section "Building a 5-Volt Regulator."

2. Connect TINI's serial 0 port to your computer's serial port using a 9-pin straight-through serial cable (don't use a "null-modem" cable).

3. Launch the special utility provided in the TINIOS package called JavaKit. JavaKit is a simple terminal application that allows communication to TINI using the serial port. It also has the capability to download the special TINI OS files. The JavaKit provided in the TINIOS download is for PCs. A special converted version of JavaKit for Mac users is located in the TINI folder that you downloaded from the companion web site.

   - **Windows Users:** Follow the instructions in the ReadMe.txt file in the TINIOS download. Open a command window and change directories to the TINI OS folder. Type **java -classpath bin\tini.jar;%CLASSPATH% JavaKit -400 -flash 40** for an 80C400-based TINI or **java -classpath bin\ tini.jar;%CLASSPATH% JavaKit** for an 80C390-based TINI.

   - **Mac Users:** Open a Terminal window and change directories to the JavaKitMac folder. Type **JavaKitMac390** for a '390-based TINI or **JavaKitMac400** for a '400-based TINI.

4. Select your serial port from JavaKit's pull-down menu, and then open the connection to TINI. Make sure the DTR Clear button is selected, and then click the Reset button to reset TINI.

5. Download the TINIOS file tini117_400.tbin using JavaKit. (If you're using a '390, download tini117-bank1to6.tbin.)

6. Clear the RAM (the "Heap").

7. Load the shell program "Slush."

8. Reboot TINI.

9. Log into TINI using JavaKit. The default user is "root" and the password is "tini."

10. Configure the basic network communication settings:

- IP Address
- Subnet Mask

11. Reboot TINI

Tip

Make sure you choose an IP address supported by your network; otherwise you won't be able to connect to TINI. Refer to the "Static and Dynamic IP Addresses" and "Routers and Firewall" sidebars in Chapter 12 for a refresher.

Once you think you're ready and have a serial cable, follow the steps in the README.txt file to install the OS. When you choose an IP address for your TINI, you have the option of using DHCP (if your network supports it). To make the setup easier, use a fixed IP address and subnet mask, because this will be the only way to communicate with TINI once you remove the serial cable. Write down the IP address during installation just in case you forget what it is later!

Caution

Double-check the power requirement for your TINI board, including the polarity of the power connector. Connecting the wrong voltage or polarity will permanently damage your TINI.

Tip

When entering file path names that contain spaces, put the entire path and file name in quotes.

Test Your Settings

Before disconnecting the serial cable or turning off power, connect TINI to your network using an appropriate Ethernet cable. Once the connection is made, the link LED on TINI should illuminate indicating the network connection is working. Open a command window on your computer and type **telnet ip_address** where ip_address is the IP address you assigned to TINI. If the setup was successful, the response from TINI will be:

```
[tims-g5:~] tbitson% Telnet 192.168.1.106
Trying 192.168.1.106...
Connected to tini6.
Escape character is '^]'.

Welcome to slush.  (Version 1.17)

tini00f6b5 login: root
tini00f6b5 password:
tini00f6b5 />
```

This indicates that you have successfully connected to TINI and your initial configuration was correct. If you get an error, go back to the serial terminal and check your settings. Do not proceed until you can communicate with TINI over the network.

FTP and Telnet

The only way to communicate with TINI is over a network or serial port connection. After setup, the serial port is disconnected (because you will use it for something else), so all you really have is the network connection.

There are two key utilities built into most operating systems called Telnet and FTP. Telnet is a network communication program that lets you type commands that get sent to TINI over the network, and then TINI sends the response back to your Telnet session. To start Telnet, open a command window and type `telnet ip_address` where `ip_address` is the address of TINI.

FTP is an acronym for File Transfer Protocol. This is how you copy your files *to* and retrieve them *from* TINI. Like Telnet, you start an FTP session by typing `ftp ip_address`. You will be prompted for your username and password. Once logged in, you send a file to TINI by typing `put local_file remote_file`, where `local_file` is the path and file name to the file you want to send, and `remote_file` is the path and file name where you want to save it on TINI. Use the `get` command to retrieve a file from TINI. To make things simpler, I usually start my FTP session from the directory that contains the files I want to send. For example, in Windows, open a command window and type:

```
>cd "C:\Documents and Settings\vak3506\My Documents\Weather
Station\TWS Install"
ftp 192.168.1.100
Connected to 192.168.1.100
220 Welcome to slush.  (Version 1.17)  Ready for user login.
User (192.168.1.100): root
331 root login allowed. Password required.
Password: tini
230 User root logged in.
put myProgram.tini /bin/myProgram.tini
get textfile.txt
bye
```

There are a few more common FTP commands you should know:

- `cd` — Change directory on the target computer.

- `lcd` — Change directory on the local computer.

- `ascii` — Send files using ascii (text) mode.

- `bin` — Send files using binary mode.

- `bye` — Quit the FTP session.

When running FTP, you can type `?` to see a list of all available commands.

Final Configuration

Once you can communicate with TINI using Telnet, there are a few more settings to configure. Log on to TINI using Telnet and set the following parameters. Some of the settings will disconnect your Telnet session, and you'll have to log back in:

- **Name:** By default, TINI has an odd name like tini00f6b5. TINI uses this name at the command prompt. Change your TINI's name to something simple, like "tini1". Use the slush command `hostname` followed by the new name, such as `hostname tini1`. The name change may not take effect until you reboot TINI.

- **Time Zone:** TINI needs to know your time zone for its internal real-time clock. Set this using the `date` command followed by your time zone such as PST, MST, CST, and so on. For example, to set your time zone to Mountain Standard Time, type **date MST**. You can also specify an offset in hours, such as `date -7`.

- **Gateway:** TINI can connect to the Internet to do many things, but it needs to know how to get there. The gateway address is the IP address of your router or Internet connection interface, like a cable modem. Use the slush command `ipconfig` to set this: `ipconfig -g ip_address` where `ip_address` is the gateway IP address.

- **DNS:** The Domain Name Server is how the Internet converts names to IP addresses. TINI needs to know the IP address of your ISP's DNS servers. Usually there are two or more, in case one fails. Your ISP provides these addresses. Set these also using the `ipconfig` command: `ipconfig -p ip_address1 -s ipaddress2` where `ip_address1` and `ip_address2` are two of your ISP's DNS servers.

To double-check your settings, type **ipconfig** at the prompt. Your current settings will be displayed. Here's an example of my settings:

```
Hostname           : tini6.
Current IPv4 addr.: 192.168.1.106/24 (255.255.255.0) (active)
Current IPv6 addr.: fe80:0:0:0:260:35ff:fe00:f6b5/64 (active)
Default IPv4 GW   : 192.168.1.1
Ethernet Address  : 00:60:35:00:f6:b5
Primary DNS       : 68.87.66.196
Secondary DNS     : 68.87.64.196
DNS Timeout       : 0 (ms)
DHCP Server       :
DHCP Enabled      : false
Mailhost          :
Restore From Flash: Not Committed
tini6 />
```

Once you have your TINI configured, test your connection to the Internet by trying to set TINI's date and time. Type **date -n** and TINI will try to connect to an Internet time server and set the date. After it is complete, type **date** (without the -n) to see the local time:

```
tini6 /> date -n
Synchronizing time to server: clock.isc.org/204.152.184.72
```

```
New system time: 12 May 2006 21:03:16 GMT

tini6 /> date
Fri May 12 14:03:26 -7 2006
```

If you're not going to connect TINI to the Internet, you need to set the time manually. Type **help date** to see how to do this.

Now is a great time for you to play with your new TINI. Type **help** to see a list of the various Slush commands (or see the sidebar "TINI Command List") and look through the TINI Specification Guide. Try some of the commands to see what they do. If you have ever worked with UNIX, you'll find that Slush is in many ways a subset of a UNIX shell.

TINI Command List

```
addc          append        arp           cat
cd            chmod         chown         copy
cp            date          del           df
dir           downserver    echo          ftp
gc            genlog        help          history
hostname      ipconfig      java          kill
ls            md            mkdir         move
mv            netstat       nslookup      passwd
ping          ps            pwd           rd
reboot        rm            rmdir         sendmail
setenv        source        startserver   stats
stopserver    su            touch         useradd
userdel       wall          wd            who
whoami
```

When you are done experimenting with TINI, type **bye** at the slush command prompt to quit your Telnet session and then close your Telnet window (Mac users can quit Terminal). Disconnect the cables and power from TINI. The next step is packaging your TINI in a protective enclosure.

Package Your TINI

The TINI module and socket are bare circuit cards. They need to be protected in an enclosure to prevent damage. In the same enclosure, you could also install the power supply and the barometer module; then the only equipment you need to connect is the weather equipment that is mounted outdoors.

TINI requires 5 volts and if you are also using the Hobby-Boards barometer, it requires about 14V to operate. Because you may not have room for two separate power supplies, you can build a regulator that converts the 14 volts to regulated 5 volts to power your TINI from the same power supply.

Building a 5-Volt Regulator

It is fairly simple to build your own voltage regulator from scratch. Figure 20-5 shows the schematic of a 5-volt regulator board and Table 20-2 lists the required parts. This is a simple circuit and can be assembled on a small piece of perf board using point-to-point wiring. The optional 1N5608 TVS diode provides some protection for TINI against power surges and spikes, and if your power is clean or you plan on running your weather station from an Uninterruptible Power Supply (UPS), then it can be eliminated.

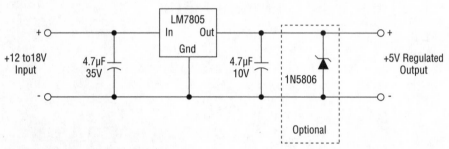

FIGURE 20-5: 5-volt regulator schematic.

Table 20-2 5-Volt Regulator Parts List		
Qty	**Description**	**Source**
2	4.7uF 35V Electrolytic Capacitor	Radio Shack 272-1027
1	LM7805 or LM340T5 5V Regulator	Radio Shack 276-1770
1	1N5908 5V TVS Diode (optional)	Digikey 1N5908RL4OSCT-ND
1	Heat Sink for LM7805 (optional)	Radio Shack 276-1363

After assembling the regulator board, double-check your work and then apply 12 to 14 volts DC to the input, paying close attention to the polarity. Using a digital voltmeter, measure the output voltage. It should be 5.0 +/− 0.15 volts. Touch the regulator IC to make sure it isn't getting hot, a sure sign something is wired wrong. Figure 20-6 shows the completed board. Later in this project when the regulator board is connected to your TINI, check if the LM7805 is running hot. If it is, you may want to add a heat sink as listed in the parts list.

FIGURE 20-6: Completed 5-Volt regulator on perf board.

Final Assembly

A good option for a TINI enclosure is an outdoor sprinkler timer box, like the one you read about in Chapter 11. The sprinkler timer box has several nice features such as an AC outlet for the power supply and has enough room to house the TINI, the 5-volt regulator board, and at least one 1-Wire module. With the cover closed, it provides a weather-resistant enclosure for your equipment. Figure 20-7 shows an example installation that houses everything except for the pole or roof mounted sensors. It includes the X10 FireCracker and relay module.

The only external connections to your TINI weather station are AC power, a network connection, and the 1-Wire bus. You can also add the BetaBrite sign from Chapter 15 and the X10 FireCracker module from Chapter 16. Figure 20-8 shows a block diagram of the complete system.

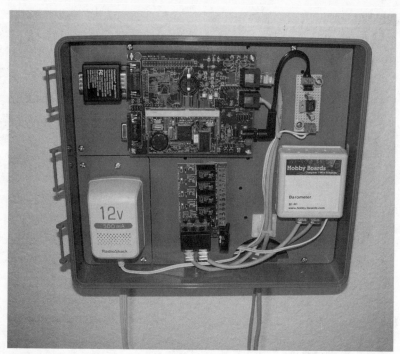

FIGURE 20-7: The TINI weather station in a sprinkler timer box with the cover removed.

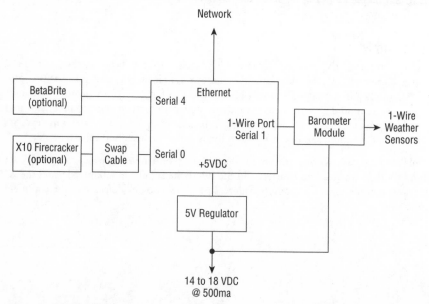

FIGURE 20-8: Block diagram of the complete TINI weather station.

As mentioned earlier, it is preferable to mount TINI indoors; however, the entire system can be mounted outdoors. The enclosure should be mounted in a location that protects it from direct exposure, such as under an eave or patio cover. If you do mount it in an unprotected area, you should add ventilation holes in the bottom of the box to keep the electronics from overheating.

Once the TINI weather station is built and connected to your network, you are ready to install the TiniWeatherServer software.

TiniWeatherServer

The software used to collect weather data is called TiniWeatherServer, or TWS for short. TWS is loaded into TINI's memory and run from TINI's Slush shell. TWS is configured using a text preferences file that contains the 1-Wire device IDs and other settings like those that were contained in the user constants in the SimpleWeather project.

What Is TWS?

TiniWeatherServer is very similar to SimpleWeather in many respects, and in fact is the predecessor to ETWS and SimpleWeather. It has been optimized to run in TINI's limited memory.

What Does It Do?

TWS collects weather data and serves up a web page just like SimpleWeather. Both use almost the same web page code. TWS also supports posting data to the Weather Underground. Here's a list of some of TWS's features:

- Collects weather data once a minute
- Builds a web page every 10 minutes
- Displays 24-hour weather data charts in your web browser
- Logs up to 2 days of weather data
- Automatically sets TINI's clock to Internet time every 24 hours
- Logs errors to a file that can be viewed via your web browser
- Supports a BetaBrite LED sign
- Weather and error log can be e-mailed every night
- Hardware watchdog can be enabled to keep TWS running

As you can see, TWS has a few features that SimpleWeather doesn't. TWS also has comprehensive weather sensor error checking. If a bad reading occurs, the data does not get averaged and the data from the previous measurement is used. If you get more than 10 errors (configurable) in a 24-hour period, that sensor is disabled.

There are many other aspects of TWS, but rather than try to go through them all, the source code is provided for you to look through and hack to your needs. But before you start changing code, get your TINI up and running first.

What Doesn't It Do?

Because TINI has only 1 MB of memory available for both your program and your data, it can only store two days' worth of data: the current day and the previous day. The previous day's data automatically gets deleted at midnight; otherwise TINI would run out of memory and crash.

In its present state, TWS doesn't support the relay board, X10, or the moisture sensor. You can add these in a similar fashion to the way you added them to SimpleWeather. In fact, you'll find that TWS has a main loop just like SimpleWeather.

TWS was originally designed to be a self-contained outdoor weather station. One of the design flaws in TWS is that instead of each sensor being a separate object, they have all been combined into a single WeatherStation object. This makes it a bit harder to add multiple sensors of the same type. This was fixed in SimpleWeather, but not in TWS (at least not yet).

In order to fit the Java Virtual Machine into the limited space on TINI, the Dallas/Maxim folks had to leave out some of the Java libraries. One of those was the Math library. This means that the method used in SimpleWeather for averaging wind direction doesn't work. This is where I could use some of your help: TWS needs a way to average wind direction. It can't take up much memory and can't use any trigonometric functions. If you come up with a good solution, email it to me!

Now that you know what TWS can and can't do, finish getting TWS loaded and your TINI running.

Setting Up TWS

Even though TINI doesn't have a hard drive, it still has a file system that lets you create directories and files similar to your desktop PC. By default, the TINIOS creates two directories: etc and tiniext. The etc directory contains a couple of system files. The tiniext directory is used to add extensions and won't be used in this project.

Configure the Prefs File

Locate the prefs.ini file in the TWS Install folder you downloaded. Launch your favorite text editor and open prefs.ini. You will see a long list of settings, similar to this:

```
# TiniWeatherServer Preferences File
#
#    Sample Settings File    05-14-06
#
#
#
Prefs_Version                = 4.0
#
# Serial Numbers of 1-wire devices
DS1820_Temp_Sensor           = 0000000000000000
```

```
DS2423_Wind_Speed_Counter     = 0000000000000000
DS2450_Wind_Direction_A_to_D  = 0000000000000000
DS2438_Humidity_Sensor_A_to_D = 0000000000000000
DS2423_Rain_Counter           = 0000000000000000
DS2423_Lightning_Counter      = 0000000000000000
Baro_Pressure_Sensor          = 0000000000000000
TAI8570_Writer_Device         = 0000000000000000
#
# Temperature Units - Degrees Fahrenheit [F] or Celsius [C]
Temperature_Units             = F
# Speed Units - Miles/Hour [MPH] or Kilometers/Hour [KPH]
Speed_Units                   = MPH
# Rain Units - Inches [In] or Centimeters [cm]
Rain_Units                    = In
# Barometer Units - inHg [inHg] or millibars [mb]
Barometer_Units               = inHg
...
```

Leave most of the settings with their default values for now, but the device serial numbers will need to be changed. Open your SimpleWeather project and copy and paste your device serial numbers from the SimpleWeather.java file into the corresponding value in the prefs.ini file. If you do not have a particular device, leave it set to all zeros. Save your prefs.ini file, making sure it is saved as plain text.

Create the TWS Directories

For this project, you are going create two more directories in TINI to keep the TWS files organized. Follow these steps:

1. Log into TINI using Telnet.

2. Type **cd /** to make sure you're at the root level.

3. Type **mkdir bin** (short for binary, a UNIX-like location to store binary files) to create a directory to put the TWS executable.

4. Type **mkdir web**. This is where TWS will build the web page (like the Web folder in SimpleWeather), and where the charting files will be copied.

5. Type **ls -l** to see a listing of the files and directories. You will see something similar to this:

```
tini6 /etc> ls -l
total 6
drwxr-x    1 root     admin       4 May 12 17:54 .
drwxr-x    1 root     admin       4 May 12 17:54 ..
drwxr -    1 root     admin       0 May 13 21:36 web
drwxr -    1 root     admin       0 May 13 21:36 bin
-rwx - -   1 root     admin     225 May 12 17:54 .startup
-rwxr -    1 root     admin     101 May 12 17:54 passwd
tini6 /etc>
```

Install the TWS Files

Leaving your Telnet window open, open a second command window and follow these steps to copy TWS and support files to your TINI:

1. Change your current directory to the TWS Install folder.

2. FTP **prefs.ini** to TINI's root directory.

3. FTP **tws.tini** to TINI's /bin directory.

4. FTP the following files to TINI's web directory:

   - allChart.html

   - tempChart.html

   - humChart.html

   - baroChart.html

   - windChart.html

   - rainChart.html

   - lightChart.html

   - dateLineApp.jar

5. Quit the FTP program.

Launch TWS

In the Telnet window, start TWS by typing

```
java /bin/tws.tini &
```

TWS should start up and display the version number and current time:

```
tini6 /> java /bin/tws.tini &
Starting TWS Vers 4.0.0
TiniWeatherServer Local Time = Sat May 20 07:25:42 -7 2006
tini6 /> TWS Vers 4.0.0 Started
Starting Web Server
```

Watch the "activity" LED on TINI (D1 on the '390, DS1 on the '400). After about 30 seconds it will start blinking on and off once a second. This indicates that TWS is running. Open your web browser and attempt to connect to TINI by typing **192.168.1.00/index.html**, replacing the IP address with your TINI's IP address. You will most likely get a "404 - File Not Found" error indicating that you have successfully connected to TINI, but TWS has not yet created the web page.

TWS only updates the web page every 10 minutes on the 10-minute mark, like SimpleWeather. Check TINI's time using the date command in the Telnet session. After the 10-minute mark, wait another 30 seconds (so TWS can collect the data and build the web page) and try your web browser again. This time you should see the weather web page as shown in Figure 20-9 (you may not see the picture at this time).

If you can't get TWS to run or you don't get a web page, look at the output in the Telnet window to see if there are any error messages. You can also start up TWS in debug mode, which displays comprehensive program execution information. To use debug mode, start TWS by typing `java /bin/tws.tini -d &`.

Current Weather Conditions
5/20/06 3:40 PM

Located in my back yard
Weather data updated every 10 Minutes

Temperature				
Current	Today's High	Today's Low	Trend	24 Hr
79.7 °F	79.7 °F	79.7 °F	- °F/Hr	- °F

Wind				
Current	Current Peak	Today's High	Trend	24 Hr
0.0 MPH SSE	0.0 MPH	0.0 MPH	- MPH/Hr	- MPH

Relative Humidity				
Current	Today's High	Today's Low	Trend	24 Hr
29.0 %	29.0 %	29.0 %	- %/Hr	- %

Dewpoint				
Current	Today's High	Today's Low	Trend	24 Hr
45.1 °F	45.1 °F	45.1 °F	- °F/Hr	- °F

Barometric Pressure				
Current	Today's High	Today's Low	Trend	24 Hr
29.89 inHg	29.89 inHg	29.89 inHg	- inHg/Hr	- inHg

Rainfall				
Year to Date	Since Midnight	-	Rate	24 Hr
0.00 In	0.00	-	0.00 In/Hr	- In

Lightning Activity				
Current	Today's High	-	Trend	24 Hr
0 spm	0 spm	-	0 spm	- spm

View: Weather Plots

Weather data collected using a 1-Wire Weather Station and TiniWeatherServer 4.0
TINI Free Ram = 550K   Server Uptime: 0 Days, 0 Hours, 0 Mins   2 Hits since 11/01/06   © 2006 by T Bitson

This weather data is provided for entertainment purposes only.
Data is believed to be accurate at the station location.
Do not use this data for applications that could jeopardize someone's safety.

FIGURE 20-9: TiniWeatherServer's web page. Look familiar?

Exploring the Settings

Once things seem to be working, you can start exploring some of TWS's options in the preferences file. Remember, TINI has very limited memory. If you enable all of TINI's features, you risk running out of memory.

Calibrations

The Calibrations section is where you enter the calibration data for some of the sensors, like the user constants in SimpleWeather. Copy and paste the calibration values from SimpleWeather into your `prefs.ini` file:

```
# Calibrations
North_Adjust                   = 0
Rain_Counter_Zero              = 0
Humidity_Slope                 = 1.0
Humidity_Intercept             = 0.0
Barometric_Pressure_Slope      = 1.1698
Barometric_Pressure_Intercept  = 20.632
```

When copying your values from SimpleWeather, the names may not be exactly the same. In SimpleWeather I used the terms gain and offset, whereas in TWS they are called slope and intercept, respectively.

System Settings

The System Settings enable or disable several of TWS's features:

```
# System Settings
Errors_Before_Device_Disable  = 10
Watchdog_Timer_Enable         = false
Weather_Data_Logging          = false
Weather_Data_Charting         = true
Web_Server_Logging            = false
Web_Log_Filename              = /web/web.log
Error_Log_Enable              = false
Error_Log_Filename            = /web/error.log
Mail_Error_Log                = false
#
# Address to mail weather log to
MailTo_Address                = user@mailserver.com
MailFrom_Address              = tini@mailserver.com
```

TWS will log up to two days of weather data if you set `Weather_Data_Logging` to true. If set, TWS will display a link on the weather page for viewing the logs.

By default, errors are displayed in the Telnet session. However, once you log out, any error messages generated are sent to oblivion. To enable logging of error messages, set `Error_Log_Enable` to `true`, and all errors will be logged to the file specified in `Error_Log_Filename`. This allows you to check the operation of TINI to make sure something hasn't failed. However, suppose one of your sensors goes bad and you don't check the log for a few days? The error log keeps filling up, and eventually will cause TINI to crash. To work around this potential problem, TWS can email the log file to you every night, deleting the file after it's mailed. To enable mailing of the error log, follow these steps:

1. Set `Mail_Error_Log` to true.

2. Set the `MailTo` and `MailFrom` addresses to valid email addresses.

3. Set the `mailhost` in TINIOS using the `ipconfig -h` command to your mail server, such as `ipconfig -h smtp.yourmailserver.net`. Your ISP provides your mail host information. See the TINI Specification Guide for more information on setting up your mail host.

Once the log mailing is set up, TWS will also email the weather logs right to your inbox, so you can save your weather history.

You can also generate weather data charts, right in your browser window. Simply set `Weather_Data_Charting` to `true` (the default setting) and TWS will save special log files that the included KavaChart engine can read to display your weather data graphically.

Web Settings

There are more than a dozen settings for the web server:

```
# Web Settings
Web_Server_Root_Directory    = /web/
Web_Server_Default_Page      = index.html
Web_Server_Port              = 80
Weather_Page_Path            = /web/
Weather_Page_Filename        = index.html
Weather_Page_Title           = <font size="4"><center>Weather Conditions
Weather_Page_Logo            = <img src="http://mywebpage.com/…"
Weather_Page_Subtitle        = <font size="3"><center>Located in …"
Weather_Page_Footer1         = This weather data is …
Weather_Page_Footer2         = Additional data…
Post_to_Wunderground         = false
WU_Username                  = KAZOROV3
WU_Password                  = password
WU_Banner                    = <a href=\"http://www.wunderground.com/…
Web_Counter_Start_Date       = 12/11/05
Web_Counter_Start_Count      = 0
```

The web server uses the `Web_Server_Root_Directory` as the top directory where files are served. This directory and all directories below it can be accessed via the server. The `Web_Server_Default_Page` is the file that the web server returns if the requester (the person asking for the web page) doesn't specify a file name. The `Weather_Page_Filename` is the name of the actual weather page. By having a different default page and weather pages, you can use your TINI web server as the default "home page" and any requests for anything but the weather pages can be redirected to another site.

There are several headers and footers to display weather page information, very similar to SimpleWeather. You can use HTML tags to modify the way the text is displayed.

Also part of the web settings is the Weather Underground information. To post your weather data, set the Post_to_Wunderground line to true, and enter your username, password, and the banner text supplied when you signed up. If you use the Weather Underground to collect your data, I suggest you turn off the system setting for data logging. This will conserve memory on TINI.

Because TINI doesn't have enough memory to store the banner or logo picture, you will have to store it on the Internet and reference the URL in the prefs file. Many ISPs offer small amounts of online storage for personal web pages at little or no charge. Even if you don't use it for a web site, you can copy your banner picture to the site and use the URL supplied by your ISP to access the banner, such as <img src="http://66.3.24.245/banner.gif">. The default image presently in the prefs file is a link to a picture of a WS-1 on my web site.

Additional Settings

There are many additional settings for TWS. For more information on the various TWS settings, look at the ReadMe.txt file located in the TWS Install folder.

Expanding Your Project

Using TINI as your weather station computer offers you numerous options. The following sections provide a sampling of some of the things you can add.

Add Lightning Protection

The TINI's 1-Wire port is built right on the TINI board. Should an ESD or lightning-induced surge zap your 1-Wire interface, you'll have to replace your TINI module. In Chapter 14, you learned how to add lightning and ESD protection to your 1-Wire bus. Adding this inexpensive protection is a must when using TINI to run your weather station!

Part of the concern is grounding. Because TINI is powered from an isolated power supply, the 1-Wire bus ground side is not connected to earth ground. This can cause several problems, because the bus can "float" to several volts above ground, lowering the immunity to noise and other interference. By adding the lightning protector and attaching the ground lead to a good earth ground, this problem is eliminated.

 Warning If you decide not to use the lightning protector, you must connect the 1-Wire bus low side or TINI's power return to earth ground.

Add a BetaBrite

TINI has two serial ports plus the 1-Wire port. This allows you to add additional serial devices to TINI. Built into the TWS code is a version of the BetaBrite code that you added to

SimpleWeather. This version uses the special "string mode" that updates the display without the flashing experienced in SimpleWeather.

To use a BetaBrite, connect it to the Serial4 port and modify the `prefs.ini` file to enable it. There are also settings that let you adjust the temperatures or humidity levels so that the corresponding value on the sign turns red or green:

```
# BetaBrite LED Sign
Enable_BetaBrite                = false
BetaBrite_Serial_Port           = serial4
Temperature_Hi_Red              = 85
Temperature_Lo_Green            = 70
Humidty_Hi_Green                = 85
Humidity_Lo_Red                 = 15
```

Add X10 or the Relay Board

The X10 class from Chapter 16 or the relay board from Chapter 18 can be added to TWS, the same way you added it to SimpleWeather. Remember, you'll need to add the class variable, define the serial port, and if you're using X10, add the X10 method. Versions of both of these classes that have been modified slightly to work in TWS are included in the TWS400 Source in the optional classes folder.

Once you have added the X10 or relay class, you'll have to edit the main loop inside `TiniWeatherServer.java` to add the rules. Unlike SimpleWeather, the weather variables are not referenced in `TiniWeatherServer.java`, so you'll have to access the variables using the WeatherCruncher class, such as:

```
weatherCruncher.temperature
weatherCruncher.windSpeed
weatherCruncher.humidity
```

Look through the WeatherCruncher class to see what other variables are available to build your rules. Then in `TiniWeatherServer.java`'s `mainLoop()`, add your rules:

```
// main program loop: Start the various servers, get the weather
public void mainLoop()
{
...
  // *** here's the call that actually gets the weather data ***
  weatherCruncher.getWeather();

  // if data logging is enabled, log measured data
  if (prefs.dataLoggingEnabled)
    todaysLog.logData(lock, weatherCruncher);

  // if charting is enabled, update chart files
  if (prefs.chartingEnabled)
    chart.log(lock, weatherCruncher);
```

```
// Add X10 and/or Relay Rules Here

// if time = 10 minute, crunch weather data & update web page
...
```

The X10 FireCracker can be connected directly to serial port 4. If you're using the BetaBrite on serial port 4, it can't be shared. You will need to purchase or build a 9-pin male-to-male null modem swap cable to connect the X10 FireCracker to serial port 0.

Building the Source Code

Once you make any changes to TWS, you will have to recompile the code, convert it to TINI format, and FTP it to TINI. This is not a trivial process and a lot can go wrong. Warning: getting your code to build the first time may make you lose sleep at night.

Here are the steps to build the code. If things don't go right, check your work and refer to the TINI Specification Guide and the docs folder in the TINIOS download for additional help.

1. Make your code changes. TWS is provided as a NetBeans project for editing purposes only; you can't run TWS in the NetBeans IDE. Make your changes in the NetBeans editor. You will need to add two libraries to the TWS project in order to build the files so you can check your syntax. Right-click your project and select Properties ➪ Libraries ➪ Add Jar/Folder. Then add these two libraries:

   ▪ Add the `owapi_dependencies_TINI.jar` file located in the TINIOS folder you downloaded earlier. It is located in the bin directory.

   ▪ Add the `tiniclasses.jar` file located in the same TINIOS folder.

2. Although you can't run the TWS project, you can check your code syntax by selecting Build ➪ Build Main Project from the NetBeans menu. Make sure to correct all errors before continuing.

3. Open a command window (or Terminal window for Mac users). Change directories to one directory *above* the source code directory. This should be the NetBeans TWS project directory.

4. To compile and build TINI code, you normally have to type in long cryptic lines into the command window. Instead, in the NetBeans project folder are two build scripts that will compile the source code, run the BuildDependency utility (which adds the 1-Wire libraries and converts the class files to a single ".tini" file), and FTP the resulting `tws.tini` file to your TINI.

5. To run the command file, Windows users type `build ip_address` or Mac users type **./buildMac ip_address**, where the `ip_address` is your TINI's IP address.

6. You will see some activity in your command window as the build script does its job:

```
[tims-g5:~/TWS400b3 Source] tbitson% buildMac 192.168.1.106
Syntax: build ip_address
IP = 192.168.1.106

Compiling...

Building....

Build successful
BuildDependency clean-up complete.
FTPing file to tini at 192.168.1.106
Connected to tini6.
220 Welcome to slush.  (Version 1.17)  Ready for user login.
331 Password Required for root
230 User root logged in.
Remote system type is UNIX.
Using binary mode to transfer files.
200 Type set to Binary
local: bin/tws.tini remote: bin/tws.tini
229 Entering Extended Passive Mode (|||52963|)
150 BINARY connection open, putting bin/tws.tini
100% |**************************************| 75006       206.72
KB/s    00:00
226 Closing data connection.
75006 bytes sent in 00:03 (19.23 KB/s)
221 Goodbye.
done!
```

7. When you're done, your new tws.tini file should be in your TINI's bin directory. If you get any error messages, go back and check your work. Note that the root user's password is used in the script. If you have changed the default password for TINI, you will need to modify the script.

8. Open a Telnet session to TINI and stop the current TWS (if it is running):

   a. Type **ps** to get a listing of the current running processes. You will see a list similar to this:

   ```
   tini3 /> ps
   3 processes
   1: Java GC (Owner root)
   2: init (Owner root)
   3: /bin/tws.tini (Owner root)
   tini3 />
   ```

   b. Find the process number that corresponds to tws.tini. In this case it is '3'.

   c. Type **kill id** where id is the tws.tini process number from step b.

9. Restart the new tws.tini by typing **java /bin/tws.tini &**.

10. With a little luck, and a whole lot of patience (not to mention your new skills), your new TWS program should start running.

Go Almost Wireless with WiFi

One of the features of the consumer weather stations reviewed in Chapter 2 was the ability to transmit the weather data wirelessly to an indoor console. This has several advantages, and if you have been yearning to have a wireless weather station, you can have one. Well, sort of.

The basic TINI weather station has three connections: power, network, and the 1-Wire cable to the outdoor weather equipment. If TINI is mounted inside a weatherproof enclosure near your weather equipment, then really all that is left are two cables: power and network. Suppose you could get rid of the network cable. Then all that remains is power, and AC power may be available or easily run to your weather station.

To lose the network cable, you will need to switch from a wired cable to WiFi, the wireless network standard. WiFi equipment has come down in price considerably over the past year, and WiFi can be added to your weather station for less than $200. You need two parts: the WiFi base station (also called a router or access point) and a wireless network client adapter for your weather station.

WiFi base stations come in many configurations and speeds. Stop by your local computer store or do a little online shopping to see what's available. The most common base stations are for wireless standards 802.11b or 802.11g. By the time you read this, the newer 802.11n standard may be available. Any of these will work fine for the TINI weather station because you really don't need high speeds like you do for online gaming or downloading video. If you already have WiFi in your home or business, you are halfway there.

To connect the weather station, you will need a small WiFi client adapter that will fit in your TINI weather station's enclosure. Several are available to choose from, and you can start by looking at the LinkSys WGA54G Wireless Game Adapter or Apple Computer's Airport Express. Both units convert the wireless signals to a standard Ethernet connector that is then connected to TINI. Figure 20-10 shows the Airport Express mounted inside the sprinkler timer box with an older TINI '390-based weather station. The entire box can then be mounted near the outdoor weather equipment, only requiring AC power.

Apple's Airport Express is easier to install, because the power supply is built into the adapter itself. However, it is a little harder to configure because it can be used as both a base station and a client. The LinkSys model requires only a few settings to identify which WiFi network to use. The Airport Express can be configured from either a PC or a Mac, whereas the LinkSys game adapter requires a Windows-based PC for initial setup.

Configure TINI's Startup File

Once your TINI weather station is configured and you're happy with the way it is running, you can modify TINI's startup file to automatically run TWS when power is applied or if it is rebooted. You can also disable TINI's serial server process to free up the console serial port.

When you were configuring TINI using JavaKit, you were communicating with the serial server. If you have added (or will add) the BetaBrite sign, or are using serial port 0 for anything else, you will need to disable the serial server so that the serial port is available for your use.

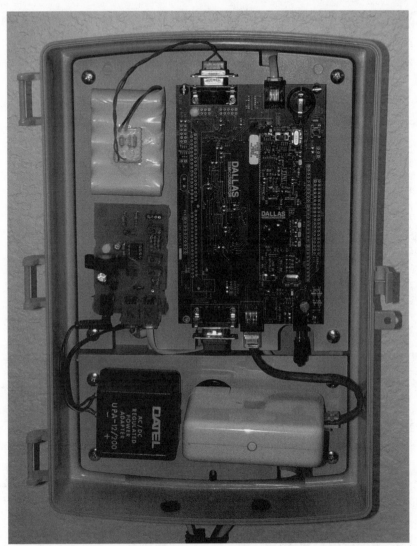

FIGURE 20-10: The TINI weather station with the Airport Express WiFi adapter.

Here's how to modify TINI's startup file:

1. Open a command window and change to your TINI development folder.

2. Open an FTP session to TINI.

3. Set the file transfer mode to plain text mode by typing **acsii**.

4. Get the startup file by typing `get /etc/.startup startup`. This will copy TINI's startup file to your TINI development folder, changing the name from .startup to startup, because some operating systems hide files that start with a period.

5. Open the startup file with a text editor. You should see this:

```
########
#Autogen'd slush startup file
setenv FTPServer enable
setenv TelnetServer enable
setenv SerialServer enable
##
#Add user calls to setenv here:
##
initializeNetwork
########
#Add other user additions here:
```

6. Change the line that enables the serial server to disable:

```
setenv SerialServer disable
```

7. At the end of the file add a new line to run TWS:

```
#Add other user additions here:
java /bin/tws.tini
```

8. Save the modified file (make sure it's plain text).

9. Copy the modified file back to TINI in the FTP session by typing `put startup /etc/.startup`, which renames the file back to .startup.

10. Quit the FTP session by typing `bye`.

11. Reboot TINI. You can do this by cycling power or typing `reboot` in a Telnet session.

Wait while TINI reboots, and after about 60 seconds you should see the activity light start blinking, indicating that TWS is running. Now, if power is lost or if you reboot TINI, TWS will automatically start up.

Convert SimpleWeather

Back when you started SimpleWeather, you used a couple of Java classes that are deprecated. Here's why: with just a few minor changes, SimpleWeather can be re-built to run on TINI. I'll leave this task up to you, but here are a few pointers:

- Use the existing build scripts; just modify the source directory.

- The wind direction averaging routine won't compile, so you'll need to remove it and come up with an alternative method. TWS simply uses the latest value without averaging.

- SimpleWeather uses the SimpleDataFormat class to convert the Date object to a String value, which isn't supported in TINIOS. Copy the date formatting routine from TWS.

- TINIOS doesn't like packages. Remove the package statement at the beginning of each java file.

- Change the serial ports to match TINI's.

- You obviously won't be able to use the user interface from Chapter 19!

- You will need to add the class containers for the relay card and the moisture sensor in the build script.

If this list seems scary, don't worry. It's really not that bad and you should be able to make the changes in just a couple of evenings. But if you have a custom version of SimpleWeather and want its capability, this is definitely worth the effort.

Wrap Up

If you have TiniWeatherServer running on TINI, you should feel proud of your accomplishment. Not only have you built a weather station, you now have built your own weather station computer. Your new TINI-based weather station chugs along, collecting weather data and serving web pages and posting it to the Weather Underground. Pretty awesome!

You have really just scratched the surface on what TINI can do. Continue to explore the various TWS settings, then start looking through the code to see what you can hack. Don't forget *The TINI Specification and Developer's Guide;* it has additional ideas to expand your project.

Looking back, SimpleWeather provided the fundamentals to show you how to develop code and troubleshoot problems and you developed a good understanding on how 1-Wire works. Now TWS takes you a step further. But don't throw away SimpleWeather. You'll find it a valuable resource should you have to test your weather station. Plus, you can use it to develop new code and try it out before moving it over to TINI.

Carry Your Weather Station with You

Consider this scenario: You decide to go for a full-day hike. You start off in the morning when it is cool, and as the day progresses it starts to warm up. By early afternoon, it's quite hot so you find a nice shady spot to take a break and eat lunch. On the way back, you swear it must be 100 degrees with at least 80% humidity! Once you get home and start thinking about your day, you wonder: was it really as hot as it seemed?

There have been many times I wished I could carry a small weather station with me. I don't necessarily need to know the temperature at the time, but if I could just log it and check it later, that would be cool. Well, the 1-Wire folks thought that would be cool too.

Take a temperature sensor, a humidity sensor, add a clock, a battery, and some memory. Package it in a can about the size of four dimes stacked on top of each other, and you have a Hygrochron: a complete temperature and humidity logger in an iButton package.

Project Overview

This project is considerably different than the rest of the projects in this book. It uses new software, new hardware, and about the only similarity is that it uses 1-Wire and measures weather. But it provides a significant capability to log temperature and humidity easily.

What's a Hygrochron?

The Hygrochron is a self-sufficient, rugged system that measures and records temperature and humidity. It takes these measurements at a user-defined interval, and stores the results in its built-in 8KB of battery-backed-up memory.

The measurement interval can be set for as little as one second or up to 273 hours, allowing you to record weather data for an hour or for many years. During setup, you can decide to use normal 8-bit resolution or the higher 12-bit resolution for the recorded data. Each measurement is time-stamped and stored in the Hygrochron memory.

Like all 1-Wire devices, the Hygrochron has a unique serial number and only requires one connection and ground to read and write to the device. When not being used as a logging device, it can act as a 1-Wire temperature and humidity sensor, similar to the ones you built in Part II.

As shown in Figure 21-1, the Hygrochron is packaged in the standard iButton F5 "microcan." Notice the small hole in the center? This is the vent where the humidity sensor detects moisture. It is protected with a special hydrophobic filter that allows moisture in the air to pass through but blocks dust, dirt, and water. Also shown is one of the many iButton holders available. The one shown allows you to attach the Hygrochron to your key chain or clip it to your belt loop.

Dallas/Maxim also makes a temperature-only version called a Thermochron. The Thermochron is about half the price of the Hygrochron. If all you need is temperature logging, this may be a better option for you.

FIGURE 21-1: The Hygrochron temperature and humidity logger.

iButton Sockets

To connect your 1-Wire bus to an iButton, you need an iButton socket. Dallas/Maxim offers several different iButton sockets for different applications. Figure 21-2 shows two different sockets: the snap-in Blue Dot receptor and a slide-on side connector. Each of these sockets provides the connection to the 1-Wire DQ connection and ground.

FIGURE 21-2: iButton sockets.

Building the Project

There really isn't too much to build for the hardware other than buying the iButton, the socket, and connecting it into your 1-Wire bus.

Buying a Hygrochron

As I'm writing this, the only supplier for the Hygrochron is Dallas/Maxim at its online store. Dallas/Maxim charges a hefty price for shipping, so if you only order a single Hygrochron it can be a bit expensive. Maybe by the time you read this, other online electronics suppliers like Digikey may be stocking these parts.

The full Hygrochron part number is DS1923-F5. You can visit the Dallas/Maxim web site at www.maxim-ic.com and download the full data sheet and check prices.

Choosing a Socket

In Figure 21-2 you saw several iButton sockets and the Dallas/Maxim web site offers a few more. If you are going to be connecting and disconnecting your Hygrochron iButton to perform logging, then the Blue Dot receptor, part number DS1402D, is probably your best bet. The iButton simply snaps in to the connector. The other end terminates in a standard RJ-11 connector that will plug into your existing DS9097U 1-Wire serial port adapter (or TINI).

Connecting It to Your 1-Wire Bus

If you're using the DS1402D iButton connector, simply plug in the socket to your 1-Wire adapter, plug the adapter into your serial port, and snap in your Hygrochron iButton.

Because SimpleWeather requires exclusive use of the 1-Wire bus, you can't run the software for the Hygrochron and SimpleWeather at the same time. If you have a 1-Wire weather station connected to your 1-Wire serial port adapter, then you either have to unplug your weather station to connect the socket or make a Y-adapter to connect it to your bus. If you have a second 1-Wire adapter and a spare serial port, you can use it to set up a separate 1-Wire bus to communicate with the Hygrochron so you can leave SimpleWeather running.

The OneWireViewer

I've been holding out on you: buried in the Java 1-Wire API code is a folder of example code, and within the code is a cool utility called the OneWireViewer.

What Is It?

The OneWireViewer utility allows you to search for and communicate with *any* device connected to your 1-Wire bus. Devices like the temperature sensor show current readings and display plots of the collected data. It is a great tool for checking out each device on your bus, playing with its settings, or even reading and writing to its memory (if it has memory, like the DS2423 counter device).

The latest version of the OneWireViewer utility supports the Hygrochron, allowing you to check the current settings, configure a new data collection, set parameters, and more. You simply start the OneWireViewer, select your Hygrochron, and you're ready to configure your device or read back the data, all using a simple Graphical User Interface.

Running It

Before you run the OneWireViewer, make sure your 1-Wire adapter is connected to your computer, and the iButton socket is plugged into the 1-Wire adapter.

You have several options for running OneWireViewer. The most unique and easiest way to run it is using the Java Web Start version. Point your browser to www.maxim-ic.com/products/ibutton/software/1wire/OneWireViewer.cfm and click the Launch the OneWireViewer button. This will launch the OneWireViewer directly from the Internet. Note that you must have at least the Java SDK installed on the machine you're running it from.

You can also run the pre-built version from a command window. Look in the Tools ➪ owapi_1_10 ➪ examples ➪ OneWireViewer folder you downloaded. Open a command window, change to the OneWireViewer directory, and type **run.bat**. Mac users need to type **./runMac**.

The third option would be to run it inside the NetBeans IDE. I've converted the OneWireViewer example code to a NetBeans project, located in the Chapter 21 source folder.

Once you have OneWireViewer running, a dialog will prompt for the serial port you have your 1-Wire adapter connected to and the adapter type, as shown in Figure 21-3. Next you will be prompted to select a polling rate for how often the viewer program updates each device. Select 10 seconds. Click Next and you will be prompted for the search mode type. Because you're not using 1-Wire tagging (a subject for another book), just make sure Show Normal Devices is selected and click Finish.

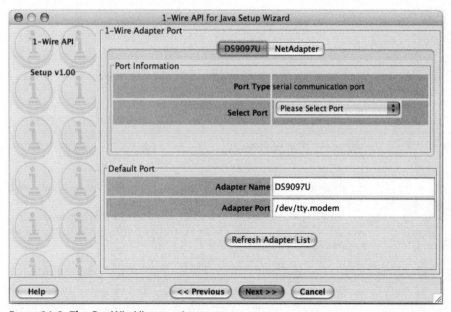

FIGURE 21-3: The OneWireViewer setup screen.

Select a Device

Once you are finished setting up your configuration, the device selection window will be displayed, as shown in Figure 21-4. 1-Wire devices that are present on the bus are shown in the Device List window. If you select a device (by clicking it), the corresponding "snap-in" window will be displayed on the left side of the window.

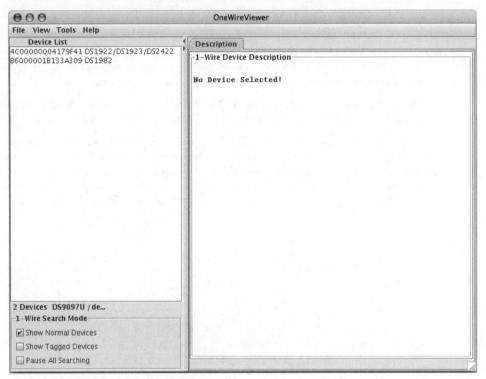

FIGURE 21-4: The device selection window.

If you have not already done so, insert your Hygrochron into the iButton socket and within a few seconds, your iButton will show up in the viewer's device list. Click the Hygrochron device in the list and the main window will provide a brief description of the Hygrochron as shown in Figure 21-5. Notice that along the top of the window there are several tabs that allow you to select different windows of information:

- The Temperature tab displays the current temperature as sensed by the Hygrochron. The graph shows the measurement history, as sampled by the OneWireViewer. This data is not the data recorded by the iButton, but simply a series of measurements taken since the window was opened.

- The Humidity tab is similar to the Temperature tab, except it displays the current humidity and history plot.

- The Clock tab allows you to set the Hygrochron's internal clock to your PC's time, or stop the clock completely, conserving battery life.

- The Memory tab allows you to view and manipulate the internal memory of the Hygrochron.

- The File tab provides an interface to the device memory as a file system, allowing you to format the memory and add directories like your PC. You won't need this to log the weather data, and storing files on your Hygrochron will limit the logging capability.

- The Mission tab is the main window where you set up and monitor the Hygrochron's data logging.

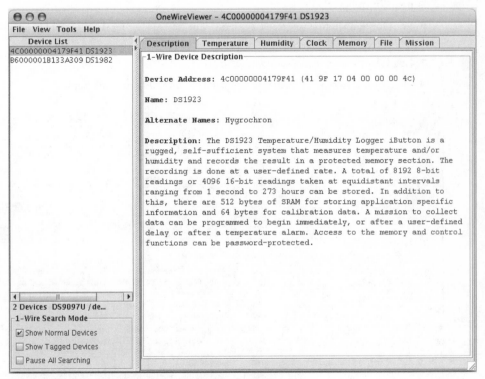

FIGURE 21-5: The Hygrochron main configuration window.

Running a Mission

The process of setting up and running a data logging activity is called a mission. Each mission can be configured differently with many options.

Setting Up a Mission

Click the Mission tab and within a few seconds, you will see the current status of your Hygrochron. If your device is new, there won't be any mission data loaded. Referring to Figure 21-6, the Mission tab displays the 16 parameters of a mission.

When defining your mission, you have several choices:

- Sampling Rate in seconds

- Start Delay in minutes

- Once the memory becomes full, allow new measurements to overwrite the oldest measurement (rollover)

- Wait for temperature to exceed a certain value before logging starts (Start Up on Temperature Alarm or SUTA)

- Log values that exceed a set point (alarms)

- High or low resolution for temperature and/or humidity measurements

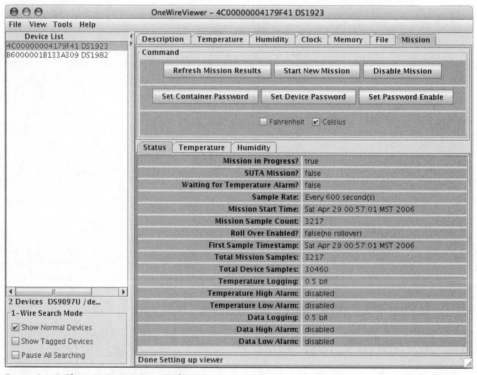

FIGURE 21-6: The mission status window.

To start your first mission, click the Start New Mission button to display the Initialize New Mission window shown in Figure 21-7. For your test mission, set the following:

- Enable Synchronize Clock to set the Hygrochron's clock to your PC's time.

- Set Sampling Rate to 10 seconds.

- Set Start Delay to 0.

- Disable the Enable Rollover option. Enable this option with caution: if you enable rollover, your Hygrochron will continue making measurements until its battery is depleted. Granted, this could take years, but there's no sense in wasting your device's life.

- Enable Sampling for both the Temperature and Humidity channels.

- Set the resolution to 0.5 for temperature and humidity (low resolution). Use this setting to maximize the number of samples you can store.

- Make sure that both the temperature and humidity Enable Alarms check boxes are unchecked.

Click OK to start your new mission and return to the status page. Wait a few seconds and then click the Refresh Mission Results button to check the status of your mission. Now you can disconnect your Hygrochron from the iButton socket and collect data (leave the OneWireViewer running for now).

Try blowing gently toward the humidity sensor's vent to change the humidity. Try varying the temperature by setting it in the freezer or out in the sun. The Hygrochron is fairly rugged, so it can withstand most environments. It is, however, subject to damage if submerged in water or zapped with static electricity. Dallas/Maxim specifies the maximum operating temperature is from −20°C to +85°C. For more information on the safe operating limits, review the data sheet.

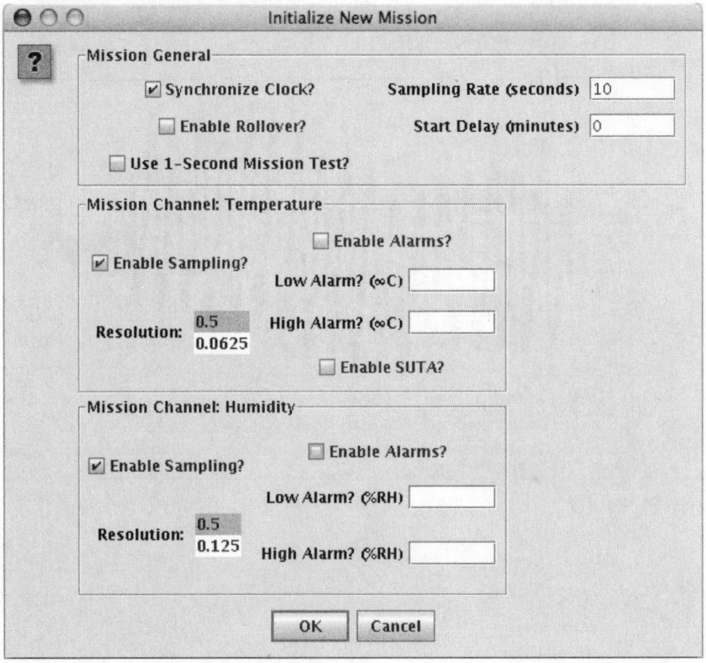

FIGURE 21-7: The mission setup window.

Reviewing the Mission Data

After you have collected data for a while, reinsert your Hygrochron into the iButton socket, select your device, and click the Mission tab. The display will update showing you how many samples were collected and the mission status.

In the Mission Status panel (about halfway down the window), click the Temperature tab to view the results of your Mission. Click the Fahrenheit or Celsius check boxes to display in the units you desire. Figure 21-8 shows the results of a week-long sample mission in my backyard. You can see near the end of the data that I brought it inside for a day before retrieving the data.

After you've viewed the temperature data, click the Humidity tab to view the humidity data logged. This device is sensitive enough that just holding it in your hand will cause a change in humidity.

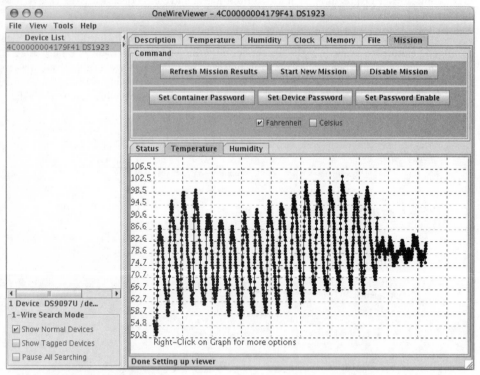

FIGURE 21-8: The mission results.

Saving the Mission

The OneWireViewer has limited means of saving your mission results. One way is to select either the temperature or humidity plot, then right-click on the plot to display a pop-up menu that provides a couple of ways to copy the data to the clipboard. You can then paste it into a text file or into an Excel spreadsheet.

Password-Protecting Your Data

A friend of mine ships biological material that must be maintained at certain temperatures. Federal guidelines require that the temperature is recorded in a secure fashion. To accomplish this, he includes a Hygrochron in the shipment and has the receiver review the mission data to ensure that the shipment was kept within the required limits during transport. To ensure that no one has tampered with the Hygrochron, he uses the Hygrochron's password feature.

The password feature allows you to set the read and/or write capability with two different passwords. The read/write password allows full control to review the data and start a new mission. The person that is responsible for the mission results typically uses this password.

The Hygrochron also has a read-only password. This lets you give the read password to someone else, allowing them to read the data but not erase it or start a new mission, providing a secure method to record the data. Because each Hygrochron has a unique serial number, you can detect if someone swapped devices in an attempt to tamper with the data.

Unless you need the security of password-protected data, I suggest you leave this feature disabled for now. If you should accidentally forget your password, I don't think there is a way to recover it.

Wrap Up

You can find many uses for the Hygrochron, and you can use the device for more than just logging temperatures. Many appliances and equipment generate heat when they are on. By attaching a Hygrochron to the equipment, you now have a way to log run time. For example, Figure 21-9 shows a plot of the run time for my air conditioner from noon one day to noon the next, generated in Excel. The data set that goes from 0 to 100 is an Excel calculation whose results are 100 if the A/C is considered on. You can see that the A/C runs considerably more during the day. The average duty cycle in this example calculates to 37.7%. You can check run times on water heaters, pool pumps, refrigerators, and well, just about any device that gets warm when on.

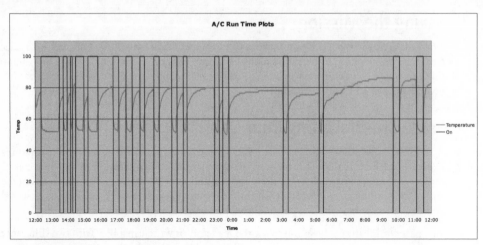

FIGURE 21-9: 24-hour air conditioner run time plot.

One of the limitations of the OneWireViewer is the way the mission results are saved. In this chapter's source code folder from the companion web site you will find an example project for building your own Hygrochron program. The example code sets a mission, reads the status, and saves the data to your hard drive. The data file is saved inside the project folder as a text file, so you can import it to Excel or another application. You can use this code as-is, or as the basis for a full-featured Hygrochron mission setup and data retrieval program.

This chapter has just covered the basics of using the Hygrochron. To learn more, download the DS1923 data sheet from the Dallas/Maxim web site at `www.maxim-ic.com/ quick_view2.cfm/qv_pk/4379`.

Wrap Up

They say all good things must come to an end, and unfortunately, we are reaching the end of our Weather Toys road. The goal of this book was to introduce you to the world of weather measurement, and to show you how to build your own weather equipment. More importantly, I wanted to provide you with the necessary tools and information to unleash your creative power to modify and expand the projects into more than just a simple weather station. I hope I have accomplished this goal.

You now have all the hardware and software tools needed to extend and hack your weather measurement system in unlimited ways, far more than you could with a pre-packaged consumer weather station. I'm sure you'll agree, knowing the intimate details of how the hardware and software work make hacking easier, providing more possibilities.

Looking Back

Let's take a quick review of each of the parts of this book and what you accomplished in each section. I'm also going to try to present a high-level look at what was done to provide the bigger picture. This may help you see options and give you ideas to continue expanding your weather station and software.

Part I

In Part I, you learned about various weather measurement techniques and how some of them evolved. Several weather sensors were presented as well as a quick look at how they operate. Having the appropriate sensor is really the key to making measurements. Every day, manufacturers are designing and producing new sensors for many aspects of our world, far more than just weather.

Although this book has just focused on weather, there are many other aspects of your environment you can add to your arsenal of sensors. For example, you could add a carbon monoxide sensor to monitor the levels in your home for your family's safety, or a Ph sensor to monitor your swimming pool. There are sensors for just about anything you can imagine. One place to start your research is *Sensors Magazine* at www.sensorsmag.com.

In this part, you also learned about 1-Wire: what it is and the fundamentals of its operation. Now that you've had a chance to use 1-Wire, the basics are actually pretty simple. But as you explore the capabilities and expand your system, you may want to learn more. Dallas/Maxim offers many application notes and tutorials on 1-Wire on its web site at www.maxim-ic.com/1-Wire.cfm.

Part II

This section of the book was the heart of the weather station project. You learned how to build the six primary sensors: temperature, humidity, wind, barometric pressure, rain, and lightning (although some may argue lightning isn't a primary sensor). Each of the modules was centered on a specific sensor for measuring the parameter. Some of the sensors output a voltage (humidity and pressure), some output a pulse train or frequency (wind speed, rain, lightning), and others output parallel data (wind direction, but read as a voltage).

Using the basic circuits presented, you can modify the designs (or in many cases use as-is) to measure voltage, current, or resistance that most sensors output. You can also detect switch closures and frequency outputs.

Look at a simple example: with just minor modifications, the humidity circuit can be modified to detect a switch closure instead of a humidity sensor. Using inexpensive alarm system parts, you can install a switch to detect whether a door or window is opened or closed. You can use this information in the smart thermostat project as part of your rules to decide when to activate the heating or cooling. Another variation is to use an alarm system motion detector. Most have simple switch outputs, letting you detect if someone is in the room and using that information in your rules.

Part II finished up with the installation of the weather station, showing a few options for mounting the weather equipment. Here's where the ease of working with 1-Wire became evident; all that was needed was a simple twisted-pair cable to connect all the weather equipment.

Part III

In the third part of the book, you were presented with 10 more projects to expand your weather station. The projects touched upon three main areas:

- **Sharing your weather data.** Whether building your own weather web server or posting it to the online weather services, sharing your weather data with others is a fun and rewarding task. Many small communities would love to have an "official" town weather station.

- **Making your weather data work for you.** There were several projects that showed how to control appliances and equipment with your weather data. Whether you're just turning on a fan when it gets hot, or building a full-featured smart home thermostat, you have the capability to *do* something with the weather data. You can adapt the designs presented to control just about anything you can think of.

- **Using a stand-alone computer.** One of the major projects in this section was switching your weather station to TINI. By using a separate computer, you have built a complete stand-alone system with far more capability than any consumer weather station. Exploring TINI and its capabilities will provide many hours of hacking fun.

Looking Forward

There are still many options for your weather station and your fun shouldn't stop at the end of this chapter. Here are a few pointers on things to explore and ideas to hack:

- Look through the 1-Wire example code in the Java OneWireAPI download. There are more than a dozen examples that show you how to use some of the other 1-Wire devices.

- If you built TINI, look through the TINI example code in the TINIOS folder. Although most are not weather related, you can learn how to do many things with TINI.

- One of the features in the consumer weather stations you didn't add is weather forecasting. It would be fun to add this capability to SimpleWeather or TWS. Looking at pressure trends, wind, and temperature you could build in some forecasting capability. Search the web to see what information you can find and then modify your software projects.

- Many PC operating systems have built-in speech capability, or it can be easily added, providing a great hacking capability. Imagine your weather station announcing the current weather or storm warnings.

- If you have a second home such as a cabin in the mountains, the projects presented provide a great way to monitor your second home's environmental conditions. With a bit of hacking, you could have your remote weather station call and connect to the Internet using a modem and send you a status report. Or combine it with the speech project and have it call you directly ("Hi, this is your cabin in the woods and it is currently 23 degrees…").

- There are a handful of additional weather parameters you can calculate using the information you already have. With a few software mods, you could calculate wind chill, heat index, degrees-days, and more.

- Look at other 1-Wire devices to see what capabilities they have. For example, the DS2890 1-Wire Potentiometer could be connected to your home stereo, providing the capability to remotely control the volume over the 1-Wire bus.

Chapter 2 looked at consumer weather stations to get some ideas for designing the 1-Wire weather station. If you run out of ideas, you can always check out their latest features. You can also look at some of the high-end commercial weather stations to see what they offer.

Movin' On…

As you continue to work with your weather station, you have a couple of invaluable resources. The companion web site at www.weathertoys.net provides a forum to users to post pictures of their projects and installations, along with links to their web sites. You'll also find code updates, notes on my latest projects, and updates to this book. You can also access the software downloads and tools at www.wiley.com/go/extremetech.

The 1-Wire Weather Mailing List is the best forum to discuss new ideas, ask for help on your problems, and learn what other 1-Wire weather enthusiasts are doing. This mailing list has been around in one form or another for almost 10 years, and the archives are full of great ideas, actual construction links, and problem resolution histories. You can subscribe to this list at `www.buoy.com/mailman/listinfo/weather`.

Dallas/Maxim offers several online discussion groups for its products. There is one for 1-Wire and iButtons, and another for TINI users. Check them both out at `discuss.dalsemi.com`.

Of course, don't forget to search the Internet. Typing the phrase "1-wire weather" returns almost 100,000 results. There are hundreds of links to user's weather stations, 1-Wire projects, software, and, well, just about anything you can think of.

You've covered a lot in this book, from humidity sensors to Hygrochrons, and as you have read, there are still many possibilities left to explore. During the research for this book, I have found several new and exciting directions for 1-Wire and weather measurement. Who knows, maybe you'll see *Weather Toys II* in your favorite bookstore someday.

Additional Resources

I n this appendix, you'll find several resources to help you continue experimenting with and expanding your weather station. There are a few listings for hardware suppliers, 1-Wire information, and discussion groups.

If you're not sure where to start, check with the 1-Wire weather discussion group to see what the latest topic is. This discussion group has been in existence in one form or another for almost 10 years.

Also check Hobby-Boards and AAG to see if there are any new or improved 1-Wire weather sensors. The Dallas/Maxim web site is also a great source of information on 1-Wire. Now that you know how to program 1-Wire using Java, there is a lot of sample code on both the Dallas/Maxim and AAG sites.

General Links

Thermocouples

Omega Inc.: `www.omega.com`

Thermistors

US Sensors: `www.ussensor.com`

Temp Sensors

National Semiconductor: `www.national.com/pf/LM/LM34.html`

Online Electronic Component Dealers

Digikey: `http://digikey.com`

Mouser Electronics: `www.mouser.com/`

Jameco Electronics: `http://jameco.com`

1-Wire

1-Wire Information

Guidelines for Reliable 1-Wire Networks, Maxim IC Nov 2001:

www.maxim-ic.com/appnotes.cfm/appnote_number/148

Dallas/Maxim 1-Wire Software Dev Site: www.maxim-ic.com/products/ibutton/software/sdk/sdks.cfm

Dallas/Maxim OneWireViewer: www.maxim-ic.com/products/ibutton/software/1wire/OneWireViewer.cfm

1-Wire Parts and Weather Sensors

Hobby-Boards: www.hobby-boards.com

AAG Electronica: www.aag.com.mx

Texas Weather Instruments: www.texas-weather.com/

1-Wire Weather Discussion Groups

bouy.com 1-Wire Mailing List: www.buoy.com/mailman/listinfo/weather

Dallas/Maxim 1-Wire Discussion Board: http://discuss.dalsemi.com/

1-Wire Weather Station Software

1-Wire WSI Software by AAG, Freeware, Platform: Windows

www.aagelectronica.com/aag/index.html

WServer by Arnie Henriksen, Freeware/Shareware Version, Platform: Windows

http://henriksens.net/1-wire/

OneWireWeather by Simon Melhuish, Freeware, Platform: Linux, RISC OS

http://oww.sourceforge.net/

WeatherDisplay by Brian Hamilton, $70, Platform: Windows, Linux

www.weather-display.com/index.php

X10

X10 Modules and Controllers

X10.com: www.x10.com/automation/index.html

Smarthome: www.smarthome.com/

X10 Software

X10.com: www.x10.com/automation/index.html

Smarthome: www.smarthome.com/

2-Phase Couplers

Smart Home: www.smarthome.com/4816a2.html

Smart Home: www.smarthome.com/4816H.HTML

Build Your Own: www.x10ideas.com/articles/displayx10article.asp?
articleid=25&title=Build%20Your%20Own%20Phase%20Coupler

TINI

TINI Software

TINI Software Download Page: http://files.dalsemi.com/tini/index.html

TINI Getting Started Guide: www.maxim-ic.com/products/tini/devguide.cfm

TINI Modules

Dallas/Maxim: www.maxim-ic.com/TINIplatform.cfm

Systronix: http://systronix.com/

EMACS Inc.: http://emacinc.com/som/som_select_guide.htm

TINI Discussion Groups

Dallas/Maxim TBI Discussion Board: http://discuss.dalsemi.com/

Accessories

BetaBrite

BetaBrite Web Site: www.betabrite.com

Alpha Protocol Programming Guide: www.ams-i.com/Pages/97088061.htm

Surge Suppressors

APC Serial Port Surge Suppressor:
www.apcc.com/products/family/index.cfm?id=145

Solar Radiation Shields

R.M. Young: www.youngusa.com/

CWOP: www.wxqa.com/shields.html

Installing and Using the ExtremeTech Weather Server

If the SimpleWeather project isn't for you, because you just aren't interested in software development or you just want to focus on the hardware, then the ExtremeTech Weather Server may be for you.

The ExtremeTech Weather Server, or ETWS, was going to be the software project for this book rather than SimpleWeather. Between the user interface and the program structure, it ended up being too much code to try to include and still have room for the hardware projects. So SimpleWeather was extracted from ETWS and re-written for simplicity (hence "SimpleWeather") and ETWS is included as a bonus.

If you're using SimpleWeather, a few features of ETWS can be added to it, such as the 1-Wire device error checking, the emailing of the logs, and more. There are also a few features that SimpleWeather has that are not in ETWS, such as the relay control code and X10. Because you have the source code for both, features can be swapped between the two.

To install and configure ETWS, you will need a 1-Wire serial port adapter and at least one 1-Wire weather sensor. If you haven't built any of the weather sensors yet, go back to Part II and complete at least one of the weather sensor projects.

About ETWS

The ExtremeTech Weather Server isn't an industrial-strength, full-featured weather program. Rather, it is meant as sample code for you to hack or use in your software project. The advantage over SimpleWeather is that it is ready to run without typing any code. It has a simple user interface and a preferences window so user-specific settings (the stuff that's hard-coded into SimpleWeather) can be entered without having to recompile. You also get all the source code as a NetBeans project, so if you want to hack the code, you're ready to go.

If you are a novice programmer who plans on modifying the code and haven't completed SimpleWeather, then I suggest you skip ETWS for now. Instead, build the SimpleWeather project to learn the basics and how the software fits together. ETWS uses a similar structure, and learning SimpleWeather will help you get started. Many of the settings are discussed in much more detail in each of the chapters than is presented here. Once you've mastered SimpleWeather, moving to ETWS is a snap.

Installation

ETWS is a Java program, so you will need to install the Java SDK along with the necessary Java libraries to access 1-Wire devices and your computer's serial port. The following steps detail what needs to be installed and where.

Setting Up Your Computer

1. If you are just starting out, follow the instructions in Chapter 4 and work up to the section titled "Building the SimpleWeather Framework" (if you've already completed Chapter 4, then skip this step). This will do the following:

   ■ Install the Java SDK and Java Docs.

   ■ Set the environment variables (Windows only).

   ■ Install the NetBeans IDE so you can modify ETWS. If you're an experienced programmer and already have a Java development environment, you can skip this step.

   ■ Install the RxTx Java Library.

   ■ Install the 1-Wire API.

2. Pick a location on your hard drive where you have the necessary privileges to read and write (such as your documents folder), and create a new folder. Name it "Weather Toys." This will be where the ETWS files get installed (you can change it later if you want to). If you have been using SimpleWeather, you already have this folder.

3. Choose a second location on your hard drive and create a new folder. For convenience reasons, it should be in the same folder that ETWS is in. Name the new folder "Web" for now. This will be the root folder for the web server. If you have been using SimpleWeather, you already have this folder.

Download ETWS

4. Download the ETWS.zip file from the companion web site. Unzip it into your Weather Toys folder created in step 2.

5. Inside the ETWS folder, you'll find two more folders. One is named ETWS 1.0 Source and contains the source code. The other folder, named ETWS 1.0 (the latest version as I'm writing this), contains an executable version of ETWS.

6. Open the ETWS Source folder. This is the project folder, and most of the files and folders here are used to build ETWS. Here's a quick synopsis of the files and folders:

   ■ **nbproject:** This is where NetBeans keeps its project files. Don't modify these files.

   ■ **build:** When NetBeans builds the project, this is where the class files are stored.

   ■ **images:** The banner and about box images are stored here. You can modify these images for your application.

   ■ **src:** This is where the Java source code for the project is kept. Use the NetBeans IDE to edit these files.

   ■ **test:** The folder used to hold test cases. This folder isn't used in this project.

   ■ **manifest.mf:** The NetBeans manifest template file for building the stand-alone jar file. Don't modify this file unless you know what you are doing!

   ■ **dist:** This is the Java distribution folder, where the double-clickable program is saved.

7. Close the ETWS Source window and open ETWS 1.0. These are pre-built files you use to run ETWS. Notice the file `ReadMe.doc`. This file contains the latest update information and additional configuration information.

8. Inside ETWS you'll find a folder named "Web." Copy the contents of this Web folder to the Web folder you created in step 3.

Launching ETWS and Configuring the Basics

9. If you have not already done so, connect the 1-Wire serial port adapter and connect your 1-Wire weather sensors.

10. Start ETWS by double-clicking the ETWS.jar icon. This will start up the program.

11. The first time ETWS is launched, you will be prompted for several items as ETWS builds the default preferences. First, ETWS needs to know the location of the Web folder created in step 3. Navigate to your Web folder in the dialog and click Choose.

12. Next you will be presented with a dialog box showing the available serial ports. Select the serial port where the 1-Wire adapter is connected and click OK.

13. ETWS will build a default set of preferences and will display the preferences dialog box with the 1-Wire Device tab as shown in Figure B-1.

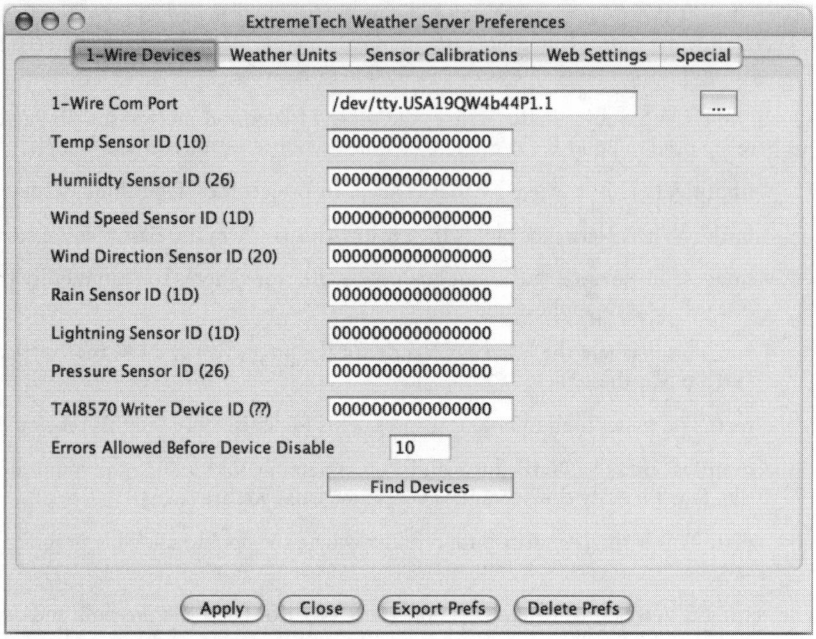

FIGURE B-1: ETWS preferences main tab.

14. The Preferences 1-Wire Devices tab is where the 1-Wire weather sensor serial numbers are entered. If you already know the serial numbers, you can enter them directly. If you need to search your 1-Wire bus, click the Find Devices button to display a dialog that will show all devices found on the selected serial port as shown in Figure B-2. Although there isn't a menu, you can still copy and paste from the search dialog to the preferences window using the standard copy and paste keyboard shortcuts. If you don't have a particular sensor, leave the field all (16) zeros to disable that sensor. ETWS also supports the AAG TAI8570 Pressure Sensor. Enter the TAI8570's reader address in the Pressure Sensor field and the writer in the TAI8571 Writer Device field.

15. Once the device serial numbers have been entered, click the Apply button to save the changes.

16. Click the Weather Units tab to see the available weather units as shown in Figure B-3. Select your preferred units for wind speed, temperature, pressure, and rain. To change the way time is displayed in the web page, modify the Date Format String as specified in the JavaDocs SimpleDateFormat class.

```
○ ○ ○            Search for Devices
     Searching...
     Adapter: DS9097U
     Port: /dev/tty.USA19QW4b44P1.1

     3A00000000CE6920
     78000800E97BA210
     550008000881E510
     3C00000073B6F426
     BE0000007416F526
     C20000000143E01D
     6F00000005B63E1D

     Done
```

       [Close]   [Rescan]

FIGURE B-2: Finding your 1-Wire devices.

```
● ○ ○          ExtremeTech Weather Server Preferences
   1-Wire Devices   Weather Units   Sensor Calibrations   Web Settings   Special

   Temperature Units          [ F        ⬍ ]

   Wind Speed Units           [ MPH      ⬍ ]

   Barometric Pressure Units  [ inHg     ⬍ ]

   Rain Units                 [ In       ⬍ ]

   Date Format String         [ MM/dd/yy h:mm a zzz        ]
```

       (Apply) (Close) (Export Prefs) (Delete Prefs)

FIGURE B-3: Weather Units tab.

17. If you made any changes, click Apply.

18. Click the Sensor Calibrations tab to display the various values that can be applied to calibrate a sensor's measurement. As shown in Figure B-4, you can set the scale factor and intercept (also called gain and offset, respectively) for the barometer and humidity sensor. There is also an offset field to set the rain counter to zero, and a field to enter an integer to adjust the wind vane for proper north orientation. If you have already built and calibrated one of the sensors, enter the calibration values and click Apply.

19. Click the Web Settings tab to view the various setting for the web server. These were automatically configured for you with the default directory you specified during initial startup. Leave these settings at their default values for now. Figure B-5 shows the setting configured with my defaults.

20. Click the Special tab to see the remaining settings available as shown in Figure B-6. For now, leave these settings at their default.

21. When you're done viewing the special settings, click Close.

22. Once the preferences window is closed, the main status window will begin to collect and display the current weather measurements based on the sensors you have. You'll see something similar to Figure B-7.

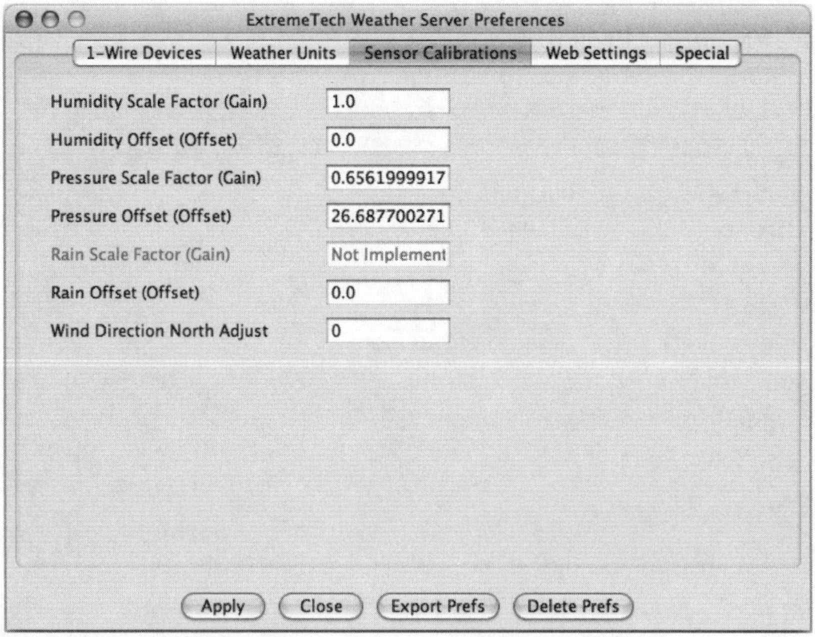

FIGURE B-4: Sensor calibrations.

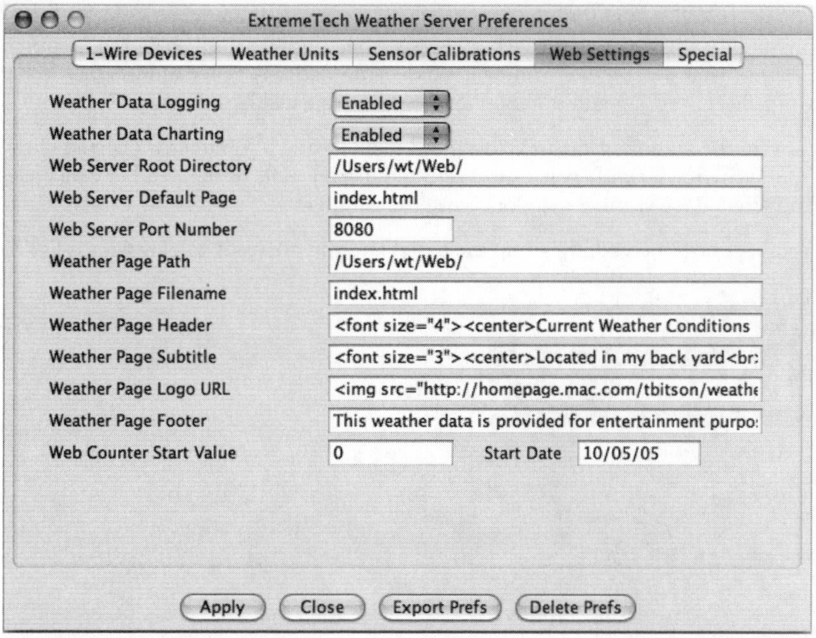

FIGURE B-5: Web Settings tab.

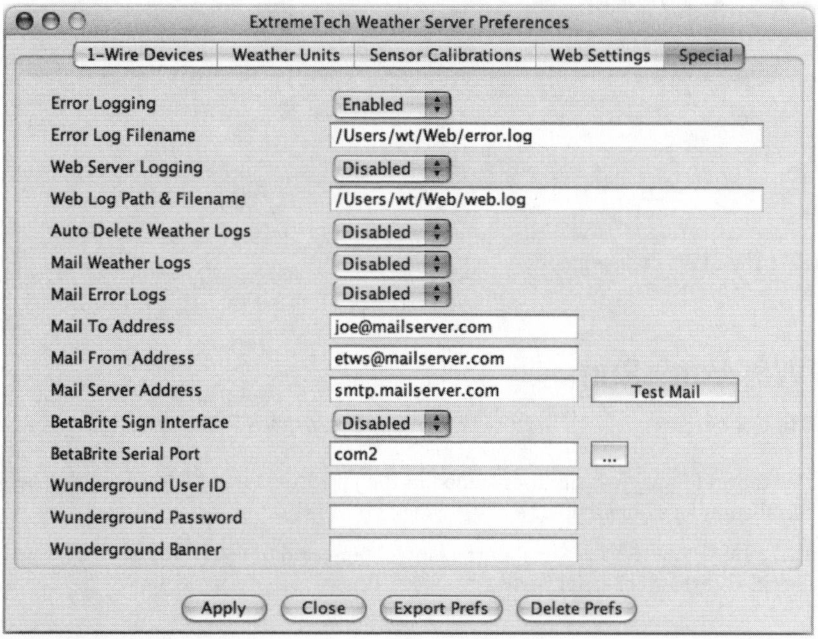

FIGURE B-6: Special settings.

23. Allow ETWS to run until the time reaches an even 10-minute mark (11:20, 3:30, and so on) and then navigate to the web folder specified during setup. Open the `index.html` file with your web browser. You will see the automatically generated weather web page. If you've been working with SimpleWeather, this page will look familiar.

24. Toward the bottom of the web page, find the link titled "Weather Plots" and click it. Your web browser will open a new page and display plots of the weather data collected so far. When done viewing the plots, close the window.

25. You have now successfully set up the basics and you are ready to play with ETWS.

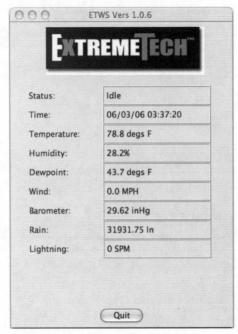

FIGURE B-7: The ETWS main window.

If ETWS Won't Run

Several things can prevent ETWS from running. Here's a quick list to check:

■ Are the additional Java libraries loaded in the Java Run Time folder? Double-check the installation listed in Chapter 4.

■ Is the DS9097U 1-Wire serial port adapter connected to your serial port? Have you specified the correct serial port?

- Try running the jar file from the command line as presented in Chapter 19's sidebar titled "What Do You Do if Your Jar File Won't Run?"

- As a last resort, try rebuilding the ETWS program using the steps described later in the section "Building ETWS."

Additional Settings and Setup

Once the basic configuration is running, you can try experimenting with some of the settings in the Web and Special tabs.

Web Settings

Referring to Figure B-5, there are many settings to configure the web page. ETWS uses the same web server as SimpleWeather. Review Chapter 12 for a discussion on configuring the web server, and accessing the web page from your network or the Internet.

The header and footer accept HTML tags to format the text displayed. You can change the font size, color, and layout. The Weather Pager Logo URL setting allows you to specify the location of a logo or banner picture. This can be a local file or a URL on a remote computer. The default is a URL to my web site that shows a WS-1 weather station.

Special Settings

Most of the special settings are unique to ETWS and are not included in SimpleWeather, except for the BetaBrite option. Here's a quick look at these features:

- If Error Logging is enabled, program and 1-Wire errors are written to the log file specified in the Error Log Filename. The default is `error.log`. Because this folder is located within the Web folder, it can be accessed via the Internet to view the file remotely.

- If you want to log web access, you can enable Web Server Logging. This will log all web requests to the file specified in the Web Log Path and Filename field.

- If Auto Delete Weather Logs is enabled, ETWS will only keep two weather logs: the current and previous days. Older logs will be automatically deleted to prevent your hard drive from filling up.

- If the Mail Weather Logs and Mail Error Logs options are enabled, ETWS will email the weather or error logs to the address specified in the Mail To Address field. This is useful for unattended weather stations. Make sure to set the Mail From Address and Mail Server Address fields to valid addresses and click Apply. Clicking the Test Mail button will send a test message using the current settings so you can make sure the configuration is correct.

- The BetaBrite settings allow the use of a BetaBrite LED sign to display weather data. See Chapter 15 for more information.

- ETWS supports posting data to the Weather Underground. Enter the user ID and password provided by the Weather Underground in the appropriate fields in the preferences window Special tab. If you want to display the banner in your web page, paste the banner text the Weather Underground supplied into the Weather Underground Banner field.

Weather Plots

The weather plots use the free KavaChart AlaCarte demo charting engine and simple HTML pages to display the weather data graphically. The demo version of KavaChart is a great tool, but if you click on the plot, you are redirected to the KavaChart home page. I suggest considering the purchase of the KavaChart Timeseries/Date applet for full functionality.

The files copied to your Web folder in step 8 are HTML files that contain the KavaChart code and parameters that build the charts. These files can be modified to tailor the charts to your liking. You may want to download the KavaChart AlaCart user's guide at www.ve.com/kavachart/tryit_download_reg.html.

Known Bugs

ETWS is not without bugs; lots of bugs. Here's a list of a few of the issues at the time this appendix was written:

- After the initial configuration, changing any device serial number requires you to quit and restart ETWS for the changes to take place.

- After enabling the weather logs, quit and restart ETWS. There is a bug that may cause ETWS to crash.

- During startup, sometimes ETWS can't open the serial port and quits. If this happens, just restart ETWS. If this happens every time, make sure you don't already have a copy of ETWS running.

- If you click the Close button in the preferences window without clicking Apply, you'll lose all changes.

- If things really start acting up, your preferences file may have become corrupted. Click the Delete Prefs button in the preferences window. Caution: as soon as you do this ETWS will quit and you'll have to reconfigure everything from scratch.

- There are many other subtle bugs. Check the ReadMe file for the latest updates and issues.

Building ETWS

If you make changes to the ETWS code, you will need to rebuild the ETWS.jar file. Here are the basic steps required:

1. Launch NetBeans and open the ETWS project located in the ETWS Source folder.

2. You may get a dialog stating the there are "Reference Problems." Follow the dialog's instructions to open the Resolve Reference Problems dialog. Select each of the references and click Resolve. Navigate to where the additional Java libraries were installed in Chapter 4. Resolve all references before continuing.

3. Edit the code. You're on your own here.

4. Run the changes inside the NetBeans environment by selecting Run ➪ Run Main Project (F6). If the code compiles and runs, great! If not, troubleshoot and debug your code changes until they work.

5. Once you have the code running the way you want, quit the running version and in NetBeans, select Build ➪ Build Main Project (F11). This will create the new Jar file in the dist folder in your project folder. The Jar file can be moved to a new location on your hard drive, such as replacing the old ETWS.jar file in your Weather Toys folder. The logo and other images are stored in the images folder. You'll need to copy this to the new location also.

To Learn More

The best place to learn more about ETWS is the source code itself. This is where you'll learn how things are done. Please keep in mind as you look through this code that there are better ways to code this project. The code here was intentionally kept as simple as possible so that it would be easy to follow (although not as easy as SimpleWeather).

The included ReadMe file will contain the latest comments and updates. Be sure to look through this file. Also, check the companion web site for updates, bug fixes, and work-arounds.

Index

How to take it to the Extreme.

If you enjoyed this book, there are many others like it for you. From *Podcasting* to *Hacking Firefox*, ExtremeTech books can fulfill your urge to hack, tweak and modify, providing the tech tips and tricks readers need to get the most out of their hi-tech lives.